NESTING BIRDS

THE BREEDING HABITS OF SOUTHERN AFRICAN BIRDS

NESTING BIRDS

THE BREEDING HABITS OF SOUTHERN AFRICAN BIRDS

PETER STEYN

CONSULTANT: DR MORNÉ A. DU PLESSIS

FERNWOOD
PRESS

THE SPONSORS

JOHN VOELCKER BIRD BOOK FUND

Fernwood Press gratefully acknowledges the financial contribution of the major sponsors above as well as Rhône-Poulenc Agrichem who, by generously supporting this publication, are expressing their concern for the birdlife of southern Africa.

For Andrew,
Linda and Susan,
who have left the nest
to establish their own
breeding territories

Fernwood Press
P O Box 15344
8018 Vlaeberg

Registration no. 90/04463/07

First published 1996

Edited by Leni Martin
Design and DTP by Neville Poulter Design, Cape Town
Production control by Abdul Latief (Bunny) Gallie
Dustjacket designed by Willem Jordaan, Hermanus
Map by Pam Eloff, Cape Town
Index by Leonie Twentyman-Jones

Reproduction by Unifoto (Pty) Ltd, Cape Town
Printed and bound by Tien Wah Press (Pte) Ltd, Singapore
Collectors' and Sponsors' Editions bound by Peter Carstens, Johannesburg

ISBN 1 874950 15 6 (Standard Edition)
ISBN 1 874950 16 4 (Collectors' Edition)
ISBN 1 874950 17 2 (Sponsors' Edition)

Half-title page: Whiskered Tern chicks; Frontispiece: Kittlitz's Plover
Sponsors' page: Pale Chanting Goshawk; Page 6: Spotted Eagle Owl; Page 7: Whiskered Tern

CONTENTS

7 Preface

INTRODUCTION
8 Observing birds
9 Social organisation and breeding systems
13 The breeding cycle
21 The main nest types

25 THE OSTRICH
26 THE JACKASS PENGUIN
27 GREBES
29 ALBATROSSES, PETRELS, SHEARWATERS, PRIONS AND SKUAS
31 PELICANS
32 GANNETS
33 CORMORANTS AND THE DARTER
35 HERONS, EGRETS AND BITTERNS
38 THE HAMERKOP
40 STORKS
43 IBISES AND THE AFRICAN SPOONBILL
45 FLAMINGOS
47 DUCKS AND GEESE
50 THE SECRETARY BIRD
51 VULTURES
54 KITES
56 THE CUCKOO HAWK AND THE BAT HAWK
57 EAGLES
62 BUZZARDS
63 SPARROWHAWKS AND GOSHAWKS
67 HARRIERS
68 THE GYMNOGENE AND THE OSPREY
69 FALCONS AND KESTRELS
71 FRANCOLINS, QUAILS, GUINEAFOWLS AND BUTTONQUAILS
73 CRANES
75 RAILS, CRAKES, FLUFFTAILS, GALLINULES, MOORHENS AND COOTS
78 THE AFRICAN FINFOOT
79 BUSTARDS AND KORHAANS
80 JACANAS
82 SNIPES
83 THE AFRICAN BLACK OYSTERCATCHER
84 PLOVERS
86 THE AVOCET AND THE BLACK-WINGED STILT
87 DIKKOPS
88 COURSERS
89 PRATINCOLES
91 GULLS
94 TERNS AND THE AFRICAN SKIMMER
98 SANDGROUSE
100 PIGEONS AND DOVES
102 PARROTS, THE ROSE-RINGED PARAKEET AND LOVEBIRDS
103 LOURIES
104 CUCKOOS
109 COUCALS
111 OWLS
115 NIGHTJARS
117 SWIFTS AND SPINETAILS
120 MOUSEBIRDS
121 THE NARINA TROGON
122 KINGFISHERS
124 BEE-EATERS
126 ROLLERS
128 THE HOOPOE AND WOODHOOPOES
130 HORNBILLS
135 BARBETS
136 HONEYGUIDES
137 WOODPECKERS AND THE RED-THROATED WRYNECK
140 THE AFRICAN BROADBILL
140 THE ANGOLA PITTA
142 LARKS AND FINCH-LARKS
145 SWALLOWS AND MARTINS
149 CUCKOO-SHRIKES
150 DRONGOS
150 ORIOLES
151 CROWS AND RAVENS
152 TITS
153 PENDULINE TITS
155 THE SPOTTED CREEPER
156 BABBLERS
157 THE BOULDER CHAT AND ROCKJUMPERS
158 BULBULS, THE BUSH BLACKCAP AND THE YELLOW-SPOTTED NICATOR
161 THRUSHES
163 CHATS
166 ROBINS AND PALM THRUSHES
169 WARBLERS
175 CISTICOLAS
177 PRINIAS AND PRINIA-LIKE WARBLERS
179 FLYCATCHERS AND BATISES
183 WAGTAILS
184 PIPITS AND LONGCLAWS
185 SHRIKES
189 STARLINGS
192 OXPECKERS
193 SUGARBIRDS
194 SUNBIRDS
200 WHITE-EYES
201 THE RED-BILLED BUFFALO WEAVER, THE WHITE-BROWED SPARROW-WEAVER AND THE SOCIABLE WEAVER
205 SPARROWS AND THE SCALY-FEATHERED FINCH
207 WEAVERS
213 QUELEAS, BISHOPS AND WIDOWS
215 WAXBILLS, MANNIKINS AND OTHER ESTRILDID FINCHES
218 WHYDAHS, WIDOW-FINCHES AND THE CUCKOO FINCH
220 CANARIES AND BUNTINGS

224 Map of southern Africa
225 Acknowledgements
225 Photographic credits
226 Glossary
226 Useful addresses
227 Bibliography
232 Index
237 List of subscribers

PREFACE

THIS BOOK WAS ORIGINALLY INTENDED to be a photographic guide to the nests of southern African birds, but as I began to put it together I realised that our birds' breeding behaviour in general is so fascinating that to include it could only enhance the whole project. Thus the concept became enlarged, developing from a basic field guide to nests into an overview of the nesting habits of the various groups of birds that breed in our region.

Writing as an ordinary birdwatcher for other birdwatchers, in this book I seek to communicate, in a style as free of jargon as possible, some of my own admiration and enthusiasm for the way in which birds nest and behave while breeding. References, many of which are articles in scientific journals (see pages 227 to 231), provide the skeleton on which the flesh of the text is moulded. The flesh itself derives from stimulating contact with colleagues as well as from personal field experience.

My birdwatching career, like that of many of my contemporaries, began when as a schoolboy I collected my first egg. It was October 1949, and the Moorhen's nest on the Cape Flats near Cape Town contained eight eggs, of which I took one. It was meticulously blown and stored in cotton wool, and my first notes were carefully written up. Little did I know it then, but that incident was the beginning of a lifetime fascination for birds, and their breeding habits in particular. Photography succeeded egg-collecting, but there can be no doubt that my youthful oological pursuits taught me to observe the habits of birds and make notes on what I saw. Thus it was an easy transition to continue making detailed observations while watching and photographing birds at their nests. The years I spent in Zimbabwe, from 1961 to 1977, were some of the most productive for me and many of the observations I made then, as well as later ones, are incorporated in the text.

One of the things that my birdwatching experiences have taught me is that it is essential to authenticate identifications and not jump to conclusions. This was illustrated when, in 1971, I consulted the Nest Record Card collection which, under the aegis of what was then the Southern African Ornithological Society, is a source of information on which most standard works draw. After examining 16 cards purporting to be nest records of Brown Snake Eagles, I found only two that were unquestionable. On another occasion I discovered two eggs in the nest of a Heuglin's Robin, one pale and the other dark chocolate. It was tempting to record the dark egg as that of a Red-chested Cuckoo, but I was uneasy about the similarity in the size and shape of the two eggs. My caution was justified and Heuglin's Robin chicks emerged from both.

Southern Africa has produced ornithologists of the highest calibre, and while writing this book I have been repeatedly impressed by the contributions made to our knowledge by dedicated observers working on their own and usually without any research grant. The doyen of field workers in southern Africa is undoubtedly C.J. Skead, whose meticulous surveys of species as diverse as the Cattle Egret, Helmeted Guineafowl, Hoopoe and Cape Penduline Tit are landmark studies. The late Leslie Brown's work on birds of prey is well known, but he also studied the breeding habits of flamingos in East Africa. Val Gargett's research on Black Eagles and Peter Mundy's on vultures have earned respect far beyond southern Africa. There have been many other contributions to the understanding of our birds: Carl Vernon's work on Red-billed Helmet Shrikes and their parasitism by Thick-billed Cuckoos, Morné du Plessis' investigation of Red-billed Woodhoopoes, Warwick Tarboton's research into polyandry in African Jacanas, Richard Brooke's studies of swifts, and many more. Yet numerous gaps in our knowledge remain. Often it is the common and accessible species which require study, and opportunities to do so are there for the taking. Even a few basic observations would greatly expand our knowledge of many of our local species.

In this book southern Africa is regarded as the region lying south of the Kunene, Okavango and Zambezi river systems. In order to avoid an unduly fragmented text I have not slavishly followed taxonomic conventions, but have placed certain families in a single account for convenience. Thus the Painted Snipe and Ethiopian Snipe are discussed together, as are the Boulder Chat and the rockjumpers, and the Cuckoo Finch and the whydahs and widow-finches. Sometimes it has been convenient to refer to a large group under a single 'umbrella' name, for example waxbills, twinspots, firefinches and mannikins are estrildid finches, or simply estrildids.

The English names follow those in the latest (1993) edition of *Roberts' Birds of southern Africa*, with occasional exceptions where alternative names are preferred. However, I have chosen to retain the use of hyphens; thus Racket-tailed Roller, not Rackettailed Roller.

With regard to the arrangement of the text, each group account follows a basic outline: introductory comments, territorial/courtship behaviour, breeding season, nest, eggs, incubation period, nestling period and post-nestling dependence. However, emphasis is placed on different aspects of breeding throughout the text, depending on their interest value and the amount of information that is available. Inevitably, read as whole, certain details may seem repetitive, but it is assumed that most birdwatchers will consult the various accounts for information relevant to what they have seen, or hope to see. If they experience even a little of the enjoyment that I have had in watching nesting birds over a period of nearly 50 years, then they will be well rewarded.

PETER STEYN
CAPE TOWN, MARCH 1996.

INTRODUCTION

WHY DO WE STUDY THE BREEDING HABITS OF BIRDS? The short answer is, to be able to conserve them. The information gathered by both professional and amateur ornithologists provides a valuable resource for research which can be used directly in the conservation of birds. A striking example is the rare Blue Swallow in Mpumalanga, where declining numbers alerted ornithologists to the loss of nesting sites, mainly as a result of the spread of commercial timber plantations into the swallows' former grassland habitats. Conservation measures have been implemented, with the result that not only the Blue Swallow but also a spectrum of plants and other animals have benefited, even though only belatedly and on a limited scale.

On a global scale, another success story resulted from research based on egg collections which established that the thinning of raptor eggshells was directly attributable to the widespread use of DDT. This led to the banning of the poison in many countries and the subsequent recovery of populations of species such as the Peregrine Falcon. Although egg-collecting is now illegal in most countries, in the past it did serve a purpose in that it resulted in some major collections (such as that housed in the Western Foundation for Vertebrate Zoology in Los Angeles) that provided important research material. It also proved to be a starting point for some prominent ornithologists.

Despite all the research that has been done over the past few decades, we still know surprisingly little about how a large number of our species breed. For even some of the common ones we do not have the basic details of incubation and nestling periods. The information that we do have about the birds of this region is stored in the Nest Record Card Scheme of BirdLife South Africa (formerly the Southern African Ornithological Society), which is administered by the Avian Demography Unit (ADU). The ADU welcomes contributions to the Nest Record Card Scheme and through its publication *Bird Numbers* it provides stimulating feedback to its contributors, most of whom are amateur birdwatchers. The addresses of BirdLife South Africa and its branches, as well as that of the ADU, are listed on page 226.

OBSERVING BIRDS

At first finding nests may seem to be a matter of chance, but by studying birds' habits one soon learns to read the tell-tale signs that they are breeding. A bird carrying nesting material or food is an obvious clue, but an observer should also watch out for birds feigning injury or carrying a faecal sac or, more specifically, a prinia with its tail bent from sitting in the confined space of its nest. The hardest nest to locate is always the first one of a particular species but, with experience as the best teacher, one soon develops a 'feel' for that species' nesting situation.

Ingenuity is sometimes required in order to examine nests. An invaluable aid for looking into high nests or ones in thorn trees is an extendible pole with a mirror attached to its end. The nest-holes of species such as barbets and woodpeckers can be inspected by means of a dentist's mirror and a bulb connected to a battery by a piece of flex. This apparatus can be adapted into a 'ripariosope' which allows an observer to peer down a long tunnel in a bank into the nest of, for example, a bee-eater. One researcher studying South African Cliff Swallows designed his own 'hirundoscope' which enables him to examine the contents of the swallows' enclosed nests. If one has access to the technology of fibre optics then some sophisticated endoscopic devices can be improvised.

Photographing nests is both useful for record purposes and enjoyable as a hobby, and I have found that a macro lens gives the best close-up results; my favourites are a 50-mm and a 105-mm. A small electronic flash is needed to illuminate nests in dark situations and, for extreme close-up work, a ring flash is invaluable.

Whether inspecting a nest or photographing it, one should always keep disturbance to a minimum. The golden rule is that the welfare of the birds is of paramount importance, and common sense should be exercised to avoid distressing either the parents or the chicks. The main dangers include the parents deserting the nest, eggs or young, and the resultant exposure of the eggs or young to cold or to heat (the latter is often more harmful), as well as the possibility of exposing well-hidden nests to predators, particularly avian ones. Birds are more likely to desert the nest at some stages of the nesting cycle than at others, and extra caution should be exercised when they are nest-building, egg-laying or incubating. Some species are also more sensitive than others: certain cisticolas may desert their nests if disturbed while they are constructing them, and the Bateleur, sandgrouse and nightjars are prone to abandoning their clutches in the early stages of incubation.

Probably the greatest threat of all is disturbance to birds which breed in colonies. Although it may appear chaotic, a colony has its own system of regulation under

A mirror attached to the end of an extendible pole makes it possible to see into a nest in a thorn tree.

which the members in it stick to the rules. Once an observer approaches too closely, or even intrudes within the perimeter of the colony, then serious disruption can occur. Nestlings which have been displaced are viciously pecked, sometimes killed, and even if they are not attacked, they often have difficulty in returning to their nests. In a gull colony exposed eggs are broken open and eaten by neighbours. The ADU advises that certain species should be monitored only by trained ornithologists and such species include a number of colonial breeders, notably the Jackass Penguin, White and Pink-backed Pelicans, the Bald Ibis and the Cape Vulture.

When looking for a nest or examining it, take care not to trample the surrounding vegetation, thus exposing the nest to predation, and if cover has to be moved aside it must be replaced exactly as it was found. As few people as possible, preferably a single observer, should visit the nest, and they should approach slowly and casually, allowing the parent bird to slip away safely. If the adult is startled at the nest it is more likely to desert, or it may displace or damage the eggs or chicks as it panics. Try not to flush away adults at dusk as they may not have time to return to the nest before nightfall. The visit should be kept as brief as possible, and once the observer has gathered all the information necessary he should move right away, allowing the parent to return in peace. A nest containing large young should be approached cautiously and inspected from a distance so that the nestlings do not depart prematurely; once they have 'exploded' from the nest it is very difficult to replace them. If they are replaced, the best ploy is to cover them with something such as a hat, wait until they settle and then remove the cover slowly and carefully.

The best place to observe birds is on one's own doorstep and, since the most common species are often neglected as subjects of study, much valuable information can be gleaned. Many species can be encouraged to nest in a garden by providing them with areas of undisturbed 'wild' vegetation and accessible water and food. Hole-nesting birds may be attracted to either manufactured or improvised nest-boxes, the species encouraged being regulated by the size of the entrance hole. Birds that excavate their own nest-holes can be lured into the garden by putting up sisal stems for them, or suitably positioned logs. In my garden in Bulawayo in Zimbabwe a strategically placed box on a water tower was used by a pair of Barn Owls and provided valuable observations on their nesting behaviour.

Throughout the world many birds display some or other remarkable breeding habits. The megapodes of Australasia, whose eggs may be incubated in a vast compost heap or in soil heated by volcanic activity, and the bower birds of the same region which decorate display areas to attract mates are just two examples. Africa, too, has its share of species with unusual habits or nests, and in the southern part of the continent we are fortunate to be able to study the likes of the Hamerkop and its extraordinarily large and complex nest structure; the Palm Swift which 'glues' its eggs to its vertical pad of a nest; and, not least, the Sociable Weaver whose communal 'apartment block' is the largest constructed nest in the world.

1

2

3

1 *A sisal stem attached to a tree in a suburban garden has been readily accepted by a Black-collared Barbet.* ***2*** *A removable panel at the front of a nest-box facilitates inspections throughout the nesting cycle of a pair of Greater Blue-eared Starlings, and with minimum disturbance to the birds.* ***3*** *When observing birds one should always be prepared for the unexpected, and authenticate observations whenever possible. The unmatched egg in a White-throated Robin's nest is almost certainly that of a Red-chested Cuckoo and it was probably intended for a Boulder Chat's nest, where it would match. Unfortunately the nest was robbed by a predator and the identity of the odd egg could not be established beyond doubt.*

Social Organisation and Breeding Systems

A comparative study of different groups of birds soon reveals that their social organisation and the breeding systems they follow are as diverse as their appearance and song. Most defend territories, but some more vigorously than others; some breed solitarily whereas others enjoy the benefits of colonial breeding; some species maintain long-term pair bonds but others mate with more than one partner even within a season; and some care for their young alone whereas others are assisted by 'helpers'. Whichever course the birds may follow, it is the one that is intended to maximise their breeding potential.

Maintaining a territory

A territory is any defended area, and in ornithological terms it may be one which is occupied by a pair all year and includes its food supply, or it may be no more than an area in which the pair remains temporarily to breed. At times it may not even be connected with breeding, but is merely the area around a food source; for example, sunbirds may defend a good nectar supply. Nomadic species such as some of the larks are not strongly territorial when breeding.

The size of a territory varies enormously, from pecking distance in species that breed colonially to an area as large as 250 square kilometres in the case of the Martial Eagle. In some circumstances the possession of a territory can be a matter of life or death. In Britain it was found that Tawny Owls with territories have a much better chance of survival in winter than those without them. The importance of a territory is demonstrated by the fact that as soon as one becomes vacant it is immediately filled by a 'floater', a bird that did not previously hold one.

An ideal territory should have nesting sites that are suitable for the species concerned and a food supply that will support the breeding pair and its young. With no more than these basic essentials a male can set about attracting a mate, but there is evidence that males with quality 'real estate' are more successful than those without. It has been observed, for example, that a male Pin-tailed Whydah with a water supply

1 *The Karoo Korhaan is inconspicuous in its Karoo habitat, but calls loudly and regularly to ensure that its presence in its territory is known.* **2** *A Fiscal Shrike advertises its territory merely by perching in a prominent position, but it may also sing from this perch.*

in his territory was able to attract more females than one that did not have this bonus.

Although birds may drive off other species which have similar habits and may compete for a food resource, they usually defend their territories against other members of their own species. They do so by means of song, by adopting certain aggressive postures and by being highly visible. Song is an essential part of territorial advertisement for many species, and its intensity increases with the onset of the breeding season. It may be delivered statically from a perch, in which case it is often loud and monotonous, as performed by most cuckoos, or it may be associated with aerial displays. The Fan-tailed Cisticola, for example, emits a monotonous series of notes as it rises and dips above open grassland, and the Crowned Eagle calls as it performs an undulating flight over its forest territory. Sometimes song is combined with a mechanical sound such as the wing-rattling of the Clapper Lark

Certain stereotyped postures, such as pointing the bill towards the intruder and sleeking the plumage, clearly indicate aggression. Many others are detailed in the following accounts. Simply being visible enables some species to maintain a territory. For example, the Fiscal Shrike sits prominently at the top of a tree or bush, and the Fish Eagle perches for most of the day, its white foreparts signalling possession of the surrounding area to potential rivals a long distance away.

If an intruder persists in its attempts to enter a territory, the holder may attack it physically as a last resort. Walt Whitman's description of an encounter between a pair of Bald Eagles is often quoted as an illustration of courtship behaviour:

'The rushing amorous contact high in space together,
The clinching interlocking claws ...'

In fact the confrontation is not amorous, as Whitman suggests, but aggressive.

Breeding in colonies

Most birds breed solitarily within a demarcated territory. However, some species – penguins, pelicans, gannets, some storks, flamingos, the Cape Vulture, avocets, stilts, pratincoles, some swifts, some bee-eaters, some swallows and martins, some starlings, most weavers, queleas and some bishops – breed colonially in tightly packed groups that range from a few dozen pairs to several million birds in the case of Adélie Penguins in Antarctica. Sometimes the colony merges into a communal nest, the most notable example being that of the Sociable Weaver. In other cases, in which pairs of the same species may have been mutually attracted to an area of suitable habitat, they nest close to one another but not in a colony in the true sense. Several pairs of Cape Canaries in a clump of pine trees around a farmhouse or Yellow-rumped Widows in an area of rank vegetation can, for example, be described as being loosely colonial.

Colonies may comprise different species, such as herons, egrets, ibises, cormorants and spoonbills which all nest in the same reedbed. On Robben Island off Cape Town Hartlaub's Gulls and Swift Terns breed together, as do Kelp Gulls and Caspian Terns elsewhere in the south-western Cape. In mixed colonies such as these the purpose of the association is to derive maximum benefits from colonial breeding.

Why do birds breed colonially? In some cases, especially amongst seabirds, colonial breeding is necessary because there is a shortage of suitable nest sites. For example there are very few islands off the southern African coast on which Cape Gannets can breed. The colony also serves as an information centre; birds which have fed successfully return to the same feeding area and are followed by others. Birds in a colony enjoy the benefit of a warning system, in that a predator is likely to be seen in good time by some of the colony members and they raise the alarm. The whole group can then mob the intruder, often actually attacking it, and thus effectively deter it. Even if a predator succeeds in entering a colony, the number of birds it may kill would be small relative to the size of the colony. The 'swamping' effect of many birds breeding in unison is evident in a Red-billed Quelea colony where even heavy predation makes little impact on its vast numbers.

Most seabirds, including the Cape Gannet, breed colonially. One reason is the shortage of suitable nesting sites along the coast.

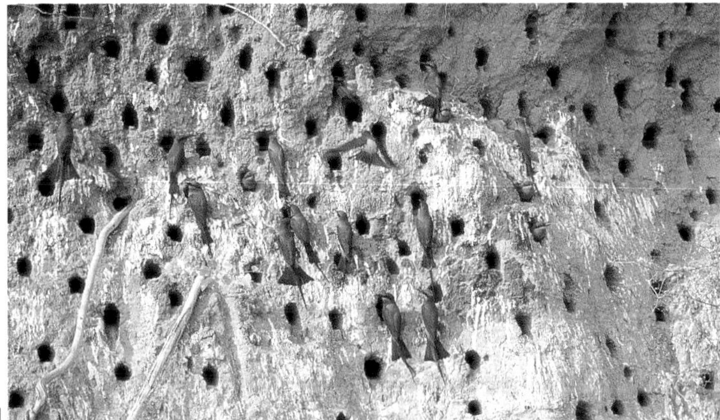

1 *A colony of colourful and graceful Carmine Bee-eaters is a spectacular sight, with as many as a thousand nest-holes sometimes honeycombing a riverbank.*
2 *Sacred Ibises often breed in mixed colonies, sharing a reedbed with species such as African Spoonbills and egrets.*

Among the disadvantages of colonial breeding are included a shortage of space and the depletion of food resources and nesting material. The latter was in such demand in a colony of Adélie Penguins in Antarctica that coloured stones placed on a single nest were soon distributed among neighbouring nests over a wide area as a result of theft. Another drawback of colonies is that disease can spread like wildfire, often with disastrous results. Sometimes when a colony has been seriously depleted, the birds that do return to breed choose certain sites first, thus illustrating that prime positions are in demand.

Colonial species are among the most fascinating of all birds to study, and observers who do so enjoy the added advantage of being able to watch many nests at the same time and marvel at the birds' ability not only to find their nests, but also to recognise their own chicks. Research on species as varied as Cape Gannets, gulls and White-fronted Bee-eaters has revealed insights into a complex array of behavioural patterns or social systems. The frenetic activity of a gannet colony and the spectacular beauty of nesting Carmine Bee-eaters are just two examples of the attractions of birds that breed in large numbers.

Mating systems

Some of the mating practices of birds would raise the eyebrows of a board of censors. Ducks and bee-eaters commit rape, European Starlings – among others – are guilty of infidelity, and polygamous behaviour is by no means rare.

Monogamy, however, is the most common mating system amongst birds. The monogamous relationship may be long-term or it may last for only one breeding season. There are even cases of 'serial monogamy' – in which a bird has a series of monogamous relationships – within a single breeding season. A life-long pair bond is maintained by certain seabirds such as albatrosses as well as many of the larger birds of prey, and even by a large number of smaller species. The advantage of a long-term association is that it allows the birds to dispense with the energy-sapping practices of extended courtship rituals, getting to know each other and then establishing new nest sites.

Sometimes monogamy is 'imposed' on a species by the very nature of its breeding behaviour. Gulls nesting colonially dare not leave their eggs unattended because their neighbours would eat them, so a pair has to share duties at the nest on a monogamous basis in order to breed successfully. Similarly, species that have to forage over long distances need to be monogamous, so that one partner can stay at the nest.

Polygyny is a system in which the male has several mates, either successively or simultaneously. Successive polygyny is more usual and is practised by species such as

Albatrosses are monogamous and maintain a life-long pair bond. A Yellow-nosed Albatross pair strengthens the bond by taking turns to preen each other.

1 *Painted Snipes are polyandrous, and the female (standing behind) has the more colourful plumage.* **2** *The Red Bishop is polygynous and the male, conspicuous in his breeding garb, will mate successively with several females attracted to the nests he has built.*

the weavers, in which the male courts one female and as soon as she has accepted his nest and laid in it, he sets about attracting another. The male Pin-tailed Whydah, on the other hand, practises simultaneous polygyny, protecting a 'harem' of females.

Polyandry, in which the female has several mates, is much rarer. It may be sequential (involving a series of successive mates) or, infrequently, simultaneous (in which the female may mate with each of her several mates more than once in a season). The female may be larger than her mate, as in the case of the African Jacana, or more colourful, as is the Painted Snipe. Some buttonquails are also polyandrous and, as is the case in most southern African species in which this mating system occurs, only the males incubate and care for the young. The Pale Chanting Goshawk is an exception, as the female of the polyandrous trio does most of the incubating.

Co-operative breeding

Although only about 3 per cent of the world's birds have been found to breed co-operatively – most of which are passerines – in southern Africa at least 42 species have been recorded with helpers at the nest. These include kingfishers, bee-eaters, the Red-billed Woodhoopoe, barbets, babblers, helmet shrikes, starlings, oxpeckers and Sociable Weavers. As research progresses other species, even unlikely ones such as the Spike-heeled Lark, continue to be added to the list.

Co-operative breeding involves helpers which are usually, but not necessarily, related to the breeding pair. The duties they assist with range from nest-building to feeding the chicks, with the latter being the most common form of co-operative behaviour. The number of helpers varies from a single bird to as many as nine in the case of the Sociable Weaver.

The Pied Kingfisher, White-fronted Bee-eater, Red-billed Woodhoopoe, helmet shrikes and the Sociable Weaver all display a range of different co-operative breeding strategies, and why they do so is only one of several aspects of this phenomenon which require further research. According to one theory, known as kin selection, the helper is related to the parents and enhances their productivity by assisting them in their breeding attempt. However, helping does not necessarily increase productivity and there are sometimes other explanations for co-operative behaviour, such as that of the Red-billed Woodhoopoe (see page 129).

Whatever the explanation may be, there can be no doubt that helpers assisting the parent birds during the nesting period, and often long after it, is one of the most interesting aspects of avian behaviour.

The Karoo Robin is the only robin in southern Africa known to have helpers at the nest.

THE BREEDING CYCLE

The timing of the breeding season

Although it can be stated as a general rule that most bird species in southern Africa lay their eggs in spring and summer – between September and March – there are many exceptions. In the winter-rainfall region of the south-western Cape breeding usually begins after the winter rains, from August onwards, although in some species it starts as early as July. The Cape Sugarbird and the Orange-breasted Sunbird breed even earlier, during the winter months, to coincide with the flowering of the plants on which they feed. In order to cope with the severe prevailing conditions they insulate their nests well and situate them where they are protected from the elements.

Further north, in the summer-rainfall regions, breeding usually coincides with the onset of the rains from October onwards. Mostly the birds anticipate the increased food supply resulting from the rain, but species such as hornbills and swallows depend on the rainfall for the mud with which they seal their nests or build them. Grassland species such as widows, which depend on grass cover and seeds for nesting sites and food, delay their breeding until late summer, but each season differs depending on when the rains occur and how much falls.

In some habitats, notably miombo woodland, breeding coincides with the spring flush of leaves. The msasa trees in large tracts of Zimbabwe break out into a riot of red and orange leaves in September as the temperature increases prior to the rainy season. In association with these phenomena, swarms of insects emerge and proliferate. Many birds begin to breed at this time and continue into November, but the duration of the season for a particular species depends on its feeding niche.

There are exceptions to the above pattern, including large birds of prey which breed in winter. In the Matobo Hills in Zimbabwe the Black Eagle breeds in cool, dry conditions which contrast with the cold and wet weather experienced by Black Eagles in the south-western Cape at the same time of year. The food supply during the nestling period and particularly when the eaglet leaves the nest is the important common factor, and in both areas prey would be more plentiful in spring.

In the equable tropical and subtropical regions of Africa birds may breed at any time of year, although there are usually peak periods. In arid habitats, too, species such as the Double-banded Courser breed at any time of year, and by laying a single egg and breeding several times within the year it is able to cope with the marginal and harsh environment in which it lives. Often in arid regions breeding is opportunistic after rainfall, although a minimum of 25 millimetres is required as a trigger. The response to rain is dramatic and birds may start nesting within a week. As a rule insectivorous species breed sooner than granivorous ones, as the latter need to wait for the development of grasses. If rainfall is substantial, several consecutive broods are raised while the food supply lasts. Nesting may cease as suddenly as it began, and larks and other nomadic species move on to new areas.

The single most important factor regulating the timing of breeding is food supply. An extreme example demonstrating this is the Sooty Falcon which nests in late summer in the inhospitable Sahara Desert, when temperatures reach 42 °C in the shade. This time of year is peak migration season, and the migrant birds can be preyed on to provide food for the nestlings.

At high latitudes day length affects the timing of the breeding season, which is often synchronised within a short period. However, in southern Africa, except perhaps in the extreme south of the region, this is not a major influence.

Courtship

There are three main reasons for courtship: to acquire the right mate, to synchronise breeding activity, and to assess the quality of a partner. It is necessary that when a bird mates it does so with another of its own species and this is achieved by certain ritualised displays which are unique to that species. Then, in order to mate successfully, the pair must be at the same reproductive stage. Nest-building is an integral part of courtship, and the fact that Brubru Shrikes sometimes destroy completed nests suggests that the breeding condition of the pair may not have been synchronised and that they need to build again to rectify the situation. A male's quality as a mate is often assessed by his ability to provide food, so courtship feeding indicates his fitness for this task. Terns and most birds of prey are amongst those species notable for courtship feeding, which also serves the purpose of enabling the female to accumulate the reserves required for egg-laying.

1

2

1*, *2 *Aigrettes in the case of the Yellow-billed Egret and the striking combination of a bare yellow patch and bizarre black wattles in the case of the Wattled Starling are designed to attract prospective partners.*

1

2

1 *A male Maccoa Duck courts a female with his vibrating trumpeting display. Unlike most other southern African ducks, he acquires a bright breeding plumage.*
2 *The essential culmination of courtship: Black-necked Grebes mate on their nest.*

Courtship behaviour varies considerably but, in general, species which maintain a permanent pair bond are the least demonstrative, and in some cases very little in the way of courtship seems to take place. A pair maintains its bond in various ways, for example by antiphonal duetting or mutual preening, or merely by perching and roosting together. Species which are more energetic in their courtship may perform either perched or aerial displays, sometimes both, and the performances range from the ordinary (such as the high flights of some larks) to the bizarre (the tumbling aerobatics of the Red-crested Korhaan). The same display or calling behaviour may serve as territorial advertisement.

Many courting birds acquire special plumes or coloration at the onset of the breeding season. Egrets grow aigrettes or plumes, several weavers assume black masks, and widows acquire long tails. In the case of the polygynous Long-tailed Whydah it has been shown that the males with the longest tails attract the most mates. As a rule, when a special breeding plumage is assumed, the males are colourful and the females, which need to be camouflaged to minimise predation, are drab. The Painted Snipe is an exception, and in this species the polyandrous female is the more colourful of the pair.

Sexual dimorphism is not restricted to differences in plumage. Male birds are usually larger than the females, but in birds of prey the reverse is true. The main reason for this appears to be that the male needs to be more agile in order to catch prey for the female, as she does most or all of the incubation and remains with the chicks for the first half of the nestling period. In vultures, on the other hand, the male and female share parental duties equally and are the same size. A difference in bill sizes is noticeable in some species that feed together as a pair all year. The male Scimitar-billed Woodhoopoe, for example, has a larger bill than the female, so that when the pair feeds together each can utilise a different spectrum of food and thus avoid competition.

Nest-building

Not all birds construct a nest, so the term 'nest-building' should be interpreted to include choosing a site. The selection of a good place to breed has survival value, and many birds will return to the same site year after year. Even within colonies certain situations are preferred to others and are competed for. Likewise, suitable nest-holes in trees are always at a premium, and competition for them is intense. In arid regions nests are often placed against a bush and on its south side so that they are shaded for most of the day. Sometimes certain species nest in association with other animals, such as Blue Waxbills which may breed near a wasp nest, or weavers which construct their nests on branches around an eagle's eyrie. One reason for associations such as these may be that they provide protection against predators, but there may also be other, less obvious, explanations.

The nest site may be selected and the nest built by the female alone, or with the co-operation of the male. In groups such as the weavers, bishops and widows the male builds the nest, or several of them, and the female chooses one and lines

1

2

1 *A Greater Striped Swallow collects mud for its nest. The timing of the breeding season of this and other swallows depends on the arrival of the rains.*
2 *A Common Waxbill carries a straw, which it characteristically grasps at one end, to incorporate in its nest.*

1 *Black Eagles may use the same nest season after season, and as they add to it each year it becomes more and more bulky.* **2** *The phenomenon that some birds breed in association with other animals is illustrated by a Collared Sunbird's nest next to that of paper wasps. Protection against predation has been suggested as a possible reason.* **3** *Many birds of arid regions, such as the Black-eared Finch-lark, place their nests so that they are shaded for the greater part of the day. A small, straggly bush provides not only shade, but also camouflage for the two chicks.*

it. Selecting the nest site and offering nesting material form an integral part of courtship for many species.

There is considerable variety in the types of nests built, ranging from a simple scrape on the ground to the complex structures built by the Hamerkop, penduline tits and weavers. But whether simple or complex, the nest always serves its purpose as a place in which the eggs may be hatched and, usually, the nestlings may be raised. The cost of nest-building in terms of energy can be substantial. For example, the male Spotted-backed Weaver may collect between 600 and 800 pieces of material and fly 20 kilometres in the course of building a single nest. Moreover, as he builds an average of seven nests in a season, the actual distance covered is 140 kilometres. In his case nest-building is an essential part of courtship, but the energy expended by many other species on nest construction is no less impressive. It is no exaggeration to say that the nest is the focal point of a bird's life.

Eggs

Often described as nature's perfect package, an egg encapsulates a bird's genetic future. Energy is required to produce eggs, and courtship feeding – in those species which practise it – serves the dual purpose of establishing the male's fitness as a partner and of building up the female's reserves. As a rule eggs form overnight and are laid in the early morning so that the female does not have to carry additional weight during the day. In the case of smaller species they are generally laid at daily intervals.

Eggs vary considerably in terms of colour, shape and size, and the colour and markings of many render them exquisite. Even a plain white dove's egg can be beautiful, for when it is fresh it has a translucent pink tinge. The scrolls and glossiness of jacana eggs make them particularly attractive, the latter also giving them a 'wet look' that conceals them on their rudimentary nest in an aquatic environment. The colouring of eggs usually serves a similar purpose, and as a rule eggs in open nests, such as those of plovers, have markings to make them less visible. Eggs in enclosed nests, on the other hand, are plain and usually white. It is often thought that hole-nesting species such as kingfishers and bee-eaters lay white eggs because camouflage is unnecessary, but a far more important reason is that white eggs are conspicuous in the dark nest and the parent is less likely to damage them when moving about.

There are many exceptions to this generalisation: doves and louries, for example, should lay camouflaged eggs, but theirs are white. The eggs of ducks, francolins and even nightjars are also conspicuous, but they are effectively concealed by the cryptic coloration of the incubating parent. Another interesting aspect which is linked to the coloration of eggs is their palatability. As a rule camouflaged eggs such as those of quails, plovers and gulls are tasty while many white ones are not.

Various terms are used for describing the shape and colour of eggs and the texture of the shell. Most species lay oval eggs but louries, for example, lay round ones, swifts lay long oval ones, and plovers lay pointed, or pyriform, ones. In oval and pyriform eggs the markings are often concentrated at one end, usually the broader one. The main colour of the shell, known as the ground colour, may be plain or it may be decorated with bold overlying markings (scrolls, spots or blotches) as well as more subdued underlying markings. The texture of eggshells is usually smooth, glossy or chalky. Remarkably, it has been found that birds are able to recognise their own eggs. For example, a female Ostrich is able to reject eggs laid by other females in her nest, and some species that are parasitised by cuckoos often reject eggs that do not closely match their own.

As a general rule, the eggs of large birds are smaller relative to the birds' body size than those of small birds. Nevertheless, the mind boggles at the thought of the eggs of the extinct elephant birds of Madagascar. The largest of these birds stood 3 metres tall and weighed 450 kilograms, and they laid eggs which had a capacity of 9 litres and took 100 days to hatch! By comparison, a male Ostrich is about 2.5 metres tall but weighs only 150 kilograms, and an Ostrich egg has a capacity of about 1 litre.

The size of a clutch of eggs is the subject of many theories. It is the strategy of a bird to produce as many young as possible in its lifetime in order to pass on its genes.

However, the most productive clutch size (i.e. one which produces the largest number of surviving young per nest) is not necessarily the best means of achieving this aim, because the bird may shorten its life expectancy through overworking. Instead, an optimal clutch size (i.e. one that produces the most surviving young in the parents' lifetime) is better in the long term, despite being slightly smaller.

Birds that are short-lived produce relatively large numbers of young per clutch, in contrast to long-lived species which lay small clutches – often a single egg – but over a longer period of time. Many of the long-lived species in southern Africa, such as vultures and Ground Hornbills, are endangered or vulnerable because if the adult population decreases they cannot compensate for the loss by increasing their breeding productivity.

Some species that live in arid regions with marginal habitats, such as the Karoo Korhaan and the Double-banded Courser, lay a single egg per clutch, but may breed several times a year. Clutch sizes are generally not constant (the exceptions being found in some groups such as pigeons and doves) and can decrease or increase depending on the food supply, which is often determined by the amount of rain that has fallen. Sociable Weavers, for example, lay larger clutches after good rains, and the Barn Owl's normal clutch of 4-6 eggs may increase to 12 when a population explosion of rodents occurs. In times of abundance not only are larger clutches laid, but several consecutive broods may be raised.

The scrolls on the eggs of the Lesser Jacana make them very attractive and, together with their 'wet look' glossiness, help to camouflage them.

The incubation period

The beginning of the incubation period is marked by the parent bird consistently brooding the eggs. In smaller species this happens when the clutch is complete, or almost complete. However in some species, especially large birds of prey which lay two eggs, incubation starts when the first has been laid. As the eggs are laid several days apart, there is considerable disparity in the size of the young and this may result in Cainism (see page 19). The Sociable Weaver, which lives in a marginal environment, also begins incubation when the first egg has been laid, with the result that if the food supply is inadequate, the last-hatched chicks may die. This means of brood reduction is found in a number of species.

Most birds have a bare brood patch which comes into contact with the eggs to give them maximum warmth from the parent's body. One species that lacks a brood patch, the Cape Gannet, enfolds its single egg in its webbed feet before adopting an incubating position. Incubating birds periodically turn the eggs with their bills to facilitate an even distribution of warmth, but this is not essential for the eggs' viability, as demonstrated by Palm Swift eggs which are glued to the nest and the eggs of Three-banded Coursers which, being buried, are also immovable. The eggs of the Maccoa Duck are so large that the female rotates her position on the nest from time to time to ensure that they are all evenly covered.

Eggs are able to withstand cooling better than overheating. European Swift eggs, for example, may be left for up to 6½ hours in poor weather when the parents have to travel long distances to find food, but still remain viable. On the other hand, some species have to prevent their eggs from getting too hot, and some plovers and the Rock Pratincole and African Skimmer soak their belly feathers and wet their eggs to keep them cool in the intense heat of their exposed nesting sites. In arid regions a

1

__1__ A Blacksmith Plover relieves its mate at the well-concealed nest. They take turns to incubate throughout the day. __2__ An Avocet turns its eggs with the tip of its bill before settling to incubate.

2

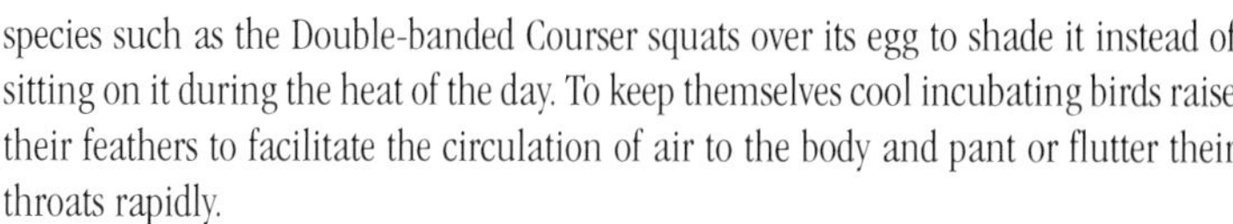

1 *In extreme heat the Double-banded Courser shades its egg rather than incubates it.* ***2*** *In the early morning a female Namaqua Sandgrouse takes over from her mate which has incubated throughout the night. She will sit for the rest of the day.*

species such as the Double-banded Courser squats over its egg to shade it instead of sitting on it during the heat of the day. To keep themselves cool incubating birds raise their feathers to facilitate the circulation of air to the body and pant or flutter their throats rapidly.

The female usually undertakes most, if not all, of the incubation, but males may also assist substantially. In harsh, arid environments both sexes may share the duty equally, changing over regularly in order to avoid undue stress on one parent. The male of some species, notably birds of prey, provides virtually all the female's food requirements during incubation and broods for short periods while she is off the nest feeding on what he has brought. In most polyandrous species, however, the female forsakes the task of incubation, leaving it to the male.

The length of the incubation period varies considerably, from as little as 11 days in the case of the Cape White-eye to about 55 days for the Bateleur (which is longer than for the Ostrich). In general it correlates to the size of the species, but there are exceptions. Victorin's Warbler, for example, incubates for a disproportionately long period of 21 days. Hole-nesters tend to incubate for longer than do species of similar size which lay in open nests. In some cuckoos partial incubation takes place in the oviduct so that their eggs are more advanced than those of the host species.

The nestling period

When the eggs hatch the brooding parent's first task is usually to take the shells and drop them away from the nest, or to eat the pieces. It has been shown experimentally that the white insides of hatched eggs attract predators, so it is important to remove them. Species which do not do so include ducks and francolins, as all their chicks hatch at the same time and vacate the nest in a group. Flamingo chicks eat pieces of eggshell as a source of calcium for their rapidly growing legs.

The chicks of many species have an egg tooth, a small sharp excrescence at the tip of the bill which they use to chip their way out of the shell. It disappears within a week of hatching, usually after a few days. Honeyguides hatch with hooks at the tips of their upper and lower mandibles and use them to kill the chicks of their foster parents. The hooks disappear once their lethal objective has been achieved.

The newly-hatched offspring fall into two main categories: altricial chicks which are naked or have a little down, and are blind, helpless and completely dependent on their parents; and precocial chicks which are covered with down, have open eyes and are able to leave the nest and feed themselves soon after hatching. The Olive Thrush produces typically altricial chicks, and the Crowned Plover precocial ones. However, the division is not always clear-cut and some species, such as birds of prey and herons, have semi-altricial chicks which are covered with down and hatch with their eyes open. Other species, such as gulls, have semi-precocial chicks which are fed in the nest until they are able to walk. African Black Oystercatcher chicks are also classified as semi-precocial as they cannot open shellfish with their undeveloped bills and need to be fed by their parents.

1 *Altricial Olive Thrush chicks beg instinctively when they feel movement or hear sound.* ***2*** *Precocial Blue Crane chicks can run and forage soon after hatching.* ***3*** *A Black Harrier chick is semi-altricial, as it is covered with down when it hatches and its eyes are open.* ***4*** *The Swift Tern's semi-precocial chick is co-ordinated but needs to be fed.*

Altricial chicks develop at a remarkable speed, and within a few days of hatching a thrush's chicks will have opened their eyes and their feather tracts will be already discernible. Precocial species also develop rapidly, and young francolins, for example, can make short flights when only ten days old.

Parental care, like incubation, may be mainly or entirely the task of the female, but even in those species where the male does not incubate he usually assists in feeding the young. As a rule most passerines share the task of feeding the chicks equally, and in birds of prey the male is the main food provider for at least the first half of the nestling period.

An altricial chick is given food either directly in its gape or by regurgitation, in which case the chick takes the regurgitated morsel from the adult's bill tip or inserts its head into the adult's throat. When the chick is still blind it is stimulated to open its gape, or 'beg', when the parent touches it or it senses the parent's movement. Once its eyes have opened the sight of the parent bringing food is enough to prompt it to open its gape. Interestingly, in some species – notably egrets and gulls – a certain colour on the parent's bill acts as a 'releaser' which prompts the chick's begging response. Once the chick is older, it can pick up food which may be regurgitated onto the nest. Birds of prey tear off small pieces of meat from a carcass and feed them one by one to their offspring. Dove and pigeon chicks are fed initially on an exudation from the wall of the parent's crop which is known as 'pigeon's milk'. Precocial chicks either find food for themselves or pick up items indicated to them by a parent.

What goes in must come out, and sanitation is an important aspect of the nestling period. In many species, especially passerines, the faeces are encapsulated in a gelatinous sac which is either carried away and dropped, or eaten at the nest by the parent. Sometimes a parent stimulates the production of a faecal sac by prodding the chick's anus with its bill. Some species, for example canaries, keep the nest clean initially but later allow faeces to collect on the nest rim. Doves and pigeons never remove the droppings, which form a rim strengthening the edge of the nest.

In some species the chicks themselves keep the nest clean. Within a few days of

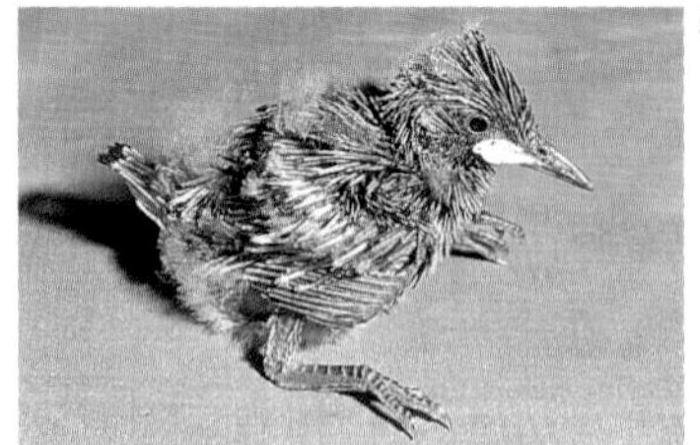

1 *An Olive Thrush deposits food directly into its chick's open gape.* ***2*** *Food is regurgitated into the bill of a young African Spoonbill.* ***3*** *Like other birds of prey, the Red-breasted Sparrowhawk tears up food for its offspring.* ***4*** *A Spike-heeled Lark keeps its nest clean by removing faeces in a gelatinous sac which it drops away from the nest.* ***5*** *In order that the feathers of the hole-nesting Hoopoe do not become fouled, they remain encased in sheaths until the chick is almost ready to leave the nest.*

hatching young birds of prey void their excreta in a strong jet over the edge of the nest, and a hornbill chick places its anus against the entrance slit and defecates accurately through it. However, other hole-nesting species merely excrete against the walls of the nest or, in the case of some kingfishers, down the tunnel so that it becomes fouled. The nests of Hoopoes and woodhoopoes are notable for the unpleasant smells that emanate from them. A feature of many hole-nesting species such as kingfishers, bee-eaters and Hoopoes is that the chicks' feathers remain encased in their sheaths until near the end of the nestling period so that they do not become fouled.

One of the most interesting features of the nestling period of some species is known as Cainism, or siblicide. Occurring mainly in pelicans, herons, egrets, birds of prey (including some owls) and the Ground Hornbill, this behaviour takes two forms: obligate siblicide in which only one chick ever survives, its sibling (sometimes there are more than one) succumbing to direct attack or starvation; and facultative siblicide where the death of siblings is neither invariable nor necessarily the result of direct and persistent attack.

Where siblicide occurs the first-hatched chick is referred to as Cain and the second-hatched as Abel. In obligate siblicide Cain, having had at least three or four days start on its sibling, relentlessly pecks Abel, and even if the female is present she does not intervene. When she feeds the chicks Cain receives all or most of the food, and Abel, gradually weakened by the vicious physical attacks and lack of food, eventually dies. The Black Eagle is a typical example of this behaviour, and there is only one positive record of two chicks having been reared together in southern Africa. At one nest the whole process was watched from start to finish and Abel was pecked 1569 times during its 72-hour life.

What makes the situation in obligate species so interesting is that Abel dies amidst an abundance of food. In facultative siblicide, on the other hand, the food supply is the decisive factor in whether or not a whole brood survives. In species where facultative siblicide occurs, for example some herons and egrets, the younger and thus weaker members of a brood may be viciously pecked by older siblings, especially at mealtimes, but their death usually results from starvation rather than from deliberate aggression as in the case of obligate siblicide.

Various explanations have been proposed for obligate siblicide. It has been argued that the second egg acts as an insurance against the failure of the first egg to hatch. This claim has been challenged – especially in the case of eagles – on the grounds that the second egg rarely serves this purpose, and eagle species that regularly lay a single egg breed just as successfully as those that lay two. However, the argument for the insurance hypothesis is that even if the second egg only occasionally serves as a back-up, its laying is justified. Another possible explanation is that by killing Abel, Cain establishes dominance and achieves a higher nestling weight, as well as enhancing its competitive abilities and eventual chance of achieving breeding status. Such interpretations are complex and involve a number of imponderables.

Anyone who studies the breeding success of birds will be struck by the high level of predation on eggs and nestlings. Snakes, monitor lizards and mammals such as baboons, monkeys and mongooses are the main predators, as well as other birds, especially species such as the Gymnogene, Gabar Goshawk, Giant Eagle Owl, Ground Hornbill and shrikes. Birds use a number of strategies in order to counteract predation, among which one of the most obvious is to avoid detection, especially when on the nest. Many species have cryptic camouflage, the most notable example being that of nightjars which remain immobile on the nest throughout the day. Many ground-nesting species such as francolins, korhaans, plovers, coursers, sandgrouse and larks have cryptic coloration, as do their chicks. In some cases, such as bishops and widows, the males are conspicuous for courtship purposes whereas the females, which are responsible for nest duties, are drab and easily overlooked.

A common strategy, especially amongst cursorial species, is to lure the predator away from the nest by distracting it. The parent feigns injury by dragging its wings along the ground or it runs off at a low crouch like a rodent, behaviour known as the 'rodent run'. White patterns on the wings often help to draw attention to birds which perform distraction displays, such as plovers and nightjars.

1

2

3

4

5

__1__ Illustrating Cainism, a Black Eagle chick attacks its weaker sibling. __2__ Both nest and well-feathered chicks of the Brubru Shrike are effectively camouflaged on a lichen-covered branch. __3__ Red-capped Lark chicks in the nest are difficult to discern in their surroundings. __4__ When a predator is too close and camouflage is no longer effective, a Fiery-necked Nightjar chick lunges with open gape. __5__ A Rufous-cheeked Nightjar performs an injury-feigning display, dragging itself along the ground and revealing its white wing patches to attract the predator's attention.

A young Spotted Eagle Owl makes itself as formidable as possible to deter a predator.

Even species which are not cursorial, such as doves and bulbuls, sometimes feign injury. Another ploy used to confuse predators is 'false brooding', in which the parent draws attention to itself and pretends to be sitting on eggs or young at a distance from the actual nest.

An alternative to distraction behaviour is threat. With the objective to appear as large and intimidating as possible, the threatening bird spreads its wings wide to reveal striking patterns, often accompanying this display with menacing noises. The Spotted Dikkop, for instance, growls and owls make snapping noises. Some hole-nesting birds emit a snake-like hiss and, in the case of the Red-throated Wryneck, even imitate the movement of a striking snake. The young of a number of species, notably coucals, some cuckoos, the Hoopoe and woodhoopoes, emit vile-smelling excreta or exudate from the preen gland when threatened.

Some species mob or directly attack a predator and strike it. Certain owls can prove particularly dangerous to observers, as the late Eric Hosking, the doyen of bird photographers, found when he lost an eye to a Tawny Owl. When a predator enters a colony of gulls or terns the concentrated mobbing behaviour of hundreds of birds can confuse and distract it to such an extent that it usually has little or no success.

Post-nestling dependence

When chicks leave the nest they are inexperienced and vulnerable to predation. In most species they need to be fed for at least a week or two, and sometimes even for months, before they are able to fend for themselves. In long-lived species, especially large birds of prey, the period of dependence may be protracted: young Crowned Eagles sometimes remain with their parents for up to a year. At the other extreme a newly-hatched Maccoa Duck is able to fend for itself and can survive without parental care. The young of some colonial species such as flamingos and Jackass Penguins form a crèche where they are looked after by certain adults, or 'nursemaids', while the rest of the flock is away feeding.

Although some larger species remain near the nest and return to it to be fed, most smaller ones move away and are difficult to keep under observation. For this reason there are few or no details on the post-nestling dependence period of many species. One thing is certain, however: even when freed from the constraints of the nest, young birds are faced with many potential threats, and the road to survival as adults is no easy one.

When a young bird leaves the nest, as this Cape White-eye has just done, it is inexperienced and extremely vulnerable to predation.

The Main Nest Types

Anyone studying birds' nests cannot help but be aware of the amazing ingenuity of birds and their ability to adjust their breeding habits to suit particular circumstances. One of the most unlikely nesting situations recorded in southern Africa was of European and Pied Starlings breeding on an offshore wreck, at least 100 metres from the coast. The energy the birds expended flying back and forth in strong winds was presumably rewarded by the security of their predator-free nest site.

Many birds have adapted to nesting on man-made structures: weavers suspend their nests from telephone wires, crows build on telegraph poles, Sociable Weavers attach their huge nests to telegraph poles and windmills, and Martial Eagles in the Karoo have adapted to nesting on power pylons. A number of other species, especially birds of prey, have also adapted to nesting on electricity pylons, and fatalities due to electrocution inevitably resulted. It is to the credit of the national supplier of electricity, Eskom, that it initiated research into the problem and has taken steps to prevent further deaths. The presence of man-made structures, including bridges, water towers and silos, across the country has enabled some species, such as swallows and swifts, to expand their ranges considerably into areas which formerly lacked nesting sites.

Birds are quick to make use of suitable artificial sites as long as they are secure, and they readily accept nest-boxes. A rusty tin lying in the veld may prove attractive to ground-nesting species, or a hollow fence post provides an alternative to a tree hole. Often birds nest near human habitation and, although they may be attracted by food and the larger trees there, they probably also derive some protection from predators.

The structure and size of nests vary considerably, as is illustrated in the summary below. Also of interest are the materials used. In some instances the use of certain plants as nesting material is their only means of seed dispersal. In southern Africa the creeper *Galium tomentosum* is particularly attractive to birds, which incorporate its seed-carrying stems into their nests and thus disperse its seeds. Similarly, the cottony seedheads of the kapok bush, or 'kapokbossie', are used to line the nest or, in the case of penduline tits, to construct it. Where available, these seedheads are selected in preference to sheep's wool which has the disadvantages that it absorbs moisture and compacts into felt, and nestlings sometimes become entangled in it.

1

2

***1** A Lesser Double-collared Sunbird's nest is made almost entirely from the creeper* Galium tomentosum. *This plant's seeds are spread only by nesting birds. **2** Nesting birds, such as the Karoo Chat, also distribute the seeds of the 'kapokbossie', as its cottony seedheads make a soft and warm lining.*

Some species, especially birds of prey, line their nests with green leaves. It was thought that this was for comfort as well as for hygienic purposes, and it has also been suggested that the lining serves to advertise occupation of a nest. Recently, however, it has been shown that the chemical properties of some green leaves reduce parasites in nests, for example in those of European Starlings. The Gabar Goshawk incorporates the nest of a social spider in its own, but it is not known for what purpose.

Most of the nests of southern African birds can be divided into 11 categories, as illustrated below. A twelfth category comprises birds which are brood parasites and build no nest at all. Some species, such as the Sociable Weaver, penduline tits and the Hamerkop, build unique nests which are described in detail in the relevant account.

Simple nests on the ground

A scrape or hollow on bare ground, lined or unlined, is used by species such as the Ostrich, the White Pelican, the Blue Crane, bustards, korhaans, oystercatchers, plovers, dikkops, coursers, pratincoles, gulls, terns, the African Skimmer, sandgrouse, some owls, and nightjars.

A sparsely lined hollow hidden in cover is the nest of some ducks, francolins, guineafowl, quails, buttonquails, some crakes and flufftails, and snipes.

1

2

***1** A Helmeted Guineafowl makes a shallow hollow in the soil. Its conspicuous eggs will be covered by the adult bird. **2** A Blacksmith Plover's nest is a sparsely lined scrape.*

Platform nests

Constructed with sticks, twigs or reeds, and with a central depression which may be lined or unlined, platform nests are placed in trees, on cliff ledges or rocky outcrops, or in reedbeds.

Ranging from the large bulky nests of storks and eagles to the flimsy nests of doves, they are constructed by the Pink-backed Pelican, cormorants, the Darter, herons, egrets, bitterns, storks, ibises, the African Spoonbill, most birds of prey, some crakes, gallinules, moorhens, pigeons, doves, and louries.

1

2

3

***1** A Namaqua Dove's nest is more durable than it appears. **2** Cattle Egrets build their platforms in reedbeds. **3** A Red-breasted Sparrowhawk on its platform of sticks.*

Nests of reeds or weeds in water

Situated in open water or among a few reeds, these nests are built by grebes, some ducks, the Red-knobbed Coot, jacanas and the Whiskered Tern. The Wattled Crane's nest is a bulky mound built up in shallow water.

__1__ A Dabchick on its floating mound of aquatic vegetation. __2__ A Whiskered Tern's nest is made of reeds.

Holes in banks or flat ground

Either solitary or in colonies, these holes are excavated by kingfishers, bee-eaters, the Ground Woodpecker, and some swallows and martins. The Grey-rumped Swallow does not excavate its own hole but nests in rodent burrows on flat ground.

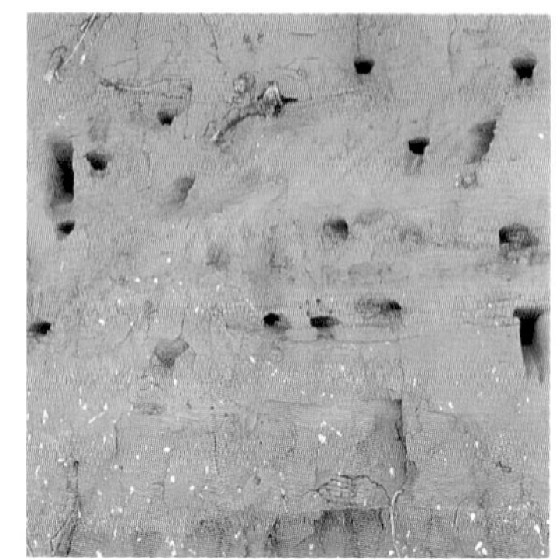

__1__ A European Bee-eater has excavated its nest in a roadside embankment. __2__ The Brown-throated Martin is a colonial species and nests in holes in banks. The tunnel to each nest may be up to 60 centimetres long.

Self-excavated tree holes

Barbets and woodpeckers are the only species which excavate their own tree holes, using their strong bills to do so. They usually carry away the wood chips and drop them some distance away to avoid attracting attention to the nest.

Their nest holes are particularly attractive to other species, which may dispossess the rightful owners, or take over disused holes.

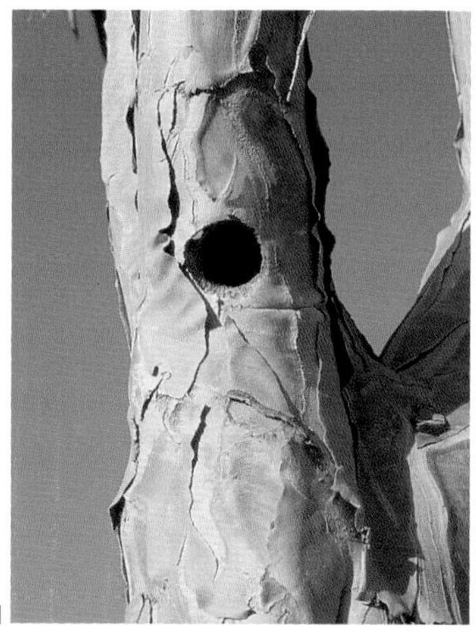

__1__ A Pied Barbet has excavated its nest in the stem of a kokerboom. __2__ A Golden-tailed Woodpecker peers out of its nest.

Tree holes

These may be either natural holes of various sizes or holes excavated by barbets or woodpeckers into which smaller species can fit. Species such as owls, kingfishers and rollers leave the cavity unlined except for wood chips or debris already there, whereas tits, flycatchers and starlings form a pad or cup of soft material inside the hole.

Hole-nesting species include some ducks, parrots, some owls, the Narina Trogon, some kingfishers, rollers, hoopoes, woodhoopoes, hornbills (which seal the entrance), the Red-throated Wryneck, tits, some flycatchers, some starlings, oxpeckers, and some sparrows.

__1__ A male Yellow-billed Hornbill perches at the entrance to his nest, a sealed hole in a dead tree. The tip of his mate's bill is just visible. __2__ A natural hole in a tree has provided a nest site for a young Hoopoe.

Nests of mud

Mud nests, made exclusively of mud pellets or of a mixture of mud and grass, are either a half-cup nest placed against a vertical surface or a hemispherical nest with a tunnel which is situated below a horizontal surface. They are made by most swallows, Rock Martins and palm thrushes.

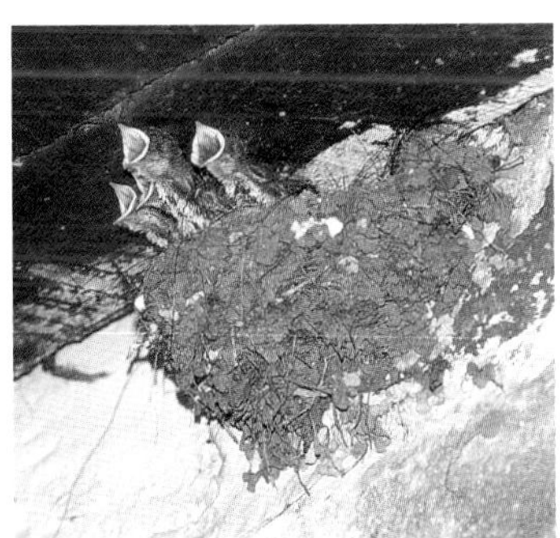

1 *The Greater Striped Swallow builds a hemispherical nest with a long tunnel.* **2** *The mud pellets in a Pearl-breasted Swallow's half-cup nest have been mixed with fibrous material.*

Cup nests

The cup-shaped nest, ranging in size from a crow's bulky structure to the tiny nest of the Karoo Eremomela, is the most common type built by birds. The nests vary considerably in depth, lining and external finish, and whereas some – like those of thrushes – are simple and functional, others – built by some flycatchers – are exquisitely decorated with lichens. They may be situated on the ground, in hollows or in trees and bushes, or they may be woven between upright reed stems.

Mousebirds, larks, cuckoo-shrikes, drongos, orioles, crows, babblers, rock-jumpers, bulbuls, thrushes, robins, chats, some warblers, flycatchers, wagtails, pipits, longclaws, shrikes, some starlings, sugarbirds, white-eyes, canaries and buntings all make cup-shaped nests.

1 *The simple cup nest of an Olive Thrush is neatly finished.* **2** *A Spike-heeled Lark's nest is set into the ground, and a rim of small sticks has been built up around it.*

Upright oval nests with a side entrance

These nests have an entrance in the side near the top or, in the case of the Fan-tailed Cisticola, at the top. They are placed low down in grass or in bushes, trees or reeds. Bishops weave them between upright stems, and the African Broadbill and some sunbirds may suspend them from branches.

Oval nests are built by the African Broadbill, some warblers, apalises, cisticolas, prinias, sunbirds, the Thick-billed Weaver, queleas, bishops, and widows.

1 *The exterior of the Spotted Prinia's elongated oval nest has a texture like that of finely woven knitting.* **2** *A Scarlet-chested Sunbird's nest is suspended from a branch, with a tail hanging below that helps to make it look like debris in the tree.*

Domed nests with a side entrance

Built on a horizontal axis and with the entrance, sometimes a spout, at one end, domed nests may be placed on the ground or in bushes or trees. They are constructed by the Buff-spotted Flufftail, coucals, the Angola Pitta, the Starred Robin, the Yellow-throated Warbler, sparrow-weavers, some sparrows, the Scaly-feathered Finch and estrildid finches.

1 *A Blue-billed Firefinch peers out of its untidy, dome-shaped nest.* **2** *The Cape Sparrow builds an untidy domed nest of grass, usually siting it in a thorn tree.*

Woven kidney-shaped nests

Weavers, with the exception of the Thick-billed Weaver, make kidney-shaped nests on a horizontal axis with the entrance beneath and at one end. The entrance may be direct into the nest, a short spout or a long tube. Although the nests of all species have a similar basic structure, most may be identified by their entrance and by their texture which results from the materials used. They are placed between upright stems in reedbeds or suspended at the ends of branches.

1

2

***1** A Cape Weaver's neatly finished nest is attached to reeds. **2** The Red-headed Weaver builds a much untidier structure, with an entrance tunnel, and often suspends it from branches.*

Brood parasites

Brood parasites are birds which build no nest of their own but lay their eggs in the nests of other species. Their young are reared by the host parents. Certain parasites specialise in a single host species, whereas others may use several different hosts. The brood parasites that occur in southern Africa are cuckoos, honeyguides, the Cuckoo Finch, whydahs and widow-finches.

A Cape Wagtail, diminutive compared with its 'offspring', feeds the Red-chested Cuckoo chick it has raised.

The Ostrich

The Ostrich is the largest living bird; an adult male may attain a height of 2.5 metres and weigh up to 150 kilograms. It is also long-lived, reaching breeding age at three years and having a life expectancy of about 40 years. The southern African race is found mainly in the arid western regions, although domesticated Ostriches, which were originally interbred with north African varieties to improve feather quality, are widely farmed in the Karoo and elsewhere.

Ostriches have a communal breeding system in which the male mates with the dominant female – the major hen – as well as with a number of subordinate females. These, known as the minor hens, lay in the major hen's nest. In some cases, however, the male has only one mate. It is a complex system which needs to be investigated further.

Although breeding may occur in all months in southern Africa, it usually peaks at the onset of the rainy season in arid areas. The males come into breeding condition, their necks and legs redden, and they chase one another aggressively as they compete for dominance and the attention of females. Courtship is spectacular: the male approaches the female with feathers raised and at an exaggerated trot, then drops into a squatting position and waves his wings from side to side, flashing the white tips like a semaphore signaller waving flags. He also twists his neck from side to side with a corkscrew motion. If the female accepts his advances she squats and mating ensues.

An Ostrich nest is a simple scrape in sandy soil measuring 2 to 3 metres in diameter. The eggs, creamy in colour and glossy, have a shell that is 2 millimetres thick and deeply pitted with pores. They vary in weight from about 780 to 1580 grams and, although one has the capacity of two dozen hens' eggs, they are not large in relation to the size of the bird, being about 1.5 per cent of the weight of the female. In comparison a kiwi's egg, at 25 per cent of the adult's weight, is proportionately the largest laid by any bird.

Clutches of 16-23 eggs have been recorded in southern Africa, but an exceptional one contained 43 eggs. The relatively small size of the eggs enables the incubating bird to cover about 20 of them comfortably; in a larger clutch it is not possible for all the eggs to be incubated. Detailed observations in Kenya have revealed that in clutches of more than about 20 eggs the major hen pushed out those of minor hens to a distance of 1 to 2 metres from the nest, where they remained and failed to hatch. The eggs she pushed out were very rarely her own; apparently she was able to recognise the ones she had laid by slight differences in their smoothness and pore structure.

Eggs are laid in the late afternoon every second day, probably because of the time required to form such thick shells. The major hen lays about 8-10 eggs and up to five minor hens may contribute to the clutch. It appears that the major hen incubates during the day, sitting from about two hours after sunrise until two hours before sunset, while the rest of the incubation, including overnight, is done by the male. Incubation begins when the major hen has completed most of the clutch; minor hens apparently do not share this duty, nor do they help to care for the chicks when they hatch.

The eggs are incubated for about 42 days. Adults in the wild are remarkably alert and wary, and at any sign of danger the incubating bird drops its head flat onto the ground so that from a distance its rounded back resembles a bush or rock. It is probably from this behaviour that the belief that Ostriches bury their heads in the sand has arisen.

The chicks are delightful creatures, their striped necks and drab coloration enhancing their camouflage. Both adults care for them and lead them away from any source of danger. In addition to mammalian predators such as jackals and hyaenas, large eagles such as the Martial Eagle prey on chicks. If they are directly threatened one of the parents performs a dramatic 'broken wing' display to lure the predator away from them. Nevertheless, the loss of chicks to predators is sometimes very high, and it is not unusual for only a third of the brood to survive. In due course some broods may merge with others to form large groups of chicks, thus finding safety in numbers.

OUT IN THE MID-DAY SUN – HOW DO OSTRICH EGGS KEEP THEIR COOL?

Why are Ostrich eggs so conspicuous? If the 'rule' applies that ground-nesting birds should lay camouflaged eggs, or hide them in cover, or incubate them from the time of laying, then the Ostrich should lay brown eggs to render them less visible. Instead, its creamy coloured eggs are easily seen, especially from the air, and they are not even incubated until the major hen has almost completed the clutch.

Light-coloured Ostrich eggs, laid on bare ground, are conspicuous to predators.

Research in Kenya has established that the major predator on Ostrich eggs is the Egyptian Vulture, which was also known to destroy eggs in South Africa before it became virtually extinct there. The vulture's technique is one the few examples of tool-using in birds: it picks up a stone in its bill and throws it onto the eggs, repeating the process until an egg is broken and the contents can be eaten. To an Egyptian Vulture flying overhead an unguarded Ostrich nest is clearly visible. When the eggs were experimentally dyed brown they survived for three times longer than undyed eggs. Why then does the Ostrich lay such conspicuous eggs?

The answer seems to be that the eggs, like the adult birds, are adapted to a hot and arid environment, bearing in mind that they may not be shaded until incubation starts. Experiments have shown that the eggs' creamy colour and glossy surface prevent them from overheating. Normal Ostrich eggs exposed to the sun had a temperature 3.5 °C lower than those artificially coloured brown. Thus the creamy colour of Ostrich eggs is basically a compromise between the higher risk of predation and the danger of overheating when they are left unattended ●

In large Ostrich clutches eggs that cannot be covered by the incubating bird are pushed to one side and will fail to hatch.

The Jackass Penguin

The Jackass Penguin, confined to southern Africa and the only breeding penguin in Africa, nests in colonies on 20 offshore islands from Bird Island in Algoa Bay westwards to the coast of Namibia. Recently two mainland colonies have been established in the south-western Cape – one near Simon's Town and the other at Betty's Bay – and on Robben Island there is now a thriving breeding colony of about 2000 pairs. This last colony comprises birds that have been released by the South African National Foundation for the Conservation of Coastal Birds (SANCCOB) after rehabilitation following oil spills.

As in most other penguin species, courtship by the Jackass Penguin mainly takes the form of 'ecstatic' and 'bowing' displays. In the 'ecstatic' display, which also acts as territorial advertisement, the penguin points its head skywards, extends its flippers stiffly sideways and then throws its head back to emit the harsh braying call from which it gets its name. Another important activity is the gathering and offering of nest material by the male. Virtually anything that can be found is offered – including seaweed, feathers, pieces of wood and bones – and in common with many other penguin species the Jackass Penguin is an innate thief. This kleptomaniac behaviour must surely have evolved because of the shortage of nest material at most penguin breeding sites. In Antarctica coloured stones placed on an Adélie Penguin's nest were soon found to have been distributed throughout the colony as a result of thieving.

Ideally the Jackass Penguin prefers to excavate a burrow in which to lay its eggs, but on rocky islands this is often not possible. Even where there may have been topsoil on the rocky substrate, more often than not it was removed with the guano which was harvested from the offshore islands over the past century or so. If they cannot burrow the penguins may lay their eggs in a shallow scrape in the open or in a hollow amongst rocks. On Robben Island they nest in thickets of alien acacias which provide both cover and protection. In some colonies, for example at Lambert's Bay, sections of concrete piping with a diameter of about 25 centimetres have been set into the ground to provide the penguins with secure nest sites.

The normal clutch is two eggs. Both sexes incubate, and the sitting bird threatens intruders by twisting its head from side to side to display the bare skin on its face which would appear intimidating at the entrance to a dark burrow. The eggs hatch after about 5½ weeks and the chicks are tended for about 11 weeks. Towards the end of the nestling period chicks may join a crèche, but each recognises its own parents.

The current Jackass Penguin population, numbering 160 000 birds, is about a quarter of what it was at the beginning of this century. Factors responsible for its decline include the harvesting of penguin eggs, an increased population of Cape fur seals which have displaced penguins from traditional breeding areas, the reduction of food supply through overfishing, human disturbance during breeding, and oil spills. Although some of these threats no longer apply, the status of southern Africa's only endemic penguin remains vulnerable.

__1__ On Malgas Island, where there is no soil in which to burrow, a Jackass Penguin places its nest on the surface among rocks. __2__ A Jackass Penguin in its burrow. If threatened, it twists its head from side to side. __3__ Jackass Penguins on Dassen Island are able to excavate their burrows in soft earth.

Grebes

All three southern African grebes build floating mound nests of decomposed material such as weeds or reeds which they collect by diving. Secured by aquatic vegetation or anchored to a submerged branch, the nests may be in open water or in the shelter of reeds or overhanging bushes along the shore. Those of the Great Crested Grebe and Dabchick are usually solitary, but the Black-necked Grebe may breed in loose colonies.

Grebes are monogamous, and their courtship displays may involve paddling chases over the water, the male in pursuit of the female. In the case of the Great Crested Grebe, the birds swim rapidly towards each other with their necks held low. As they come close, they raise their heads to full height, crests erect, and flag them from side to side, each bird performing movements identical to the other's, like a face in the mirror. Either sex may initiate the display. Both birds build the nest platform and this becomes the centre of activity where mating takes place.

The clutch is usually 2-6 eggs which are white when fresh but they are soon stained brown by the wet nest vegetation. If disturbed, both the Great Crested Grebe and the Dabchick regularly cover their eggs on leaving the nest, but in the case of the Black-necked Grebe this behaviour appears to be occasional. Detailed observations of Dabchicks on the nest have revealed the egg-covering technique. The incubating bird stands and rapidly pulls vegetation over the eggs with its bill, at the same time rotating its position on the nest so that all sides are evenly covered. The bird then flattens the heaped-up material by standing or sitting on it, or by trampling over it as it leaves. On returning to the nest the Dabchick usually pops up beside it after an underwater approach, observes for a while to see if there is any danger, then jumps up onto the nest edge. It quickly uncovers the eggs with a few deft movements of the bill and settles to incubate. Both parents share incubation, and when they change over the departing bird does not cover the eggs.

When the chicks hatch they are immediately active and can swim well, their striped necks and backs concealing them effectively in vegetation when they venture from the nest. Both parents care for the brood and sometimes divide it between them, in which case each tends the same chicks until they become independent. Soon after hatching the chicks are able to ride on an adult's back, clambering onto it via the rump and being fed there by the other parent. Alternatively, they are 'parked' on the nest, which often remains the focal point of activity for some time after they have hatched.

5

1

3

6

2

4

***1**, **2** Great Crested Grebe eggs covered and uncovered. **3**, **4** The grebe uncovers its eggs before settling down to incubate. **5** A recently-hatched Dabchick clambers onto its parent's back. **6** The Black-necked Grebe often leaves its eggs uncovered.*

Albatrosses, Petrels, Shearwaters, Prions and Skuas

Although a wide variety of seabirds visits the coastal waters of southern Africa, they do not breed here. Some come from the northern hemisphere, but most inhabit the windswept southern oceans where they are supremely well adapted for survival in their extreme environment. There they breed on the remote islands of the sub-Antarctic or in Antarctica itself. Much important and revealing research has been carried out on the biology of seabirds at weather stations on Marion Island, one of the Prince Edward group, and on remote Gough Island in the southern Atlantic near Tristan da Cunha.

Birds adapted for a life at sea are clumsy on land. With a wingspan of up to 3.5 metres, the Wandering Albatross soars with consummate grace even in the most appalling conditions, but once on land it walks awkwardly, leaning well forward to maintain its balance. It lands and takes off with difficulty and, like a large aircraft, needs a long runway. Consequently it builds its nest in highlying open areas where it breeds in loose colonies. Other albatross species breed where they can get lift from the wind, and the smaller Dark-mantled Sooty Albatross nests on cliff edges where there is a constant updraft. Most albatrosses build a characteristic raised mound so that the nest resembles a miniature volcano. The areas where they breed are often boggy, or may be covered in snow, and the raised nest keeps the incubating adult dry, and in due course the chick.

The giant petrels nest in similar fashion to albatrosses, but do not make such conical nests. Most of the smaller petrels, as well as the shearwaters and prions, nest in burrows which they excavate. They often breed in huge colonies, and the night-time cacophony produced by tens of thousands of breeding petrels has to be heard to be believed. The birds visit their nests under the cover of darkness to avoid being preyed on by skuas, and it has been found that some petrels have the ability to locate their nest burrows using their sense of smell. Of the petrels which do not burrow, the Pintado Petrel nests on open cliff ledges, and the small storm petrels place their nests in crevices in rocks.

The Sub-Antarctic Skua makes a nest like a gull's (see page 91) but breeds solitarily, laying usually two or three camouflaged eggs. Both parents incubate and care for the chicks, and they are notoriously bold in defending them, dive-bombing and

2

3

4

5

__1__ Windswept Marion Island is an ideal nesting habitat for the Wandering Albatross. __2__ Dark-mantled Sooty Albatrosses breed on cliff edges where there are constant air currents. __3__ Green 'fertility patches', seen among a colony of Southern Giant Petrels, indicate the importance of seabirds in bringing nutrients to the islands. __4__ A Black-browed Albatross chick on its 'volcano' nest which keeps it above the surrounding mire. __5__ Yellow-nosed Albatrosses on their nest on Gough Island.

1

even striking humans if they come too close. Skuas are strongly territorial and it has been found that certain dominant pairs that nest closest to a food source (such as a penguin colony) are the most successful breeders.

With the exception of the skuas, the normal clutch of the species discussed above is a single egg which is white and often large relative to the size of the adult. The incubation period, even for the smaller species, is long, for reasons as yet unexplained. The nestling period is also protracted: two months in the case of the small storm petrels, and three to five months for medium-sized petrels. The young of the larger albatrosses take even longer to leave the nest. The nestling Wandering Albatross is a year old by the time it flies, so this species can breed only every second year. The long nestling period and the raising of a single chick are adaptations to a food supply that may often be erratic.

Oceanic birds' low breeding productivity is compensated by their long lifespan. The Wandering Albatross, for example, regularly attains an age of 30 years, often much more, but it is quite old – between five and ten years – by the time it breeds. Adolescent birds return to the islands where they were born and begin the breeding rituals. Several males vie for an unmated female's attention with impressive displays: they raise their huge wings and spread them forwards, extend their necks, clatter their bills and emit weird groaning noises. The female eventually chooses one of her importunate suitors and he leads her to his selected nest site. It is not surprising that, having gone to so much effort to establish a pair bond, Wandering Albatrosses pair for life.

Populations of oceanic birds that are undisturbed face no threat from their low breeding productivity, but it can have serious implications when the normal routine is upset. A classic example is the catastrophic predation on Marion Island's burrow-nesting seabirds by feral cats which have only recently been eradicated. One wonders also how much of an effect sealers and early travellers had on some Wandering Albatross colonies when they collected the birds' eggs, ate the chicks and used the wing bones of adults for pipe stems.

1 *A Soft-plumaged Petrel emerges from its burrow at night to minimise predation by skuas.* ***2*** *The Northern Giant Petrel, a solitary breeder, makes a nest similar to that of an albatross but less conical.* ***3*** *Pintado Petrels do not burrow, but nest on open cliff ledges such as those on Deception Island in Antarctica.* ***4*** *A Sub-Antarctic Skua on its nest in Antarctica, within easy reach of prey in the form of Adélie Penguin eggs and chicks.*

Pelicans

Both pelican species which occur in southern Africa, as well as elsewhere on the continent, breed colonially but each has distinct nesting habits. The White Pelican makes a rudimentary nest on the ground, often a mere scrape, while the Pink-backed Pelican constructs a substantial platform in a tree.

For White Pelicans the most important nesting requirement is a secure and inaccessible site where there are no mammalian predators. Consequently they breed on islands in lakes or pans, on an offshore island and on artificial platforms in the sea at Walvis Bay. The South African breeding population is limited to one colony of about 400 pairs on Dassen Island off the Cape west coast and another numbering approximately 2000 pairs at Lake St Lucia. In Namibia there are breeding sites at Walvis Bay, Hardap Dam and Etosha Pan, while in Botswana White Pelicans breed at Lake Ngami and Makgadikgadi Pan when conditions are suitable. Breeding appears to be linked to food supply and if this fails the attempt may be aborted. The pelicans commute over considerable distances from the colony to their feeding grounds, and a round trip of 200 kilometres daily is not unusual.

White Pelicans in breeding condition are recognisable by a knob-like swelling on the forehead, brighter facial skin (pinkish yellow in the male, bright orange in the female), a small crest of feathers on the crown and a rosy tinge to the plumage. Although pre-breeding group displays take place, there are no distinct individual courtship rituals, except for a strutting walk during which the male follows the female on a winding course before they return to the main group.

The Pink-backed Pelican breeds colonially in trees at Lake St Lucia and in the Okavango Delta in Botswana. In contrast to the White Pelican, this species advertises courtship by a striking bill-clapping display during which it throws its head back until the tip of its bill is upside down over its tail. The bill is held open to reveal the red interior and is then clapped once or several times. New nests are constructed with dead or leafy branches broken off from surrounding trees, or old ones from the previous season may be repaired.

1

Both White and Pink-backed Pelicans usually hatch out two young but only one survives because the smaller chick is attacked and killed by its older sibling. The chick is fed on food regurgitated into the pouch, a rather vigorous and uncomfortable process when a larger chick inserts its head right into the adult's gullet. Older chicks in White Pelican colonies form groups or 'pods', but the parents can distinguish their own chick and the chick its parents. From the age of 70 days Pink-backed Pelican chicks make practice flights to nearby branches or neighbouring trees, then gradually extend their range. They continue to be fed by their parents, even if they have travelled a considerable distance from the nest site.

Both pelican species are wary and sensitive to disturbance while breeding and for this reason, as well as because both are classified as rare in South Africa, it is essential that existing breeding sites are rigorously protected.

2

3

***1** Pink-backed Pelicans build sturdy platform nests in trees. **2** An unusually substantial White Pelican's nest. Often the nest is no more than a scrape lined with a little material. **3** A pod of young White Pelicans on Dassen Island.*

Gannets

The only breeding gannet in southern Africa is the Cape Gannet which nests on six offshore islands, all off the west coast except for a colony on Bird Island in Algoa Bay. We know that these colonies have existed for a long time, for there is a record of adult birds and eggs having been collected from Malgas Island in Saldanha Bay in 1648. When the sailing vessel *Dodington* was wrecked off Bird Island in 1755 some of the survivors supplemented their diet with gannets and their eggs.

Gannets breed in large colonies: on Bird Island there are 40 000 pairs, and on the west coast 25 000 pairs occur on Malgas Island and 7000 pairs at Lambert's Bay. For conservation reasons access to the islands is strictly controlled. The best place to observe gannets is at Lambert's Bay where the breeding island is connected to the mainland by a breakwater. An observation tower overlooks the colony, enabling visitors to watch and photograph the breeding activity – which extends from August until early March – without disturbing the birds.

Unlike the Northern Gannet which nests on the steep sides of islands and cliffs, the Cape Gannet breeds on flat islands. Another difference between the two species is that the Cape Gannet does not gather nest material from a distance. Once a nest site has been secured the birds merely scrape guano and other miscellaneous material into a mound which has a hollow on top for the egg. The size of the nest depends on the amount of guano available, and some birds may have to lay on the bare ground. Nests are close together and in three sample areas at Lambert's Bay were no more than 58, 53 and 39 centimetres apart. There may be as many as seven nests in a square metre, the greatest breeding density known for any species of gannet.

A gannet colony gives the impression of constant frenetic activity, but it is one of the best places to observe a whole range of behaviour patterns which maintain order amidst apparent chaos and ensure that injuries to adults and chicks seldom occur. It is, for example, quite remarkable how a parent returning from a foraging expedition is able to pinpoint its own nest amid the turmoil surrounding tens of thousands of others. One of the most frequently seen behaviour patterns is 'sky pointing', which indicates the gannet's intention to take off as it moves through the mass of birds to the edge of the colony. The bird points its bill skywards and spreads its wings, the amount of wing spread and vertical pointing of the head varying from low to high intensity, depending on the circumstances. The displaying gannet also moves slowly so as not to upset its neighbours, but sometimes it becomes impatient and makes a dash for the edge of the colony. Once it commits itself to running it has blown its cover, so to speak, and is viciously pecked by all the birds it passes.

A rich repertoire of courtship behaviour patterns also establishes the pair bond and ensures that strife in the tightly packed colony is kept to a minimum. An important display is 'mutual greeting' during which the male and female face each other breast to breast with their wings partially spread sideways. Then they stretch their necks upwards and shake their heads a few times, sometimes clashing bills with a knife-sharpening action, before dipping their heads over each other's backs with a bowing movement, and at the same time emitting harsh, grating calls. Once a bond is formed the pair may stay together for life. Having established a nest site, one of the pair stays on it while the other is away feeding. Sometimes these single birds perform a 'solo bow' display while waiting for the mate to return. When the female returns to the nest site the male may grasp her behind the head with his bill in a vigorous 'kiss' which often precedes copulation. The birds also preen each other, and in so doing strengthen the pair bond.

A single egg (very occasionally two) is laid and the incubating adult encases it with its webbed feet before settling into a brooding position. After a six-week incubation period the chick emerges and is then brooded on top of its parent's feet. Initially it is naked with black skin, but later develops a thick coat of white down. Once the feathers emerge the chick is sooty brown with white spots on the wing coverts. Fed by regurgitation on a rich diet of fish, it grows rapidly and leaves the colony to fend for itself once it is 14 to 15 weeks old. Adult plumage is assumed after two years and the birds first breed when three to four years old.

No account of the breeding of the Cape Gannet is complete without mention of the collecting of guano. This valuable source of fertiliser has been harvested off the southern African coast since the early 1840s. Today concessions are still granted but collection ceased on Malgas Island in 1985 when it was incorporated into the West Coast National Park. At Lambert's Bay, only 100 kilometres away, guano is still harvested and there is a striking difference in the sizes of the nests in the two areas. Nests on Malgas are significantly larger and, with a plentiful supply of guano available, the birds are easily able to rebuild the previous year's nests. The larger nests enable them to breed about 40 days earlier than birds at Lambert's Bay because they are not flooded by late winter rains. Also, if eggs are lost, Malgas birds can lay again at about the same time that the gannets at Lambert's Bay are breeding for the first time. The gannets on Malgas have the advantage of completing their breeding cycle when the food supply is at its best.

1

2

3

1 *A Cape Gannet chick is brooded on its parent's feet.* ***2*** *A 'sky-pointing' Cape Gannet works its way towards the edge of the colony before it can take off.*
3 *Cape Gannets perform the 'mutual greeting' display.*

Cormorants and the Darter

There are five cormorant species in southern Africa, of which the Cape, Bank and Crowned Cormorants are marine species, the Reed Cormorant is found on inland waters, and the White-breasted Cormorant occurs both inland and in a marine environment. The sixth species of this group, the Darter, is found only inland. They all breed colonially, the inland species sometimes in mixed 'heronries' in which, for example, Reed Cormorants, Darters, egrets, herons and ibises congregate in one reedbed. These species construct substantial basket-shaped structures of sticks or reeds in reedbeds, trees or on rocks, while the marine cormorants build mounds of seaweed and other miscellaneous material on flat ground or on rocks.

A number of breeding characteristics are common to the group, notably a brighter breeding plumage which usually includes a glossy sheen to the feathers. In some species there is spotting on the wing coverts, and the Darter acquires long plumes on its upperparts. The eye colour may change and the skin on the throat of the Cape Cormorant becomes deep orange. Courtship may involve the offering of nest material, as well as swaying movements of the head and neck which display bright colouring, such as that of the Cape Cormorant's gular patch.

The breeding season is very variable and may occur in all months in some species. Along the Namibian coast the Cape Cormorant is the most numerous bird and breeds in colonies numbering many thousands, including those on artificial platforms at Walvis Bay. Its breeding season there is triggered by the south-westerly

1 *In a Cape Cormorant colony on Dassen Island the nests are regularly spaced.* ***2*** *A Bank Cormorant's substantial nest is made entirely of seaweed and strengthened with excreta.* ***3*** *The marine Crowned Cormorant breeds mainly on offshore islands.*

coast and on offshore islands, sometimes on man-made structures and on shipwrecks. The Cape Cormorant is a prolific breeding species that congregates in huge colonies on offshore islands as well on the platforms at Walvis Bay in Namibia, where White Pelicans have been seen preying on the chicks. At Cape Point in the Cape of Good Hope Nature Reserve there is a unique colony that breeds on sheer cliffs. This cormorant's mound-shaped nest incorporates all manner of material, such as pieces of plastic, fishing net and even lengths of nylon fishing line, a real danger to both adults and chicks.

The Crowned and Reed Cormorants are very similar in appearance and were once considered to be a single species. However, the former is entirely marine and has a restricted range along the western and south-western coasts, breeding on coastal cliffs or offshore islands. The Reed Cormorant, on the other hand, nests in trees or reedbeds in its inland environment, often alongside Darters which occupy a similar niche. An interesting feature of the Crowned Cormorant's breeding biology is its strategy to reduce the size of its broods by laying eggs of different sizes, staggering the hatching times and preferentially feeding the strongest chicks. Thus in essence, the third and subsequent eggs in a clutch are smaller and hatch later than the first two eggs, and the stronger chicks are fed first by the parents so that their weaker siblings survive only if there is a plentiful food supply.

The endemic Bank Cormorant has a particularly interesting breeding biology. It occurs only along the west coast of southern Africa in association with large kelp beds where it feeds; of the total breeding population 70 per cent is found off the Namibian coast on Mercury and Ichaboe islands. When they are in breeding condition the birds develop a white patch on the rump which later disappears. They build a nest almost entirely of kelp, which the male collects by diving and then delivers to the female so that she can proceed with the construction. The whole process may take five weeks, and it has been calculated that during this time the male dives on average 238 times, for a total duration of 18 hours. The amount of seaweed he collects is substantial, and some of the heaviest nests constructed weigh up to 6 kilograms. To help bind the material and form a solid construction the birds deliberately defecate on the nest. Why does the Bank Cormorant invest so much effort in building its nest? The answer seems to be that often the only suitable sites near kelp beds are on steeply sloping rocks near the sea. The solidity of its construction and the strengthening effect of the guano ensure that the nest is able to withstand the rough surf.

winds which blow in September, causing the upwelling that brings nutrients to the surface for the fish shoals. However, if the food supply fails the whole breeding attempt is aborted and the chicks are left to die in their nests.

In all species the male collects the nest material and the female builds. Unguarded nests are the target of thieving by neighbouring birds, which steal whatever material they can. Normally cormorants lay 2-4 eggs, rarely six or seven, and the Bank Cormorant never lays more than three. The eggs, pale blue when fresh, turn whitish as incubation progresses. The chicks hatch naked and blind, but later acquire a covering of down. Both parents share incubation and both feed the chicks by regurgitation.

The nests and nest sites of several species merit individual mention. The White-breasted Cormorant breeds in trees or on rocks in lakes or dams, as well as along the

1 *White-breasted and Reed Cormorants often nest together in a mixed colony.* ***2*** *The Darter is an inland species and, like other members of this group, breeds colonially.* ***3*** *Reed Cormorants nest in reedbeds as well as in trees.*

Herons, Egrets and Bitterns

The 18 species in this group range in size from the towering Goliath Heron to the diminutive Dwarf and Little Bitterns, but the nests of all of them are basically platforms of sticks or reeds. Although usually located in trees or reedbeds, the nests are sometimes built on rocks or on the ground on islands, and in one unusual case Little Egrets reared chicks from nests on the beach in the midst of a colony of Kelp Gulls.

For most species in this group the male collects material – even stealing sticks whenever possible from unguarded nests – and brings it to the female, who builds the nest. The birds add material to the construction during incubation and, to a lesser extent, during the early part of the nestling period. The nests of the large herons end up being substantial structures, and even the apparently flimsy stick platform of a Cattle Egret may contain as many as 400 sticks, although the average is about 300. Despite the exposed position of some Cattle Egret colonies in tall trees, the nests are remarkably durable and are able to withstand strong winds.

Most species breed colonially in mixed 'heronries' which may also include cormorants, Darters, African Spoonbills, ibises and storks. No two heronries are alike, their numbers and combinations of species varying considerably, but as a rule the different species tend to associate in groups of their own kind within the colony. There is usually a high degree of breeding synchrony throughout the heronry and this has the advantage of 'flooding the market' for predators. While some losses may occur, often on the periphery of the colony, most of the progeny survive. There is also some evidence that smaller species may derive a protective advantage by breeding in proximity to larger and more aggressive species. Some members of this group, such as Goliath, Purple and Green-backed Herons, may breed either singly or in association with their own or other species, but others, such as the White-backed Night Heron, Little Bittern and Bittern, are solitary nesters.

The courtship displays of the herons and their kin are varied and elaborate, and many include both aerial and non-aerial elements and often involve harsh croaking calls. Most species, notably the egrets, acquire special breeding plumes which are used in displays, including threat. At the onset of breeding the eyes, lores, bills and legs of many species change to a brighter colour, but once nesting commences these colours begin to fade. In the Great White Egret the colour changes are particularly interesting: at the beginning of the breeding season the eyes become bright red, the lores emerald green and the bill shiny black. After about three days of courtship the eyes change back to yellow, and during the 26-day incubation period the lores fade to olive green. On the day the first chick hatches the shiny black bill begins to revert to yellow, initially at the base and then gradually towards the tip so that by the end of the breeding cycle it is almost entirely yellow again. Experiments with Cattle Egret chicks have shown that a yellow bill elicits from them the most vigorous begging response. It may be assumed that the colour change in the Great White Egret's bill back to yellow precisely when the chicks hatch is timed to provoke a similar response. In the Little Bittern a change in the colour of the bill is 'emotional' rather than chick-related, as the base of the bill flushes red in both male and female during courtship and copulation, and when the pair change over at the nest.

The clutch usually comprises three or four eggs, but may range from 2 to 6. Most species lay plain bluish green eggs, except the White-backed Night Heron and Dwarf Bittern whose eggs are very pale bluish white, and the Little Bittern which lays white eggs. Incubation usually starts when the first egg has been laid, and in the case of species such as the Purple Heron, which lays eggs two or three days apart, there is considerable variation in the size of the chicks in a brood. If there is a food shortage the smaller chicks may not survive, as they cannot compete with the vigorous begging responses of their older and stronger siblings. However, a study of the Great White Egret revealed that its chicks were extremely tolerant of each other, and there was a high breeding success.

Chicks are fed by regurgitation by both parents and, although initially helpless, they develop quickly. Their legs in particular grow rapidly, appearing disproportionately large compared with the rest of the body. If they fall out of the nest onto surrounding branches or reeds they are able to clamber back using their strong legs and their bills. Indeed, the chicks regularly leave the nest before they can fly, clambering out onto nearby branches or reeds.

1

2

3

4

1 *A solitary Purple Heron at its large platform of reeds.* ***2*** *The White-backed Night Heron, a reclusive and solitary species, hides its nest among branches on a quiet backwater.* ***3*** *The neatly constructed nest of a Green-backed Heron.* ***4*** *The bluish green eggs of the Rufous-bellied Heron are typical of most of this group.*

2

1 *A male Little Bittern at its nest among bulrushes, a favourite breeding site.* **2** *Always well-concealed among reeds, nests of the elusive Bittern are seldom found.* **3** *Grey Herons build substantial nests in reedbeds (as shown) or in tree-tops.* **4** *Although incubation has already begun, this Cattle Egret pair still add twigs to their nest.*

3

1 4

Several species merit special mention. The breeding biology of the Slaty Egret, for example, has long been a mystery, and only recently were nests of this species found in the Okavango Delta. It breeds beneath the canopy in thick stands of water figs, often in association with Rufous-bellied Herons. The nests, eggs and chicks of both species are so similar that great care is needed to ensure that a nest has been correctly identified.

The only detailed study of the Little Bittern made in southern Africa has been carried out at Rondevlei Nature Reserve in the south-western Cape. Here this secretive species nests exclusively in bulrushes near channels. The bulrushes are invasive and tend to spread rapidly, but if they were to be eradicated at Rondevlei the Little Bitterns would lose their nesting sites. However, control of the bulrushes is proving difficult, and it seems likely that the association between them and the Little Bittern will continue.

Perhaps the most elusive of all the species in this group is the Bittern. It is more often heard than seen and in the breeding season its frequently repeated booming call carries over several kilometres. The nest is well hidden in reeds or sedges and very few have ever been found in southern Africa. If disturbed at the nest, the adult freezes in a typical 'bittern' posture, its neck extended vertically so that it resembles the surrounding reeds and is almost impossible to detect. Some herons and their young also adopt this posture if disturbed, but for them the disguise is not nearly as effective as it is for the Bittern.

THE OPPORTUNISTIC BITTERN

The Dwarf Bittern, sometimes known as the Rail Heron, is interesting in that it appears almost miraculously when pans and marshy areas are seasonally inundated and, taking full advantage of optimal conditions, immediately starts to nest. An intra-African migrant, it comes from as far afield as equatorial Africa to breed opportunistically in the southern part of the continent. We have yet to discover how it locates suitable breeding areas and what happens in years when the summer rains fail.

Dwarf Bitterns are crepuscular feeders and exploit the abundant supply of frogs in inundated areas. Their nests may be spread over a large area, but are sometimes only 19 metres apart. Sometimes Green-backed Herons nest in the same areas and in the past confusion between the two species has undoubtedly resulted in erroneous breeding records. The bitterns compress their nesting cycle into the shortest possible time while conditions remain suitable. Clutches are completed quickly, the incubation period is short, and the chicks leave the nest early and gather in groups in trees where they are fed by their parents ●

The Dwarf Bittern breeds quickly in its ephemeral habitat, and its chicks are able to clamber off the nest for short periods at only seven days old.

The Hamerkop

The Hamerkop is an extraordinary bird, a taxonomic enigma that has been variously linked to herons, the Shoebill, storks and shorebirds. On anatomical and behavioural characteristics it shows little affinity to any of these, although recent findings show that it may be more closely related to shorebirds than previously thought. Whatever its true relationship may be, there is no doubt that its huge domed nest is unique.

The preferred site for the nest is the main fork of a tree, but in arid areas nests are often placed on a cliff ledge or on a prominent rock in a watercourse. Nests have also been recorded on sandbanks and in one instance on the roof of an occupied farmhouse. Frequently they are situated over water, but this is not necessarily the case; the important factor seems to be that the entrance should be inaccessible to potential predators. No particular orientation of the entrance has been determined, although nests in trees are built so that it faces in the direction towards which the tree's fork inclines. The height of the nests above the ground varies considerably: in west Africa 80 nests were between 3.6 and 14.6 metres above the ground, the average height being 9 metres.

The nest construction is complex. Initially miscellaneous materials such as sticks, reeds and grass are piled into an inverted pyramid with a hollow on top. From this base the walls are built up with sticks laid horizontally so that a deep basket is formed. Then the roofing sticks are added, some placed vertically leaning inwards while others are interlaced horizontally. Once a solid network of sticks is in place it is covered with an incredible assortment of materials. The completed roof may be as much as 1 metre thick and easily supports the weight of a heavy man. The front of the nest, where the opening is to be, is completed last. The V-shaped opening that has been left is made smaller with sticks until it has a diameter of 13 to 15 centimetres and a tunnel to the main chamber is formed. This relatively long tunnel, sloping upward from the entrance, is probably the reason for the belief among indigenous African people that the nest contains three rooms. Both the tunnel and the main nest compartment are then plastered with mud, the interior roughly but the tunnel with a smooth coating. Thus not only is the finished structure enormously strong, but a stable microclimate is created inside it. When an observer made an inspection hatch at the back of a nest he found that the birds repeatedly plastered it over with mud from the inside.

The completed nest may weigh 25 to 50 kilograms, but much heavier estimates are often given. Although its size varies depending on the amount of material on the roof, the external height is usually 1.5 to 2 metres and the depth 1.6 metres from the front entrance to the back of the nest. The spacious inside chamber is about 80 centimetres in diameter and 30 to 40 centimetres high, easily accommodating both adults and chicks. The eggs are laid in a shallow unlined depression at the bottom of it.

A considerable amount of time and energy is invested in building the nest, and it has been estimated that 8000 pieces of material are used in its construction. Both male and female build, and although the sexes are not usually identifiable, it seems that they take an almost equal share. Both find material and add it to the nest, apparently working independently except on the roof, where one bird working from

1

2

3

1 *The Hamerkop often uses large and unwieldy sticks in the construction of its nest, and its sturdy bill is a useful tool for carrying them.* ***2*** *Hamerkop nests sometimes provide shelter for creatures other than those which built them, and in this case a half-complete nest has been taken over by an African rock python.* ***3*** *In areas where there are no suitable trees, nests are often built on a cliff ledge, sometimes over water.*

WHAT CONSTITUTES A HAMERKOP'S NEST?

An account of the contents of a Hamerkop's nest was published in *Honeyguide*, the journal of the Ornithological Association of Zimbabwe, in March 1985. The nest had fallen to the ground on the property of a Mr King, who noted the following materials incorporated in it: a pan brush, a broken cassette tape, a glove, a plastic dish, a plastic cup, two peacock feathers, chicken feathers, two socks, rabbit fur, 45 rags, four mealie cobs, one piece of glass, four bits of wire, a plastic comb, one pair of underpants (male), a typewriter ribbon, a piece of leather belt, four bits of stocking, two bits of tin, two bits of foam rubber, seven bits of hose pipe, nine bits of plastic pipe (electrical), six bits of asbestos (roofing), 11 miscellaneous bones (T-bone etc.), 12 pieces of sandpaper, four lengths of insulation tape, ten plastic bags, nine pieces of paper, 56 scraps of tinfoil, six bicycle tyres, six lengths of insulating wire and approximately 100 kilograms of twigs, sticks and grass ●

The miscellaneous material added to the roof of this Hamerkop's nest is nearly a metre thick.

the inside is helped by its mate on the outside. Some of the sticks used are very large, up to 1.5 metres long, and one was found to weigh 232 grams, half the weight of the bird itself. It has been suggested that the relatively sturdy bill is adapted not only to feeding but also to lifting heavy sticks.

The main periods of building activity are for about two hours in the early morning and again for about two hours in the late afternoon. When fetching material both birds make a combined total of 50 to 60 trips per hour, and will frequently visit a good source of material, such as the sticks and other debris piled up by flood waters. The first inverted pyramid stage is completed in less than a week and the roof, started some two to three weeks later, is complete after about four weeks have elapsed. It takes about six weeks for the whole structure to be finished. However, in west Africa nests may be completed in as little as 20 to 25 days, and three or four nests may be constructed over a period of 12 months. It would seem that nest building is an important factor in maintaining the pair bond, and it is not unusual to find several nests of a single pair in a relatively small area. In fact, it is not inconceivable that nests may be a form of territorial advertisement, albeit a very time-consuming and energetically expensive one.

The inaccessibility and security of a Hamerkop's nest is more apparent than real, and it may be usurped by a variety of creatures. Perhaps the construction of several nests may be a form of insurance against losing one or more to an intruder. One of the main culprits is the Barn Owl which frequently takes over both unused and occupied nests, apparently dispossessing the Hamerkops with little difficulty. Giant Eagle Owls may take over a nest at the basket stage or lay on the roof of a completed nest. In one instance the owls were breeding on top of a nest while an Egyptian Goose incubated her eggs inside it. Knob-billed Ducks, Pygmy Geese, Grey Kestrels and Rock Pigeons have also been known to utilise Hamerkop nests for breeding. Some other inmates recorded are grey squirrels, genets, monitor lizards, snakes and, not infrequently, bees. A Hamerkop's nest should therefore always be inspected with caution!

Hamerkops use their nests for roosting, but not necessarily on a regular basis. They emerge in the morning like a cork out of a popgun and return in the late evening with an 'upward dive', folding their wings and entering like a dart. The smoothly plastered tunnel eliminates any chance of snagging as they enter. After fledging the young may return to roost in the nest for about two to four weeks before they move away.

False mounting forms part of the Hamerkop's courtship ritual.

Why does the Hamerkop build such a large nest? Although the species' breeding biology has been well studied, especially in west Africa, a really convincing answer has yet to be advanced. The reason for the bulky, enclosed structure probably embraces a number of facets, but the stable microclimate within it must surely confer considerable advantages. Although the Hamerkop's nest remains something of a mystery, there can be no doubt that it is one of the most remarkable structures to be built by any bird.

Nor is it only the Hamerkop's nest that is remarkable; the bird's courtship and social behaviour is also bizarre. The male and female will, for example, make rapid flights towards each other; bow or nod to each other; run towards one another with drooped wings and crest rising and falling; and run in circles side by side making repeated false mountings (the male standing on the female's back and vice versa). These displays are accompanied by loud, trilling *yip-purr* calls that sound maniacal. Actual copulation may occur after a succession of false mountings and usually takes place on or near the nest.

Social interactions, in which several birds participate, frequently occur and usually involve false mountings, when some birds stand on the backs of others and then change around at random. Sometimes mounted birds may even face in the wrong direction. Remarkable to watch, these gatherings are accompanied by loud, excited calling. Despite this form of social interaction, Hamerkops are considered to be monogamous (although no studies have actually involved marked birds), and there is only one record of helpers bringing sticks to a nest.

In southern Africa the breeding season is extended, lasting from July to January in South Africa, for example. In Zimbabwe breeding has been recorded in all months, but accurate laying dates can seldom be established since it is difficult to inspect nests. Clutches of 3-7 eggs appear to be standard in southern Africa, although larger ones are occasionally recorded. The eggs, laid at intervals of 24 to 72 hours, are chalky white to begin with but soon become heavily stained with mud.

Incubation appears to start with the laying of the first egg because the chicks hatch at different times. The male and female share incubation duties over a period of 28 to 32 days, but the latter does the major share during daylight. The eggs may be left unattended for quite long periods, and although they sometimes become completely cold they still hatch successfully. Both parents tend the chicks, bringing food mainly in the morning and evening. Although their visits may be infrequent, the growth rate of the nestlings indicates that they receive sufficient nourishment. Initially the chicks are covered in grey down, and their crests appear as early as the sixth day. During the nestling period, which lasts between 44 and 50 (usually 47) days, they defecate against the walls of the nest chamber which soon becomes whitewashed with excreta. After they have left the nest the young may return to roost for two to four weeks.

Storks

Eight species of storks occur in southern Africa and, with the exception of Abdim's Stork which is an intra-African migrant, all of them breed here. The Shoebill, often considered taxonomically close to storks, has yet to be authentically recorded in southern Africa, despite its inclusion in several standard reference books for the region. Of the seven breeding species, four (the White, Black, Woolly-necked and Saddlebill Storks) nest solitarily, and three (the Openbill, Marabou and Yellow-billed Storks) breed colonially. All build stick platforms which are lined with soft material such as grass, weeds or leaves. These nests range in size from the Saddlebill's enormous platform, which may be 2 metres across, to that of the Openbill which has a diameter of 45 to 75 centimetres.

In the breeding season most storks assume brighter plumage colours and the bill, legs and soft parts also brighten in some species. It is significant that the Saddlebill, which maintains a pair bond throughout the year, shows no colour changes whereas species such as the Yellow-billed Stork and Marabou, which do not remain paired, assume impressive breeding colours. The latter, distinguished by some as 'the world's ugliest bird', redeems itself somewhat during the breeding season by growing lovely white undertail coverts which are known as 'marabou down' and were once used in the millinery trade. Its gular air-sac, which it inflates, becomes bright pink and the bare skin at the base of the neck reddens.

Courtship displays among storks are complex, but the one common to most species is the 'up-down' display which is also used in greeting throughout the breeding cycle. The bill is raised to a vertical position and then dropped down towards the feet with a sideways swaying movement. The display is usually accompanied by bill clattering as well as by vocalisations which in the case of the Marabou comprise weird squealing, grunting and mooing sounds. The White Stork's version of the 'up-down' display is the best known and the most extreme: it throws its head back until the bill touches the tail, accompanying the movement with bill clattering. When not breeding storks are generally silent, except for occasional bill clattering during aggressive encounters over food.

The Saddlebill performs little in the way of display, probably because it forms a long-term pair bond. The only courtship behaviour observed is the 'flap-dash', which involves one bird, usually the male, flapping its wings as it runs away from and then back to its mate. Even in the breeding season this species is silent and no bill clattering behaviour has been noted.

When the breeding season begins the male stork returns to the nesting area or to an established nest. In species such as the White and Black Storks the female also returns and is soon accepted if she is recognised as the mate of the previous season.

1

2

3

4

__1__ The Yellow-billed Stork's plumage takes on a delicate pink hue in the breeding season. __2__ A wagon wheel placed on top of a pole provides a pair of White Storks with a platform for their bulky nest. __3__ The clutch of one of the few pairs of White Storks that breed near Bredasdorp. __4__ Woolly-necked Stork chicks lie flat in their platform nest to avoid detection.

Black Storks nest on cliff ledges as well as on ledges in quarries and abandoned mine workings.

The male Yellow-billed Stork perches in a chosen spot in the colony and advertises his presence with bouts of 'display preening' during which he performs ritualised movements without actually preening. He drives off all other birds of the same species, male and female, when they approach his chosen spot, but a reproductively 'ripe' female will persist, holding her bill open and spreading her wings in a 'balancing' display. She is eventually accepted and the birds then begin nest building. Mating usually takes place on the nest and the male clatters his bill against the female's during copulation, a characteristic feature of stork behaviour.

In storks both sexes build the nest and when it is complete the female lays a clutch of 3-5 white eggs. She and the male share the incubation and care of the chicks, which are guarded and shaded during the first few weeks of the nestling period until they are sufficiently developed to be left alone. Food is regurgitated onto the nest and the young pick it up, unlike ibis and spoonbill chicks which insert their bills into the crop of the adult.

The White Stork has become part of European folklore and is probably one of the world's best known birds because of its habit of nesting on buildings. Some nests in Europe have been used for hundreds of years; one that was in use in 1549 was still occupied in 1930. Some exceptionally large nests may measure 2.25 metres across, and one removed from a cathedral weighed 800 kilograms. Sadly, the White Stork has disappeared as a breeding species from large parts of its former range, especially in western Europe. Although a number of factors are involved, it is clear that loss of habitat is one of the main reasons for its decline.

For many years the White Stork was considered to be a Palaearctic migrant to Africa, but this view was modified in 1940 when the renowned naturalist Austin Roberts discovered a nest containing three chicks on a farm between Calitzdorp and Oudtshoorn in the Cape Province. The farmer informed him that the storks had nested there for at least seven years. Some 20 years later, in the early 1960s, a few nests were found in the Bredasdorp district, as far south as it is possible to be on the African continent. This population has apparently not increased, and two to four pairs still breed there. Interestingly, one of the chicks ringed at Bredasdorp migrated north and was recovered on the Zambian border with Tanzania, 3300 kilometres from its birthplace. The only other southern African breeding birds known are at Mossel Bay and at Tygerberg Zoo near Cape Town, where an injured captive White Stork mated with a wild bird and bred successfully. On present information there are probably fewer than ten pairs breeding in southern Africa. It has been suggested that the southern African breeding population may have lost its migratory urge because there were too few birds in the extreme south to provide the required stimulus. This, combined with the availability of food on a year-round basis, may have motivated the storks to overwinter and eventually breed.

White Stork nests in southern Africa are typical bulky stick platforms. They have been built in various kinds of trees, mainly eucalypts, but also in a rooikrans and a milkwood. One nest was on a large aloe in a family graveyard and others have been built on artificial platforms put up for them.

The Black Stork's breeding range is the most extensive of any stork, stretching through most of the Palaearctic and in Africa from Malawi south to the Western Cape. Like the White Stork it has declined in western Europe and has disappeared as a breeding species from several countries there. In the Palaearctic Black Storks breed mainly in trees and the loss of woodland has been a major factor in their decline. In southern Africa, however, nests are always on cliffs – on a ledge under an overhang, in caves, on rock columns or on the ledges of abandoned opencast mines. Sometimes the nest may be built on top of the disused one of another species such as a Hamerkop, Black Eagle or Gymnogene. Occasionally Black Storks breed near a Cape Vulture or Bald Ibis colony, or near the nest of a Black Eagle, Peregrine Falcon or Lanner Falcon, but usually their nests are solitary. In contrast to the White Stork, they prefer remote areas away from human habitation. The first recorded southern African breeding pair was found in the Bloemfontein district in 1901. In 1967 the

The Woolly-necked Stork is a rare and elusive breeding species in southern Africa. Although it is known to breed in Zululand and the lowveld of the Northern Province and Mpumalanga (formerly Transvaal), the first nest in Zimbabwe was discovered only in 1970. There are probably fewer than 30 breeding pairs in South Africa, while in Zimbabwe only two breeding sites are so far known. The nest is situated in swamp forest or riparian woodland and is usually placed on a horizontal fork at the end of a lateral branch. It is about 1 metre across and 30 centimetres deep, and during the breeding cycle the adults regularly bring green leaves as lining. They are very shy at the nest and disturbance should be kept to the minimum.

The Openbill is a colonial breeder, congregating in colonies of up to 500 pairs, and it often nests alongside cormorants, Darters, herons, ibises and African Spoonbills. It locates its nest in a tree on an island or in a partially submerged tree in a recently inundated area. In fact, its breeding appears to be linked to flooding, when its specialised diet of molluscs is most likely to be available. In the 1970s it bred opportunistically in Zimbabwe during a succession of particularly wet seasons, when its breeding population built up to unprecedented numbers and then declined again. The nest of an Openbill measures 45 to 75 centimetres in diameter and has a central depression of about 20 centimetres across which is lined with wet weed collected from the bottom of a pool. During the incubation period the nest is kept moist, either by collecting more wet weed or by regurgitating water onto the nest. Water is also regurgitated onto the chicks during the nestling period. The young are fed exclusively on molluscs.

The Saddlebill's huge solitary nest may be 2 metres across and 1.5 metres deep. Open to the sky, it is usually situated 20 to 30 metres above the ground at the top of a tall tree, but is occasionally much lower. Sometimes it is built on the disused platform of a Secretary Bird or Tawny Eagle. Both adults construct and repair the nest, which may be used for many years.

An uncommon breeding species in southern Africa, the Marabou nests in colonies, the best known of which is in the Okavango Delta in Botswana. Although elsewhere in Africa it builds its nest on cliffs, in southern Africa it has been recorded as breeding only in trees, often in baobabs or, in the Okavango, in thick stands of water figs. Marabous often nest in association with other colonial species such as Pink-backed Pelicans, cormorants, herons and ibises. They are fascinating to watch when breeding as they inflate their gular air-sacs, display the red skin at the base of their necks, clatter their bills and emit an extraordinary cacophony of sounds. In addition, when females approach males with territories they hold out their huge wings in an impressive 'balancing' display to express submission.

In the Okavango Delta Yellow-billed Storks often nest alongside Marabous, and this is the only area in southern Africa where this species is known to breed in substantial numbers. The nest, measuring 80 to 100 centimetres across, is built in a tree above water or, if it is located away from water, in a species such as a flat-topped acacia. The Yellow-billed Stork's courtship displays are much enhanced by the subtle beauty of its breeding plumage, which is suffused with pink.

South African population was estimated to be 34 pairs, but subsequently it has been found to be considerably larger, with 50 to 70 pairs north of the Vaal River alone. In addition, there are at least 100 breeding pairs in Zimbabwe.

The nest is a platform of sticks 1 to 1.5 metres across and lined with soft material. Nests in caves are often quite flat structures and some may be fairly rudimentary. Most are usually inaccessible, but one nest in a small cave on a hillside could be walked to with ease. The same site may be used year after year, and a nest in Botswana, first located in 1941, was still in use in 1978.

Black Storks are fairly vocal and emit high-pitched whistling calls which sometimes draw attention to the presence of additional birds near the nest, possibly young from the previous breeding season. In north-eastern South Africa and Zimbabwe eggs are laid in the winter months, mainly June and July, when water levels are at their lowest and isolated pools provide a supply of easily caught fish for the nestlings. Most nests are situated near water, but some may be a considerable distance from the nearest pools.

1 *A male Saddle-billed Stork with three large chicks. The nest, on top of a thorny acacia, measures 2 metres across.* ***2*** *A Marabou performs its 'wing balancing' display.* ***3*** *A Marabou's nest in the Okavango Delta.*

Ibises and the African Spoonbill

Four ibises (the Sacred, Bald, Glossy and Hadeda) and the African Spoonbill breed in southern Africa. Three of these, the Sacred and Glossy Ibises and the African Spoonbill, usually breed in mixed heronries in trees, reedbeds and sometimes on rocky islands. The Bald Ibis breeds colonially on cliffs, and the Hadeda nests solitarily in trees.

The appearance of all these species undergoes change at the onset of the breeding season. In all of them the colours of the soft parts brighten and some acquire breeding plumes. The fine black ornamental plumes of the Sacred Ibis develop a metallic sheen, its flank feathers become yellowish and its head, neck and inflatable neck sac glossy black, and a line of bare skin on the underwing turns a vivid red. The bare skin on the head of the Bald Ibis becomes bright red and its plumage much glossier, and the feathers of the Glossy Ibis develop an iridescent sheen, its facial skin turns bluish and there is a distinctive bluish white line where its face and forehead meet. The Hadeda shows least change, probably because it nests solitarily and maintains a long-term pair bond. The only significant difference in its appearance is that the red on the bill turns bright crimson. The African Spoonbill's plumage becomes an immaculate white, its soft parts brighten and a thick crest of feathers develops at the back of its head.

In all the species in this group the male fetches most of the nest material and the female builds; both share incubation and both care for the chicks, feeding them by regurgitation when they vigorously insert their bills into the adult bird's crop.

The Sacred Ibis nests in a variety of sites, in either single-species or mixed colonies. Even in the latter the birds tend to form groups of their own species and in reedbeds build their nests close together, forming rafts. Preparatory to breeding, the males gather in groups and establish pairing territories with a number of complex displays that include 'display flights' and 'sparring displays', as well as 'bill popping' in which the bird gapes, raises its head and snaps its mandibles closed. Once a mate has been acquired the pair moves to a nesting territory, usually in synchrony with all the other pairs in the area. At this stage their courtship behaviour includes a 'bowing display' during which their necks and bills are intertwined.

Some colonies contain hundreds of Sacred Ibis nests, which are constructed with sticks or reeds and lined with leaves or grass. They measure 30 to 45 centimetres across and are 20 centimetres deep. The clutch usually comprises two or three eggs – sometimes up to five – which are white with reddish spots. The Sacred Ibis differs from other ibises in that when the young leave the nest they form a crèche in which they are fed by their own parents on the basis of mutual recognition. They persistently run or fly after their parents, importuning them with frantic begging calls.

The Bald Ibis, which is endemic to South Africa, has a restricted range centred on the Drakensberg massif stretching from Kwazulu-Natal and Lesotho in the south to Pietersburg in the north. It used to occur about 1000 kilometres west of its present range, but has declined apparently as a result of overgrazing in the Eastern Cape which altered its habitat. It was also once considered good eating and at its breeding

1 

2 

3 

 4

5 

1 *A Bald Ibis on its typically flat platform of sticks on a cliff ledge.* ***2*** *The Glossy Ibis's eggs are a beautiful deep blue in colour when fresh.* ***3*** *The Hadeda builds a flimsy nest, and lays eggs that are very variable in ground colour and markings.* ***4*** *A raft of Sacred Ibis nests. Even in a mixed colony, Sacred Ibises tend to stay in groups of their own species.* ***5*** *African Spoonbills and Sacred Ibises often breed together, and their nests and eggs are similar.*

1 *African Spoonbills are monogamous and maintain the pair bond with bouts of mutual preening.* **2** *The African Spoonbill's eggs are noticeably pointed.*

BALD IBISES AND BUTTONS

In the early 1970s Peter Milstein of the Transvaal Division of Nature Conservation documented the occurrence of numerous buttons of assorted sizes beneath a Bald Ibis colony near Pietersburg. A thorough search below the nests revealed 156 buttons, but a similar search below cliffs where Bald Ibises were not nesting turned up none.

This puzzling phenomenon seemed to have only one possible explanation – that the ibises had swallowed the buttons and then excreted them undigested. When the buttons were examined closely it became clear that the chitinous remains of beetles were sticking to several of them, or were caught in their holes. As beetles form part of the normal diet of Bald Ibises, it would appear that the birds had mistaken the buttons for beetles and eaten them ●

colonies it was an easy target for hunters. Although classified as 'Threatened' in the *Red Data Book* of 1984, the breeding population was thought to be stable at 1250 pairs (possibly an underestimate). The species is now regarded as out of danger, although disturbance at colonies remains a potential threat.

Bald Ibises are gregarious and nest colonially on inaccessible ledges or in potholes on cliffs, often near a waterfall or overlooking a river. Traditional breeding cliffs are used for long periods, although individual birds may move to other colonies. The males occupy the nest sites first, and a pair may move from ledge to ledge indulging in courtship behaviour which includes mutual preening and grasping a twig and wagging it about, sometimes with both birds holding onto it. However, it has been difficult to determine courtship patterns because of the capricious behaviour of apparently mated birds which may copulate randomly with other partners. During coition the male grasps his mate's bill and shakes it vigorously. When breeding the birds are fairly vocal and emit a variety of calls, including a series of piping notes used in greeting.

The nest is a fairly flat platform of sticks 50 centimetres across and 15 centimetres deep, and its shallow depression is lined with dry grass. Some nests are rather flimsy and may disintegrate before the chicks are ready to fly. The eggs, two or three in a clutch, are pale bluish white with reddish spots.

The Glossy Ibis is southern Africa's smallest ibis and has only relatively recently become established as a breeding species; the first nests were discovered on the Witwatersrand in 1950. Since then this highly nomadic species has extended its range considerably and breeds widely as far south as the Western Cape. It normally nests in mixed heronries, favouring reedbeds although trees are also used, and unlike the Sacred Ibis does not form groups of its own species within the heronry. Instead, Glossy Ibis nests are scattered at random amongst those of other species. About 30 centimetres in diameter, they are compact structures of sticks or reeds lined with finer material. The eggs, 2-5 in a clutch, are a beautiful deep blue colour when fresh, but fade to paler blue as incubation progresses. Courtship behaviour is not well known but it includes 'mutual bowing', rubbing bills together and emitting cooing calls.

Hadedas differ from other southern African ibises in that a pair forms a long-term bond, and even at communal roosts will perch side by side. Because a pair bond is maintained, the courtship behaviour of the Hadeda is less complicated than, for example, that of the Sacred Ibis. Nevertheless, it includes head nodding, rattling bills together, offering sticks, intertwining of necks and mutual preening. Pairs return to a nest area for many years, and one pair or their successors is known to have bred in the same kloof for at least 15 years.

The solitary nests are built in trees, 1.2 to 12 metres up and usually in a horizontal fork at the end of a branch, although ones in a vertical fork have also been found. Often, but not always, over water, they are sometimes located above a dry watercourse or gully quite a distance from a river. Nor are they always built in trees: one Hadeda nest has been found on a telephone pole, possibly on the disused nest of a Black Crow, and another on the floor of a cave. The nest, which measures 25 to 45 centimetres across and 15 centimetres deep, is a relatively insubstantial structure of sticks with a shallow central depression lined with grass or lichens. The 112 sticks counted in one Hadeda nest compares with an average of 300 in a Cattle Egret's nest, and the flimsy construction of the former probably explains why eggs or chicks quite often fall out of it. The clutch is 2-4 eggs which are enormously variable both in ground colour and the amount of blotching. The usual ground colour is pale olive green and it is heavily overlaid with brown spots and blotches.

In recent decades the Hadeda has expanded its range into the drier west and, since the late 1960s, also into the south-western Cape. Factors promoting its spread are the increase in the area covered by irrigated agricultural land and in the stands of alien trees for roosting and breeding.

The African Spoonbill is nomadic and, like the Glossy Ibis and Hadeda, has recently expanded its range. The first breeding in the south-western Cape was recorded in the late 1950s and the species is now well established in the region. It nests colonially in mixed heronries – frequently in association with Sacred Ibises – but also in single-species colonies. It often breeds opportunistically, for example in flooded areas after heavy rains have fallen in the Karoo. The platform nests are built of sticks and may be located in reedbeds, in trees or on rocks on islands. They are similar in size to those of the Sacred Ibis and, as the eggs have similar red markings, care should be taken when identifying the owner of a nest where these two species are breeding close to one another. The clutch comprises 2-4 eggs.

The breeding displays of the African Spoonbill have been thoroughly studied and no fewer than 14 ritualised social displays have been described. During pair formation the males establish themselves at nest sites where they are approached by unmated females. Initially they attack the females, but the latter persist with 'display flights' (during which the wings produce a woofing sound) and 'bowing' displays until they are accepted. Once the bond is formed the pair perform other displays, notably the 'mutual greeting' during which they erect their crests, point their open bills upwards and flap their wings slowly while calling loudly. Other important displays, usually in the following sequence, are 'display shaking', 'display preening' and 'bill popping'. When mating the male steps slowly onto the female's back, grasps her bill loosely behind the spoon and nibbles it during coition.

In view of the similarities between some of the displays of the African Spoonbill and the Sacred Ibis it is interesting to note that in captivity the two species have been known to interbreed and produce young successfully.

FLAMINGOS

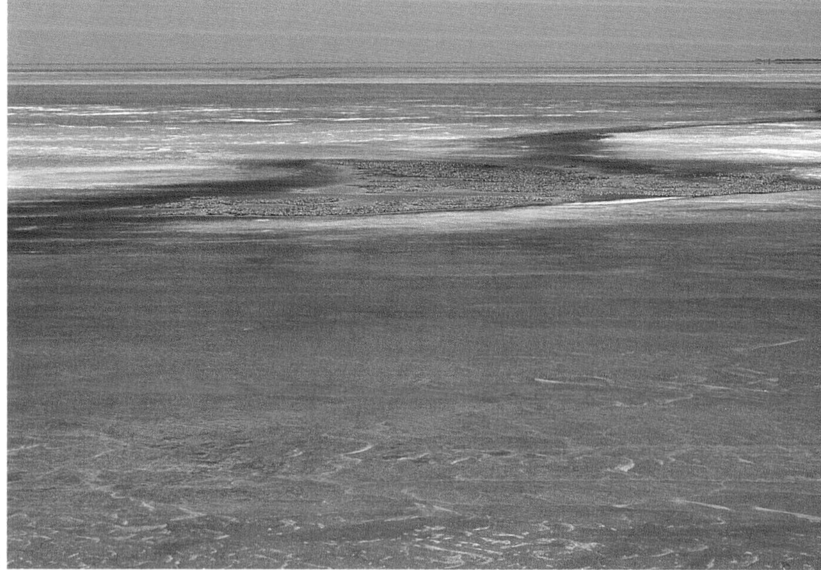

FLAMINGOS ARE SPECTACULAR BIRDS, especially when they congregate in huge numbers as they do on the lakes of the Rift Valley of east Africa. To see a breeding colony is a memorable experience but one that, for several reasons, few people have been able to enjoy. In the first place, flamingos are opportunistic and erratic breeders, and they choose to nest in inhospitable and inaccessible places. Moreover, they are prone to natural disasters such as flooding, and often abort their nesting attempts. The first detailed observations on their breeding in Africa were made in the early 1950s in the eastern part of the continent by the late Leslie Brown.

Two species of flamingos occur in Africa, the Greater and the Lesser. The latter is the most prolific flamingo species on earth, with an African population estimated at three to four million birds. A breeding colony at Lake Magadi in east Africa in 1962 numbered at least a million pairs. It is mind-boggling to consider how much food is required by so many birds. Every day they would consume 200 tonnes of food, filtering with their specialised bills the microscopic life which teems in the nutrient-rich alkaline lake.

Breeding flamingos in southern Africa have proved no less elusive than their counterparts in east Africa. The first published account of the mass breeding of Greater Flamingos related to successful nesting at Bredasdorp in the south-western Cape in 1960, when about 800 eggs were laid and approximately 350 young fledged. Prior to this there were sporadic accounts of nesting from various parts of southern Africa but no evidence of mass breeding. Subsequently both Greater and Lesser Flamingos were found to be breeding in large numbers on Etosha Pan in Namibia, and the 1971 season there was the first to be fully documented. In 1972 Greater Flamingos bred successfully at Lake St Lucia in Natal and 4000 young survived. Six years later the same species bred on a dam at Van Wyksvlei in the Karoo which was subsequently drained for irrigation purposes. The Cape Department of Nature Conservation transported the starving chicks, 623 in all, to another dam in the district where many of them were reared successfully by their parents. Recently both Greater and Lesser Flamingos have bred on Lake Makgadikgadi in Botswana where they had almost certainly been overlooked previously because of the enormous size of the pan and the remoteness of the breeding areas. It is quite probable that breeding attempts in other remote areas of southern Africa have gone unnoticed.

Robert Porter Allen, the noted authority on flamingos, has summarised the breeding requirements of flamingos as: 'a shallow lake, usually of high salinity, which is situated in an isolated area, desolate of aspect and, except for the flamingos and the fellow members of their food web, is nearly devoid of animal life'. Freedom from mammalian predators is as necessary for successful breeding in flamingos as it is in White Pelicans, and it is significant that flamingos and pelicans have been found breeding in the same areas, for example at Etosha Pan. In 1961 the Greater Flamingos at Bredasdorp attempted to breed for a second year but failed and all 120 eggs were lost. The suspected culprits were Cape clawless otters.

It appears that Etosha Pan and Makgadikgadi Pan are the only two areas in southern Africa where breeding occurs on a large scale when conditions are suitable. At Etosha it has been established that flamingos breed whenever the annual rainfall exceeds 500 millimetres. Between 1956 and 1971 breeding was recorded five times, each time in a year of above average rainfall when there were large expanses of shallow water in the normally dry pan. Flamingos compensate for their erratic breeding habits by being long-lived: it is estimated that Greater Flamingos live to an age of 30 to 50 years and Lesser Flamingos to about 20 years.

When a Greater Flamingo comes into breeding condition its red colours intensify. In the case of the Lesser Flamingo the whole plumage becomes a deep pink, and a mass of displaying birds is a scene of incredible beauty. For the observer the memory is more than visual, for the babel produced by thousands of birds is like the roar of distant surf.

At the beginning of courtship male Greater Flamingos gather in groups and, with necks extended, flag their heads from side to side, emitting at the same time loud, honking calls. They are then joined by females who behave in a similar way. Thereafter individual birds indulge in a series of ritualised displays, the most prominent of which is the 'wing salute'. With neck at full stretch, the bird flicks out its wings and holds them extended, then snaps them closed. In another, quite different, 'wing salute' the bird bows forward with lowered neck, stretches its partially extended wings forward and then snaps them closed again. These two displays serve to expose the maximum amount of black and red respectively. Another posture is the 'twist preen' during which the bird lays its head along its back and appears to preen

1 Lesser Flamingos at Makgadikgadi Pan in Botswana, one of the inhospitable areas of southern Africa where flamingos breed. ***2*** *Flamingos typically build mound nests such as these of Lesser Flamingos at Makgadikgadi Pan.* ***3*** *At Bredasdorp, however, the unsuitable substrate prevented a colony of Greater Flamingos from building more than rudimentary mounds.*

1

THE GREAT FLAMINGO TREK

Both at Etosha Pan in Namibia and at Makgadikgadi Pan in Botswana remarkable feats of survival by flamingo chicks have been observed. In years when the water in these pans rapidly recedes and dries up the young birds, not yet able to fly, are left with no choice but to march or die. The slush left by the receding alkaline water becomes a death-trap as encrustations of soda form around the chicks' lower legs, and many die pitifully as they are weighed down by these 'anklets of death'.

When Etosha Pan dried up in 1971 30 000 young Lesser Flamingos set off on a two-stage march that lasted 30 days. First they trekked north for 30 kilometres until they reached the nearest water. When this also dried up they marched westwards for another 50 kilometres until they reached the sanctuary of a good water supply. The research workers who monitored their progress estimated that approximately 27 000 young flamingos survived the great trek. Although chicks died along the way, the mortality was not as high as expected, and some of the older chicks may have been able to fly to water before the journey was completed.

At Makgadikgadi some 20 years later Lesser Flamingo chicks trekked north across the bleak pan for 80 kilometres until they reached water. The journey took about a week and it was estimated that three-quarters of the young survived.

How do the young birds know where to go and how do they survive? It appears that they are guided in the right direction by adults flying ahead of them, probably the same 'nursemaids' which supervised the original crèche at the colony. That they survive is due to the fact that they are fed along the way by their parents, usually at night. The adults commute between the marching chicks and the nearest feeding grounds, sustaining the youngsters with nutritious meals along the way. There can be no doubt that, in the annals of avian survival, the Great Flamingo Trek must be one of the most remarkable episodes ●

2

3

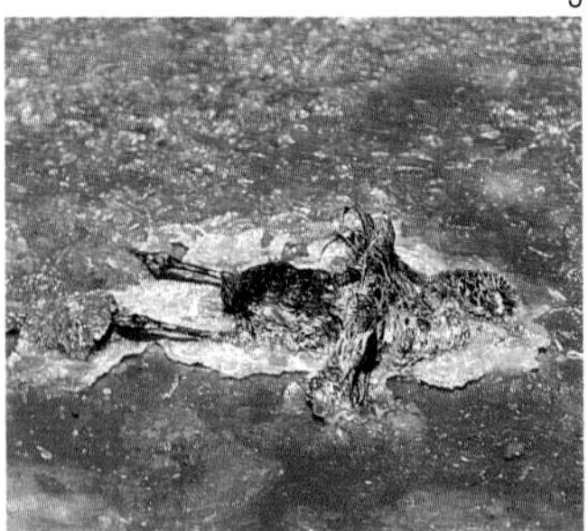

***1** A tightly packed crèche of Lesser Flamingo chicks treks across the Makgadikgadi Pan to find water. A few 'nursemaids' may be seen at the edge of the group. **2** After the gruelling 80-kilometre trek a chick slakes its thirst. **3** Some did not survive, weighed down by soda encrustations that form around the chicks' lower legs.*

at the base of a partially opened wing. In the prelude to copulation, known as 'hooking', the neck is held in the form of a crook with the bill pointing downwards. A pair will walk away from the main displaying group and the male places his hooked neck across the female's back. He then jumps onto her back, she holds his flexed legs with her wings and mating takes place. Copulation has been observed in breast-deep water, in shallow water and on dry land.

The Lesser Flamingo shares with the Greater Flamingo a number of displays such as the 'head wag', 'wing salute', 'twist preen' and 'hooking', but its behaviour is far more intensely communal. Birds flock together into small groups and these are joined by others until they form a tightly packed mass that may number thousands. On a forest of red legs the flamingos march and counter-march in what Leslie Brown called the 'communal stomp', an apt description of the incredible spectacle of this moving pink mass.

Although wonderful to watch, courtship displays are not necessarily followed by breeding. The conditions that stimulate courtship may disappear and the birds will abort their displays as suddenly as they began them. In east Africa courtship performances of massed Lesser Flamingos have been observed at sites far from their known breeding grounds.

Once the flamingos decide to breed they do not necessarily do so at the same time. Birds in a group within the colony will mate at the same time, but the different groups may stagger their breeding. Thus there will sometimes be several batches of chicks of distinct age classes, as well as birds incubating fresh eggs. As a rule the broods of late layers are most likely to fail if the water dries up too quickly.

Both male and female flamingos build the nest, the form and size of which depend on the habitat in which it is situated. Those on rocky islands may be very basic, a collection of anything available such as gravel, grass or feathers. If there is some mud or soil then a low mound may be built. On mudflats a volcano-like nest with a shallow depression on top may be constructed, its size varying according to how much mud is available. Sometimes where traditional sites are used, as at Etosha, nests may be built on the remains of those of previous years. As mud is scooped up to add to the nest a moat is dug around it, and it has been calculated that a colony numbering 500 000 pairs of Lesser Flamingos would excavate about 15 000 tonnes of mud! The raised nest structure is important in an environment of extreme heat: at Etosha Pan it was found that the nest cup was 2 to 8 °C cooler than the surrounding pan. At Lake Natron in Tanzania, probably the most severe conditions that flamingos nest in anywhere, the nest cup was 25 °C cooler than the surrounding mud.

The normal clutch is a single egg but on very rare occasions two may be laid; at one Greater Flamingo colony five nests out of 900 contained two eggs. The elongate oval egg is initially chalky white, but it becomes heavily nest-stained as incubation progresses. The yolk is a rich blood-red. Both adults incubate, the male being readily identifiable because he is taller than the female. Change-overs at the nest appear to be synchronised, with groups of birds coming in together from their feeding area to relieve their incubating mates. To settle on the nest a flamingo characteristically places one leg on the rim while spreading its wings to maintain balance. It extends its free leg backwards and shakes it vigorously, then stands on it while stretching and shaking the other leg. Both legs are then flexed and the bird settles with its 'knee' joints extending backwards well over the rim of the nest. Since the leg-shaking ritual occurs irrespective of whether nests are on hard ground or in slushy mud it is probably innate, but it has a functional value in that it prevents the egg from becoming encrusted with mud which would inhibit the hatching of the chick.

A pecking distance is maintained between nests and incubating birds in the tightly packed colony are remarkably tolerant of each other. Aggression is confined to bickering between neighbours which lunge at each other with extended necks. To express threat, for example when a returning bird passes by too close, an incubating bird adopts the 'chrysanthemum' posture by raising its back feathers. Occasionally, for no apparent reason, the whole colony will rise up, run off for a short distance, and then return to the nests. Low-flying aircraft may cause panic and research workers have found that 300 metres is the minimum height from which a colony can be inspected without causing disturbance.

The incubation period of Greater and Lesser Flamingos is close on 28 days and chick development and parental care are basically similar for both species. The swollen and rubbery coral-red legs of a newly-hatched flamingo are its most striking feature. Its down is grey but varies from light to dark among individuals. The chick is brooded by its parents and is fed in a characteristic way that remains constant irrespective of its age. Standing over the forward-facing chick, the parent rests the tip of its bill on the lower mandible of the chick's open bill and regurgitates a liquid meal. Initially newly-hatched young are fed on a special red fluid, probably an exudation of the upper digestive tract, which contains red blood corpuscles and is rich in fat and glucose.

The chick develops remarkably quickly on its nutritious diet and within a few days its red legs have darkened to grey and it is able to leave the nest. While still small it pecks at the ground and eats any pieces of eggshell it may find, thus absorbing calcium for its rapidly growing legs. Within a week or so a number of chicks form a crèche, remaining in a group that may number many thousands until, at ten to 11 weeks old, they are able to fly. They mill around all over the place, often upsetting adults that are still incubating. While the parents are away at the feeding grounds these groups are supervised by a few adults known as 'nursemaids'. Remarkably, in the midst of the mass of young birds, each chick is able to recognise its parent's voice and an adult will feed only its own offspring.

Flamingos face many difficulties in completing their breeding cycle. A change of wind may create waves that inundate the nests, as has happened at St Lucia, Etosha and Makgadikgadi. Alternatively, if a nesting area dries up the chicks may die of starvation if they cannot walk to the nearest water. A colony may fail if it becomes accessible to mammalian predators, as happened at Bredasdorp in 1961 when the water level dropped. When White Pelicans nest in association with flamingos they may inhibit breeding merely by taking over areas that are suitable for flamingos. Avian predators such as Tawny Eagles and African Fish Eagles may prey on chicks. A menace not yet encountered in southern Africa is posed by Marabou Storks. Flamingos have an absolute dread of these birds and in east Africa the mere presence of no more than 17 Marabous caused a flourishing colony of 4500 pairs of Greater Flamingos to desert.

Ducks and Geese

Sixteen species of ducks and geese breed in southern Africa. In addition to these a small introduced population of Mute Swans once bred in the southern Cape but is now considered extinct in the wild, and feral Mallards occur in small numbers. The latter should be eradicated because they interbreed with the indigenous Yellow-billed Duck to which they are closely related.

A notable feature of southern African ducks is their lack of sexual dimorphism, in contrast to those in the northern hemisphere where the males of many species assume a conspicuous nuptial plumage. Why should this be so? The most acceptable explanation, if one allows for certain anomalies, is that many northern species are regular long-distance migrants and have a relatively short period in which to perform their courtship displays. The males' bright plumage compensates for this. Southern African species, on the other hand, are either resident or nomadic and do not have to complete their courtship as quickly as their northern counterparts do. Moreover, they are exposed to greater predation pressure than are ducks in the northern hemisphere, and drab plumage is a decided advantage, as is courtship behaviour that is less intense and less elaborate.

The variety of the southern African ducks and geese is considerable and they range in size from the Spur-winged Goose – the largest males can weigh as much as 10 kilograms – to the relatively diminutive Hottentot Teal and Pygmy Goose which weigh a little less than 300 grams. The 16 species may be loosely divided into six groups: the White-faced, Fulvous and White-backed Ducks; the Egyptian Goose and South African Shelduck; the *Anas* ducks (Yellow-billed Duck, African Black Duck, Cape Teal, Hottentot Teal, Red-billed Teal and Cape Shoveller); the perching 'ducks'

1

2

3

1 *Egyptian Geese are highly adaptable in their choice of nesting site, a factor which has undoubtedly contributed to their success as a species.* ***2*** *The White-backed Duck nests on a floating mound of aquatic vegetation.* ***3*** *The female with her newly-hatched chicks. Her partner regularly assists with their care.*

***1** Even on its relatively exposed nest, an incubating Black Duck is well camouflaged. **2, 3** When it leaves the nest it covers the eggs with down that keeps them warm. Uncovered, the light-coloured eggs are an obvious target for predators. **4** When a female Cape Shoveller is frightened off the nest she defecates on the eggs, rendering them distasteful to a predator. **5** The Pygmy Goose normally nests in tree holes, but may be enticed into laying in a nest-box.*

(Knob-billed Duck, Spur-winged Goose and Pygmy Goose); the Southern Pochard; and the distinctive Maccoa Duck. The Spur-winged, Egyptian and Pygmy Geese are not true geese, but ducks that have developed certain goose-like features.

Courtship behaviour varies considerably from one species to another. Some species, usually those that maintain long-term pair bonds, have fairly simple displays while others perform a whole range of ritualised actions – as many as 15 in the case of the Yellow-billed Duck. Sometimes the courtship behaviour suggests affinities that can be used in taxonomic classification. The White-backed Duck, for example, was once placed close to the Maccoa Duck on the basis of its diving habits, but it is now grouped with the White-faced and Fulvous Ducks on other grounds, including a post-copulatory 'step dance' that it performs.

In general visual signals predominate in the courtship of the different species, but vocalisations are also important to a certain degree. Most species exhibit a range of behaviour patterns, initially in social groups and then during the formation and subsequent maintenance of the pair bond. These displays include preening movements during which the colourful speculum is revealed, swimming around the female, dipping the bill in the water, head-nodding, chasing flights and 'jump flights', in which the duck takes off and almost immediately lands.

The displays of the Maccoa Duck and Knob-billed Duck merit special mention. In the breeding season the male Maccoa Duck assumes a handsome chestnut plumage which is offset by a bright blue bill. His performances are all aquatic and include short paddling rushes, a vibrating trumpeting display during which the neck is inflated, and a posture in which the bill is partially submerged and the tail feathers are stiffly cocked. The Knob-billed Duck's displays are closely linked to its habit of perching in trees which makes it highly visible. The male positions himself so that his prospective partner sees him in profile, thus displaying the size of the knob on his bill to best advantage.

Several ducks maintain long-term pair bonds, including the African Black Duck and the South African Shelduck. The former is highly territorial, and the female will even physically help her partner to attack intruders. Most southern African ducks and geese are monogamous but there are a number of exceptions. The Knob-billed Duck may be either monogamous or polygynous, and if the latter it exhibits either simultaneous polygyny (the male having two or more mates in a 'harem') or successive polygyny (the male mating with several females one after the other). The Spur-winged Goose is also considered to be either monogamous or polygynous but this species has been little studied. The male Maccoa Duck is promiscuous, mating with a large number of females. The males of *Anas* species also often make promiscuous assaults on various females and may violently overpower, or 'rape', them.

In all species of this group copulation usually takes place in the water. The male swims up to the female while making pumping movements with his head, mounts her and grasps her behind the head with his bill. Several ducks also perform characteristic post-copulatory displays.

The breeding season varies according to region and rainfall, with the basic requirements of an adequate food supply and cover for the nests applying for most species. For example, the Knob-billed Duck, a long-distance intra-African migrant, breeds in southern Africa when pans are inundated after good rains. Because the breeding season may vary from year to year depending on rainfall, it is important that the open season for duck shooting be regulated on a flexible basis.

Most species breed solitarily, with the exception of White-backed Ducks which sometimes build their nests a few metres apart. In some species a number of nests have been found close together on an island, but the reason for this is less to do with sociability than with the safety and suitability of the site. Generally the nests are simple hollows which the female builds with material available at the chosen site and conceals in vegetation near or over water. The surrounding cover is often pulled down over the nest to form a canopy. The Hottentot Teal often nests in clumps of reeds and the African Black Duck in flood debris caught in trees on riverbanks. The White-backed Duck builds its nest in emergent vegetation in water 1 to 2 metres deep and well away from the bank, constructing a mound of aquatic vegetation and sometimes incorporating surrounding reeds. The Maccoa Duck makes a similar nest, or it may nest in a hollow under a bush near water. Sometimes the nest of another species of waterbird is used; for example, a Southern Pochard has been known to lay in the disused nest of a Red-knobbed Coot.

Several species do not conform to the above pattern. Both the Pygmy Goose and Knob-billed Duck nest regularly in holes in trees, investing a great deal of time and energy in selecting a suitable site. A female Knob-billed Duck was observed peering into a hole for 20 minutes before entering, a wise precaution in view of the possibility that a snake or mammalian predator may already have been in residence. The Spur-winged Goose also occasionally nests in holes in trees. The South African Shelduck breeds underground in the abandoned burrow of an antbear, porcupine or springhare, and in one instance laid its eggs 8 metres from the entrance. The fact that the Egyptian Goose chooses a wide variety of nesting situations explains in part its success as a species. It nests on the ground, in holes in cliffs or on cliff ledges, in holes in trees, on the nests of birds such as raptors and crows, on top of or in a Hamerkop's nest, and on buildings.

Down from their own bodies is used by many ducks to line their nests, even if they are no more than rudimentary scrapes. When a parent leaves the nest to feed it pulls some of the lining over the eggs so that they are not only kept warm by the down's excellent insulating properties, but they are also concealed from predators. The amount of down in some nests is considerable: the Hottentot Teal may deposit a layer

The Maccoa Duck lays large eggs and can only brood them successfully if she regularly rotates her position on the nest.

2.5 centimetres thick, and 3562 down feathers were found in the nest of an African Black Duck. The White-faced, Fulvous, White-backed and Maccoa Ducks do not line their nests in this way.

Most of the eggs are glossy and their colour ranges from plain cream or buffy yellow to light brown. Clutches of 6-12 eggs are normally laid, but there are often many more. Recorded clutches of 20-27 eggs for the Spur-winged Goose are undoubtedly the product of two females. Occasionally eggs are dumped at random in the nest of another member of the same species or of another one. The Maccoa Duck is often responsible for dumping, its eggs having been recorded in the nests of the Fulvous Duck, Egyptian Goose, Hottentot Teal, Southern Pochard and Red-knobbed Coot. It also dumps in the nests of other members of its own species, and it is possible that it is evolving into a regular brood parasite, like its relative the Black-headed Duck of South America. Maccoa Duck eggs are extremely large, a single one being equivalent to 16 per cent of the female's body weight, and they produce highly precocial chicks which are able to dive and feed themselves soon after hatching.

Ducks usually lay their eggs at daily intervals, and as incubation begins only once the clutch is complete the chicks all hatch at the same time. In most species only the female incubates, but male White-faced, Fulvous and White-backed Ducks also do a substantial share. Incubation spells may be very long; a female African Black Duck under observation left the nest only twice a day for periods varying from 22 to 165 minutes and, just prior to hatching, she remained on the nest continuously for two days. The female Maccoa Duck, on the other hand, leaves the nest fairly frequently for short periods of intensive feeding to make up for the energy she lost in laying her extra-large eggs. Because of their size she cannot cover the eggs adequately when incubating and ensures that the whole clutch is evenly warmed by turning round on the nest from time to time. As Maccoa Ducks feed almost entirely by diving they do not moult their body feathers, so their nests do not have a downy lining.

Incubating birds sit very tight, flying off only at the last moment and then often feigning injury. Some species, especially the Cape Shoveller, excrete foul-smelling faeces onto the eggs when surprised off the nest. Although the function of this behaviour is debatable, experiments in Europe have shown that the faeces render the eggs distasteful to mammalian predators.

The incubation period of most species lasts approximately a month and the downy ducklings are active soon after hatching. If the nest is some distance from water they are led to it by the female as soon as they are mobile. In one case a Spur-winged Goose guided her ducklings 1.5 kilometres through thick undergrowth, arriving at the dam three hours after leaving the nest. The young of tree-nesting species such as the Pygmy Goose and Knob-billed Duck, as well as those of the Egyptian Goose, have lost their fear of heights and jump to the ground in response to the female's call below. Egyptian Goose chicks have been watched jumping from the nest, their tiny wings flailing and their feet paddling like miniature air-brakes. Once all the young have been accounted for the female leads them away.

Authentic cases of adults carrying young are rare, but in Zimbabwe two White-faced Ducks flew into a house and two ducklings were found running around the lounge with them. As the bottom half of the stable door was closed and this was the only possible means of entry, the adults must have carried the ducklings, although how they did so is not known. There is also an account from Botswana of a Pygmy Goose flying with a chick on its back and dropping it as it flew past the observer. The goose flew back, retrieved the chick and then flew off with it once more on its back.

In many species the female raises the brood on her own, but the males of the Fulvous, White-faced and White-backed Ducks, the Egyptian Goose, the South African Shelduck and the Cape Teal regularly assist in caring for the young. In some other species males have occasionally been observed accompanying the female and her brood.

The Secretary Bird

The taxonomic status of the Secretary Bird remains uncertain, but it has traditionally been regarded as an unusual bird of prey and the description of it as a 'long-legged marching eagle' is as suitable as any. Its undulating display flights are reminiscent of raptors, but its running displays with wings raised are similar to those of cranes, as is its bowing behaviour used in greeting at the nest.

The nest is the focal point in the life of the Secretary Bird and it is often used as a roost outside the long breeding cycle (the nestling period is usually 80 to 90 days but may be as long as 106 days). The preferred site is a thick, flat-topped thorn tree, but where these are not available other trees such as pines are used. Nests are large, flat platforms of sticks 1 to 1.5 metres across, although some may measure 2.5 metres in diameter. They are built between 2 and 12 metres above the ground, but usually at a height of 5 metres. Both birds take part in building, sometimes carrying considerable amounts of material in their bills. Their feet, which are unsuitable for grasping, are never used to transport nest material or food. The centre of the nest is lined with dry grass, and dry dung is also frequently used. Lining continues to be added during incubation and sometimes well into the nestling period.

The nest may be used for a single breeding cycle or repeatedly for many years. A move to a new site is usually prompted by the conformation of the tree's crown changing in such a way that it can no longer adequately support the nest. Sometimes another species such as a Saddlebill Stork or a Lappet-faced Vulture may take over a disused Secretary Bird's nest.

In southern Africa breeding has been recorded in all months of the year, except in the south-western Cape where the Secretary Bird lays mainly in spring (August and September). The clutch comprises 1-3 chalky white eggs which sometimes have red blotches, and they are characteristically pointed at one end. The female undertakes most of the incubation, but is occasionally assisted by the male. His main task, however, is to bring food for his mate at the nest in typical raptor fashion. If the female detects any danger she lies flat on the nest so that she cannot be seen from below, a position frequently adopted amongst raptors. The Secretary Bird is likely to stand up and leave the nest only if an observer actually looks over its rim.

Initially the chicks are tended and guarded by the female, but her time on the nest gradually decreases and once they are a month old and have a covering of feathers on their backs she leaves them unattended for long periods to assist the male in hunting. The chicks are fed by regurgitation, usually a liquified stream of digested matter, but solid food is also disgorged onto the nest where it is gobbled up by the young. Water, too, may be transported in the adult's crop and is regurgitated to the nestlings.

At nests kept under observation in the Karoo no more than four feeds were brought to the chicks during a day, which indicates that the foraging periods are long. Contrary to popular belief, snakes are infrequently brought; instead insects are an important component of the Secretary Bird's diet. Because the birds may forage far afield, sometimes many kilometres away, they conserve time and energy by soaring up on a thermal and then gliding back to the nest. Once the chicks have been fed the adults march off again from the nest area in their seemingly ceaseless quartering of the veld.

1 *A Secretary Bird's eggs appear dwarfed on the huge platform nest atop a thorn tree.* ***2*** *The parent feeds its offspring by regurgitating digested matter, often insects, into the clamouring chicks' bills.*

Vultures

Of the eight vulture species that occur in southern Africa, seven – the Palm-nut, Bearded, Hooded, Cape, White-backed, Lappet-faced and White-headed Vultures – breed in the region. There are still occasional sightings of the eighth, the Egyptian Vulture, but the last authentic observation of it breeding was recorded in 1923. This species' decline in southern Africa is linked to the disappearance of game herds, to farmers' tendency to remove carcasses of their stock from the veld, and to direct persecution by means of poisoning and shooting. Egyptian Vultures were particularly unpopular with ostrich farmers as they would break the ostrich eggs by throwing stones on them. They also lived in association with man, thriving where conditions of poor sanitation existed and declining as these improved.

The Palm-nut Vulture is also very rare, with a small localised distribution in Zululand linked to areas where raphia palms grow. This species is unusual in that it does not share the carrion-eating habits of other vultures, feeding instead on the fruits of raphia and oil palms as well as on fish, crabs, amphibians, and the occasional bird. Indeed, it is sometimes called the Vulturine Fish Eagle and may be more closely related to fish eagles and snake eagles than vultures. Very few Palm-nut Vulture nests have been found in Zululand, where they are built in mainly raphia palms, although in one instance in a eucalypt. Elsewhere various other trees are utilised, and in southern Angola baobabs are preferred. The nest is a substantial structure of sticks measuring 60 to 90 centimetres across and 30 to 60 centimetres deep, and it is lined with dry grass and other dry material as well as dung.

The Bearded Vulture nests in potholes and caves on sheer basalt or sandstone cliffs. Its southern African range is now restricted to the Drakensberg massif, although formerly it extended almost as far south as Cape Town and coincided with that of the Egyptian Vulture. Estimates of this species' status and numbers have been vague or inaccurate in the past, but fortunately a thesis published in 1988 has separated fact from fiction, arriving at a breeding population of 203 pairs in southern Africa, of which 122 were in Lesotho. There can be no doubt that the pastoral practices in Lesotho, where there is fairly high mortality of stock and the carcasses are seldom removed, provide ideal conditions for the survival of these vultures.

Bearded Vultures lay their eggs in winter in areas where conditions are severe with some heavy snowfalls. In view of this they usually breed on cliffs at lower altitudes and select sheltered sites under an overhang, placing the large stick nest at the back of the cavity where it is protected from the elements. It closely resembles that of the Egyptian Vulture and is usually 1 metre in diameter and 50 centimetres high, although the size varies depending on the configuration of the pothole or cave. Some nests have measured 2 metres across and one was 2 metres high. The nest bowl is about 40 centimetres in diameter and is lined with an assortment of materials, mainly sheep's wool, including large pieces of fleece, as well as rope, rags, sacking, skin and fur. An interesting feature of the Bearded Vulture's breeding habits is that it rarely uses the same nest twice, irrespective of whether it bred successfully in the previous season. It is quite normal for a pair to have several nests, usually about three.

The Hooded Vulture may be termed the 'tail-end Charlie' among vultures as it hangs around on the periphery of a vulture feast. Unable to compete with the other scavengers because of its smaller size, it waits until the scrimmage is over and then picks in the crevices of the carcass using its long thin bill. Although extremely common locally in Africa, in the southern part of the continent it is rare and rather shy, seldom associating with humans. It nests solitarily and secretively, and relatively few sites have been found. Unlike most other tree-nesting vultures, it usually builds in a major fork beneath the canopy of a well-foliaged tree, sometimes taking over the nest of another bird of prey. A favourite tree is the jackal-berry and it is interesting that the tree's range and that of the Hooded Vulture coincide almost exactly in the region. The nest is relatively small, measuring about 60 centimetres across and 40 centimetres deep. It is lined with dry material such as grass which is overlaid with a thick carpet of green leaves, a unique feature amongst southern African vultures. This lining is replenished throughout the incubation period and most of the nestling period.

The Cape Vulture, a southern African endemic, has declined dramatically over the years and is currently classed as 'Vulnerable' in the *Red Data Book* for birds. When Jan van Riebeeck founded his settlement at the base of Table Mountain they soared around the cliffs above, and although many other mountains in southern Africa still bear the name 'Aasvoëlberg' or 'Aasvoëlkrans', they no longer sport the characteristic whitewash droppings that indicate an active Cape Vulture roost or breeding cliff.

All is not gloom and doom, however, as the total population numbers some 12 000 birds. Cape Vultures breed in the Magaliesberg only 40 kilometres from the centre of Pretoria, and a colony of almost 1000 pairs extends along a 5-kilometre stretch of cliff near Thabazimbi in the Northern Province. The mountain ranges north of the Vaal River remain the premier stronghold of the species, followed by the Drakensberg massif, Eastern Cape and eastern Botswana. A small population still

1

2

3

***1** Bearded Vultures place their nests in pothole caves in cliffs to shelter them from harsh winter conditions.* ***2** The Hooded Vulture is unusual among vultures in that it places its nest beneath a tree canopy and lines it with green leaves.* ***3** In Zululand the Palm-nut Vulture's bulky nest of sticks is usually built in a raphia palm.*

survives in the south-western Cape at Potberg near Bredasdorp. In Namibia a few birds maintain a precarious toe-hold on the Waterberg where there was once a stable colony, and in the midlands of Zimbabwe substantial numbers of birds roost on Wabai Hill at Shangani. In theory they could establish a viable colony but, for reasons as yet unexplained, they breed unpredictably, with sometimes only a single pair rearing young.

As the descriptive Afrikaans name *Kransaasvoël* ('cliff vulture') indicates, the Cape Vulture is dependent on cliffs for roosting and breeding. The nests are sited on ledges as much as 150 metres above the cliff base, and usually in the shade facing south or south-east. They are built close together, sometimes almost touching, and the birds are tolerant of their neighbours. A Cape Vulture nest is a compact structure of sticks and grass about 70 centimetres across and 11 centimetres high, with a cup 35 centimetres in diameter. Some nests are substantial and others are rudimentary, but in general they are prone to weathering and have to be rebuilt each year. Studies of ringed birds have shown that a pair may occupy the same site year after year.

The main threat at Cape Vulture breeding colonies is disturbance by humans. Baboons have the potential to be a serious problem but accounts of actual predation are rare and the vultures do not usually react to their presence nearby. They do, however, react with definite alarm to Black Eagles, stretching out their necks towards them and emitting a hoarse braying call. The eagles, which often breed on nearby cliffs, have been seen to snatch chicks from the nest and are justifiably regarded as a threat by the vultures.

White-backed Vultures nest in trees rather than on cliffs, forming loose colonies in a suitable tract of country where, for example, large trees line a watercourse. Usually the nests are placed at the very top of the tallest trees in the locality, and often in a thorny species. Sometimes they are built on top of a Red-billed Buffalo Weaver's nest, and recently some have been found on the superstructure of electricity pylons near Kimberley, a unique instance of vultures breeding on a man-made structure. The nest is small relative to the size of the vulture and comprises a flat platform of sticks about 65 centimetres across and 27 centimetres high that is lined with dry grass. The vultures may use the same sites year after year – for at least 15 years in one case – or they may breed opportunistically, settling in a different area to take advantage of a favourable food source such as a culling operation and then moving away again after a year or two. Mobility such as this must have considerable advantages for breeding.

The impressive Lappet-faced Vulture is found mainly in semi-arid and desert environments, particularly in the Namib Desert Park in Namibia where it breeds in harsh conditions. Because suitable large, flat-topped trees do not grow in the more arid regions, it usually builds its nest lower than do other tree-nesting vultures, at a height between 3 and 15 metres above the ground. Thorn trees are favoured, but in the mopane scrub of the Gonarezhou National Park in Zimbabwe purple-pod terminalias are used. Covering the whole top of the tree, the nest is virtually inaccessible to mammalian predators, including humans. It is a sturdy structure and huge, measuring 2.2 metres across and 70 centimetres deep, and may contain sticks 2 metres long. Initially lined with dry grass, the central depression becomes carpeted with fragmented hair-ball pellets as the breeding cycle progresses. In a remarkable case in east Africa a Lappet-faced Vulture shared its nest with a Greater Kestrel and the two birds incubated their respective clutches only 65 centimetres apart!

Despite the huge size of the nest and the considerable investment of time and energy in its construction, the vultures may frequently move to other sites and have as many as four nests clumped in their territory. Do these nests perhaps serve as a form of territorial advertisement? On the other hand, 11 pairs of Lappet-faced Vultures at a location in Zululand are known to have remained in the same nest trees for 13 years.

The Lappet-faced Vulture is classified as 'Vulnerable' in southern Africa and has suffered, along with other vultures, from the indiscriminate use of poisons, especially strychnine. It is particularly wary and sensitive to disturbance when breeding, and an egg or small chick could easily overheat if an adult is kept off the nest for any length of time.

The White-headed Vulture is a most attractive and colourful species with a regal demeanour that matches its appearance. Its large, flat platform of sticks is located at the top of a tree, between 10 and 21 metres above the ground. A variety of trees may be used, but acacias and baobabs are frequently chosen. Attachment to a nest site varies, with birds sometimes remaining at the same nest for a long period – at least 11 years in one observation – or sometimes moving regularly to an alternative site. The nest is about 1.2 metres across and 36 centimetres deep and, like that of the Lappet-faced Vulture, its lining of dry grass becomes covered with a carpet of disintegrated pellets as the nesting cycle progresses.

With certain exceptions, the breeding habits of southern African vultures are essentially similar. Unlike eagles and other birds of prey, there is virtually no difference in size between the sexes, nor in colour (except in the case of the White-headed Vulture, the male of which has dark secondaries whereas the female has white ones). This means that in most observations the role of the sexes must be assumed, although in the case of the Cape Vulture ringed birds have confirmed certain assumptions.

Most species appear to have no distinct courtship displays, except for the Palm-nut, Egyptian and Bearded Vultures whose impressive aerobatics are similar to those of some eagles. The Bearded Vulture pair may perform cartwheeling flights with interlocked talons, a rare instance where this behaviour is genuine courtship rather than an aggressive interaction. Although it is not certain whether such activities serve a courtship function, the Cape Vulture often indulges in tandem flying, when two birds fly close together, and in 'jetting', when a pair flies in at speed to land on the nest ledge. Copulation takes place on the nest or in trees near carcasses. In the case of the Cape Vulture when the male mounts the female he grasps the ruff at the back of her neck with his bill to the accompaniment of characteristic hoarse calls. The Bearded Vulture performs a pre-copulatory 'dance', and some species, especially the Hooded Vulture, indulge in mutual preening after mating.

Most vultures in southern Africa lay their eggs in the winter months between late

1

2

1 *In some localities White-backed Vultures have adapted to nesting on electricity pylons.* ***2*** *A White-backed Vulture chick on its nest, at the very top of a tall tree.*

1 *Cape Vultures protect their downy offspring from the elements on exposed ledges on which they breed.* ***2*** *Bone deformity caused by a lack of calcium in the diet will prevent this Cape Vulture chick from flying.* ***3*** *Like most other tree-nesting vultures, the White-headed Vulture places its nest right at the top of a suitable tree.* ***4*** *A Lappet-faced Vulture chick crouches low on its huge nest, which measures 2 metres across.*

May and July, but the Palm-nut Vulture lays later, in August and September. In a particular region eggs are often laid at more or less the same time, usually within a period of six to eight weeks. The usual clutch is a single egg, except in the case of the Bearded and Egyptian Vultures which regularly lay two eggs. Occasionally some single-egg species have been recorded with two eggs in the nest, but usually it is not possible to know whether they were laid by the same female. However, there is a recent record of two young being raised by a Lappet-faced Vulture pair from their own two-egg clutch.

The eggs are either plain chalky white or variably marked with red-brown. Those of the Egyptian Vulture are particularly handsome and are thus unfortunately prized by egg-collectors. The Cape Vulture lays a plain white egg which becomes nest-stained as incubation progresses, and the White-backed Vulture lays eggs that are either plain or marked. The Bearded Vulture's eggs are almost entirely rust-red, but in this case the colour is acquired through contact with the ventral plumage of the adults as incubation progresses. The adults' feathers in turn have become red 'cosmetically' through contact with iron oxide in the caves and potholes in which the birds roost. The eggs of the other species are all marked with red-brown.

The long nesting cycle of vultures is characterised by the degree to which both sexes share duties at the nest, in marked contrast to eagles for example. An equal division of labour is a strategy that compensates for the long periods spent away from the nest by one member of the pair on its often extensive forays for carrion. Both partners are involved in nest-building, the female being mainly responsible for forming the structure with material provided by the male. Both sexes share incubation, usually sitting for about 48 hours at a stretch, although a Cape Vulture has been recorded at a nest for 94 hours before being relieved. In the case of the Bearded Vulture both birds take turns during the day, but only the female broods overnight. The incubation period for most species is 54 to 58 days, except for the smaller Palm-nut and Hooded Vultures which incubate for a shorter time.

In general vulture nestlings are fed by mouth-to-mouth regurgitation, but Bearded and Egyptian Vultures bring food to the nest in their feet and tear off pieces for the young in the manner of an eagle. Initially the downy chick, which is white in some species and grey in others, is constantly attended, but once it is substantially feathered both parents may leave it alone in the nest while they forage. Cape Vultures, however, are remarkably attentive towards their young and an adult may remain with the chick even when it is well-grown. Although the Bearded Vulture usually lays two eggs it raises only a single chick. This is because the eggs are laid several days apart but incubation commences once the first egg is laid. The first-hatched chick has a size advantage over its weaker sibling which starves to death because it cannot compete for food.

The nestling period of vultures is long, up to 125 days for the Bearded and Lappet-faced Vultures, and 140 days for the Cape Vulture. Sometimes Cape Vulture chicks are prevented from flying by wing deformities from a metabolic bone disease caused by calcium deficiency. This occurs in areas where there are no longer hyaenas and other carnivores to break up large bones at carcasses. To try and rectify the situation the Vulture Study Group, an organisation which has been established to monitor and conserve vulture populations, has set up 'vulture restaurants' where bones are artifically broken up and left near the feeding place for the vultures to find. This is but one example of the effort of this group to protect vultures in general and Cape Vultures in particular.

Once the young vulture has left the nest it is by no means equipped to survive, especially as it is dominated at a carcass by adults and rarely gets its fill. It continues to be fed on or near the nest and may remain dependent on its parents for long periods, usually for four to five months in the case of the Cape Vulture, but sometimes for as long as seven months. All young vultures have a distinct juvenile plumage so that aggression from adults is avoided. Nevertheless, the juvenile from a previous breeding season may have to be chased off when its parents begin their new breeding cycle.

Vultures are long-lived birds which reproduce slowly and the mortality of young birds before reaching breeding condition is high. Under normal circumstances the survival of the species is maintained, but if excessive and unnatural mortality occurs, especially of adults, then it is difficult to halt the decline. This has been dramatically shown in the case of the Californian Condor. In southern Africa, except for the Egyptian Vulture, no species has reached such a critical stage, but there is no room for complacency.

Kites

Two kite species – the Yellow-billed and the Black-shouldered – breed in southern Africa, but they belong to different genera – *Milvus* and *Elanus* respectively – and their nesting habits are different. The Yellow-billed Kite is an intra-African migrant, arriving in the region in late July or early August and departing in March. It is usually considered to be an African race of the widely distributed Black Kite, which has seven races, but some authorities have suggested that it should be considered a species in its own right. In Africa it is certainly convenient to call it the Yellow-billed Kite to distinguish it from migrant Black Kites from the Palaearctic. These kites are probably among the most successful raptors in the world, being versatile predators, opportunistic scavengers and bold pirates.

The Yellow-billed Kite occurs throughout southern Africa and breeds as far south as the south-western Cape, but it appears to be declining as a breeding species there and in Zimbabwe. Soon after the birds arrive in their accustomed breeding area a pair establishes a territory and commences courtship. The male 'chases' the female in a slow, weaving flight to the accompaniment of their characteristic mewing and tremulous whistling calls. Copulation takes place in trees near the nest, often after the male has placed nesting material on it.

Both birds build and may use the same nest year after year, or they may move to another site nearby. They are usually solitary breeders but sometimes several pairs may form a loose colony, their nests spaced 75 to 100 metres apart in a suitable area of large trees. Various trees are used, including introduced species such as eucalypts. The nest is placed in a main fork or on a lateral branch beneath the canopy, mostly at a height of 6 to 15 metres above the ground, but sometimes much higher. It is a substantial structure of sturdy sticks 45 to 60 centimetres in diameter and 30 to 45 centimetres deep. The cup, measuring 20 to 25 centimetres across, is lined with an incredible assortment of material such as dung, sheep's wool, clods of mud, plastic bags, paper and rags, but green leaves are never used. The Yellow-billed Kite's habit of gathering all kinds of material for lining is well summed up on a nest record card in Zimbabwe which succinctly gave the contents of a nest as: 'two eggs, two blue socks, a pair of pink panties, some dung – probably cow'.

Eggs are laid in September and October throughout southern Africa, usually in clutches of two, although one and three have also been observed and there is one record of a four-egg clutch. They are white, variably marked with spots, blotches and scrolls of dark red. Some eggs are very handsomely marked whereas others may be plain. As is the case with most birds of prey, incubation begins when the first egg is laid so the young vary in size. Incubation is carried out mainly by the female and the male feeds her. Both birds are very bold in defence of the nest.

The newly-hatched downy chicks are attractive creatures, grey above and white below, with fine, hair-like down on top of the head. They are tended by the female who feeds them with prey items provided by the male, sometimes in excessive amounts. One nest with a well-grown nestling was found to contain 25 Cape mole-rats, five mice and a small snake, many of which were decomposing. The young leave the nest after about six weeks but remain dependent on their parents for about another seven. Although the Yellow-billed Kite is widespread in southern Africa, its breeding biology needs to be thoroughly studied; even the incubation period has yet to be accurately established.

In contrast, the breeding behaviour of the ubiquitous and handsome Black-shouldered Kite is well known and includes some fascinating habits. Although a population in an area may appear to be sedentary, ringed and colour-marked birds have revealed that there is in fact a considerable turnover of individuals. Not only are these

A Yellow-billed Kite chick on its nest, which is lined with miscellaneous items. A rhombic skaapsteker and a gerbil have been brought as prey but lie uneaten.

kites nomadic, but some may even make long-distance movements, and one ringed as a nestling near Pretoria was recovered $4^1/_2$ years later in Uganda, more than 3000 kilometres from its birthplace.

Black-shouldered Kites roost communally, flying in at dusk to settle in trees or reedbeds to form assemblies of 30 to 70 birds, or even as many as 150 at times. Despite this roosting sociability, individuals are highly territorial and when perched wag their tails up and down as a signal to warn off passing intruders. During the breeding season a female selects a male with the best territory, but she does not form a permanent pair bond and may move to the territory of another male the next time she breeds.

There is no distinct breeding season for Black-shouldered Kites, and they have been recorded laying in all months of the year. The only area in southern Africa where there appears to be a regular laying peak is the south-western Cape, where eggs are laid in spring and early summer, from August to November. Breeding is largely opportunistic in response to a good food supply, but it appears that the kites may be able to predict a future abundance and lay accordingly. It has been suggested that reproductive steroids in their predominantly rodent prey trigger hormonal activity in the kites in anticipation of an increase in the rodent population. During a population explosion of rodents the birds may raise several consecutive broods, and then not breed at all for a time.

Courtship consists of circling flights during which the male dives down towards the female and she turns over and briefly presents her talons. The male also flies round with dangling legs and slow, exaggerated wing-beats in what is known as a 'butterfly flight' display. Mating takes place in trees near the nest. According to one study, during the nest-building and egg-laying stages the male delivered larger prey – in the form of Angoni vlei rats as opposed to the smaller striped and multimammate mice – to the female. This may well be an important prerequisite for successful breeding.

The nests are usually built in trees, although occasionally they are also found on man-made structures such as telephone poles and electricity pylons. Thorn trees are preferred but the choice varies according to availability, and in the south-western Cape pine trees are often used. The nest is placed near the top of the tree, beneath the highest branches of the canopy, at heights varying from 75 centimetres on a low bush to 18 metres at the top of a pine. Most nests, however, are not more than 6 metres above the ground. The structure comprises a flat platform of small sticks and twigs, usually 30 centimetres across and 7 centimetres deep, and it is lined with dry grass that during the nestling period becomes carpeted with decomposed pellets that have been regurgitated by the chicks. The male and female take equal shares in building the nest and are most active in the early part of the morning, breaking off sticks and twigs with their bills and also collecting some from the ground. The nest is not substantial, and a new one is usually built for each breeding attempt. The birds complete the structure in about two weeks, although in one case the first egg was laid only 13 days after construction had begun. The more quickly a pair builds its nest, the more rapidly it can respond to prey abundance and take advantage of it.

The clutch comprises 3-6 eggs whose cream ground colour is overlaid with red-brown blotches and streaks. Usually the eggs are so well marked that the ground colour is almost obscured, but some have a well-defined red-brown cap at the broad end. They are laid at intervals of one to two days and incubation may begin when the first egg has been laid or when the clutch is nearing completion. The female does almost all the incubation but the male may assist her for brief spells. It has been possible to establish the incubation period accurately at an average of 31 days. Like the Yellow-billed Kite, these small kites are bold in defence of their nests and will attack and strike intruders, including humans.

The attractive newly-hatched chick is covered in buff down and appears to have a head too large for its body. No sibling aggression has been observed and the nestlings grow rapidly: by 21 days old they are well feathered. The emergent dorsal feathers are tipped with white, effectively camouflaging the chicks when they crouch flat in the nest should a potential predator such as a crow fly overhead. Initially the female broods and feeds them on food provided by the male, but later both birds hunt and merely deposit prey on the nest, leaving the young to feed themselves.

The nestling period varies between 30 and 38 days, averaging 35. Once the young leave the nest the male takes over the entire responsibility for feeding them, sometimes for as long as three months. In the meantime the female may pair with another male and lay a new clutch. This is not so much 'desertion' of her original mate and brood, but rather a means whereby maximum productivity can be achieved during periods of plenty.

1

2

***1** Black-shouldered Kites place their nests in thorn trees where these are available, but also nest in other trees and even on man-made structures.* ***2** Their well-marked eggs are laid at intervals of a day or two, and incubation usually begins before the clutch is complete.*

The Cuckoo Hawk and the Bat Hawk

Although somewhat loosely sharing the term hawks, the Cuckoo Hawk and Bat Hawk are quite distinct from one another and, like the kites, belong to different genera: the Cuckoo Hawk to the genus *Aviceda* and the Bat Hawk to the genus *Macheiramphus*. The former is an unaggressive species that subsists mainly on insects, chameleons and lizards and occasionally on mice, snakes and small birds. It is normally unobtrusive but during courtship becomes highly vocal, emitting a loud whistling call. Its courtship displays consist of tumbling aerobatics, during which it reveals to striking effect its chestnut underwing coverts.

The nest, one of the most unusual of all African raptor nests, is placed beneath the canopy of a leafy tree between 7 and 25 metres above the ground. Both birds take part in building it, either working independently or the male may pass twigs to the female for her to place in position. Remarkably small, the nest measures 25 to 40 centimetres in diameter, with a leaf-lined cup 10 to 15 centimetres across and 12 to 20 centimetres deep, and is only just large enough to accommodate two young at the end of the nestling period. It comprises mainly leafy twigs a few millimetres thick which are collected in the vicinity of the site in an unusual way. From a perch the bird peers about, selects a suitable twig, and then flies up and grabs it in its feet, often hanging upside down with flapping wings before snipping it off with its bill. It then rights itself dextrously, alights on a perch below the nest, transfers the twig to its bill and flies up to the nest with it. Usually the twig is snipped off quickly, but an observer watched a bird struggling for six minutes before it managed to detach its chosen branchlet. Sometimes the hawk will sidle along a branch like a parrot and snip off a twig. It has been suggested that the pronounced notch near the bill's tip gives the bill a secateur-like action that facilitates this unusual method of gathering nesting material.

According to one record, it took a Cuckoo Hawk pair 11 days to build a nest, for which 144 leafy twigs were collected. Another nest taken to pieces after a breeding attempt had failed contained 98 branchlets. The birds add leafy twigs to the nest throughout the incubation period, and to a lesser extent during the nestling period too. The leaves attached to the sticks soon die, and when they do the nest resembles leaf debris caught in the tree, or the nest of a bushbaby (galago). Usually a new nest is built each time a pair breeds, but in one observation a second clutch was laid in the same nest ten weeks after the first brood flew successfully.

Clutches of two eggs have been recorded between September and March, the eggs being white with variable dark red markings. Since the male and female birds can be identified by their respective red and yellow irises, observers have been able to determine how the sexes share duties at the nest. Both birds incubate the eggs, doing so for a period of 29 to 30 days and changing over frequently. Unlike most raptors where the male brings prey for the female and incubates for short

WHY DOES THE CUCKOO HAWK BUILD A LEAFY NEST?

Although the Cuckoo Hawk is not dashing and rapacious as are the sparrowhawks and goshawks, it does have in common with these species a barred plumage that may protect it to some degree. Any predator that would hesitate to attack a sparrowhawk or goshawk would equally think twice about attacking a Cuckoo Hawk. However, the unaggressive Cuckoo Hawk is apparently still vulnerable to predation: there are two records of adults having been killed on the nest and several which suggest that other raptors or crows have killed and eaten nestlings.

The fact that the nest looks like dry debris caught in a tree undoubtedly helps to conceal it from predators – and, incidentally, from ornithologists. One of the reasons for the scarcity of Cuckoo Hawk nest records is that the nest is seldom recognised as that of a raptor. The adults are silent and unobtrusive when breeding, although the loud begging calls of the chicks during the nestling period somewhat nullify the advantage of a camouflaged nest.

Several nest records remark on the proximity of Cuckoo Hawk nests to human habitation and have suggested that this confers a degree of protection to the breeding birds. It would be very difficult to substantiate this and it may be purely a fortuitous association. However, it would be reasonable to conclude that the unusual leafy nest of the Cuckoo Hawk does provide this vulnerable species with a measure of protection from potential predators ●

The dead leaves in a Cuckoo Hawk's nest give it the appearance of debris in the tree, helping to conceal it from predators.

1

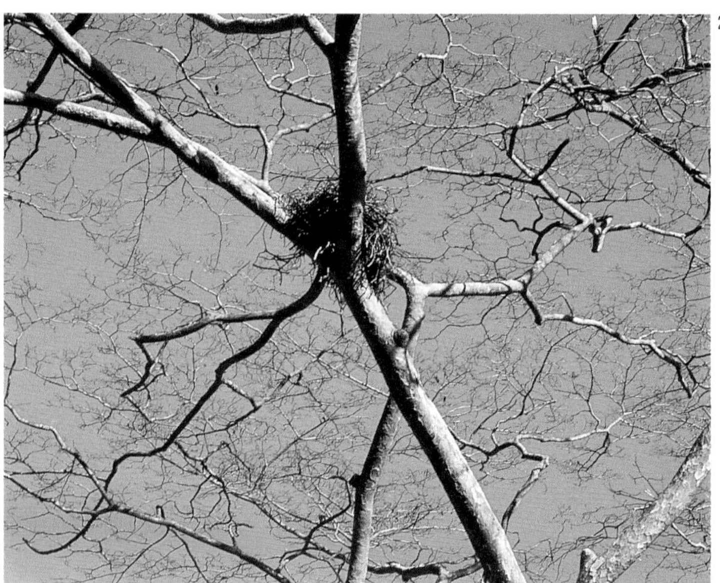

2

1 *The Bat Hawk invariably lays a single large egg.* ***2*** *A Bat Hawk's nest in a smooth-barked mountain acacia is placed on a lateral branch.*

spells while she is away feeding on it, each bird forages for itself. Male and female also share care of the chicks, taking turns to brood them while they are still small and to bring food to the nest. Initially the adult tears up larger prey items such as chameleons, but later it clings briefly to the side of the nest with prey in its bill and passes it to a chick which swallows the item whole. The food brought to nests varies, presumably depending on local availability. At one nest mostly chameleons were brought, while in another study involving several nests insects predominated, comprising 95 per cent of 206 food items identified. Most of these insects were green juvenile grasshoppers, and it appeared that the hawks caught these in preference to the brown adults which seemed to be more abundant – at least to the observer – in the area where the birds hunted.

Initially covered in white down, the chicks develop feathers rapidly from the age of 14 days, and by the time they leave the nest after 35 to 42 days they have a juvenile plumage that closely resembles that of an African Goshawk. After leaving the nest the young remain in its vicinity but do not return to it, staying with their parents in a family group for at least five weeks.

The Bat Hawk, as its name suggests, subsists mainly on bats, although it does also prey on birds. It hunts mainly at dusk and dawn, but sometimes also at night and can even catch prey when there is no moonlight. Small bats – up to 30 grams – are preferred, and these are swallowed in flight to make maximum use of the period after sunset when bats emerge.

The speed and fluency of this dashing raptor's flight not only make it an efficient predator, but also contribute to its courtship displays. These consist of spectacular aerobatics and chasing flights, during which the birds may zigzag through the branches of the nest tree before alighting. They build the nest 10 to 25 metres above the ground, characteristically placing it in a fork on a horizontal branch of a smooth-barked tree such as a baobab, a mountain acacia, a star chestnut or a eucalypt.

The nest is constructed of sticks about a centimetre thick, with finer ones used for the cup. It has a characteristic shape, being wider at the base than at the top so that it looks like an inverted basin. One typical nest measured 56 centimetres in diameter and 38 centimetres deep, with a cup 14 centimetres across. The cup is fairly deep and is sparsely lined with a few green leaves. Both birds build the nest, collecting sticks from trees by grasping them in flight with their feet and breaking them off. Nests may be re-used, and in one case a pair returned to the same nest for six consecutive years. The birds are bold in defence of the nest and may attack humans who climb the tree.

The single egg laid is generally white, although some have a few red blotches and sepia spots. It is elongate, and large for the size of the bird. The laying of a single egg is undoubtedly linked to the Bat Hawk's specialised feeding habits which would make the raising of more than one chick difficult. During the day only the female incubates, and it is not known whether the male assists at night. He spends the day in the nest tree or nearby, making no particular effort to conceal himself, and in the early morning and evening may bring material to add to the nest. The incubation period is about 48 days, as long as that of some eagles, and the nestling period is 67 days. These long incubation and brooding periods would also seem to be linked to the Bat Hawk's specialised feeding habits. The chick is covered in white down and initially it is brooded and fed by the female on food brought by the male. When it is small bats are torn up for it, but later it swallows them whole.

The Bat Hawk's main breeding season is from September to November but there are also records from other times of the year. Presumably in these cases the birds are breeding in response to seasonal prey abundance.

Eagles

Eagles are admired for their inherent majesty and power, and their nests are among the most impressive of all birds'. They may reach massive proportions: a Golden Eagle nest in Montana, USA, supported between two rock columns on a cliff, attained a height of 6.7 metres, and in southern Africa the tallest recorded nest was that of a Black Eagle which was 4.1 metres high.

Most eagles occupy their territories for long periods. Black Eagles, for example, have been known to nest in the same area for as long as 67 years, and two pairs of Tawny Eagles and their successors bred on the same farm for at least 50 years. While individual nests may be used for a long time, eagles also quite frequently move to other sites in their territory, sometimes using three or four nests over a period of years. These moves often appear to be random and unpredictable, and bear no relation to whether the previous season's breeding attempt was successful or not. Sometimes, however, they become necessary, as when a Black Eagle's nest under an overhang on a cliff runs out of headroom because of its height.

Of the 15 eagle species that breed in southern Africa, most build bulky nests of sticks which vary in size and situation from one species to another; snake eagles,

1 *A pair of Black Eagles at their sizeable nest in the Matobo Hills of Zimbabwe.*
2 *A cliff ledge is the usual site for a Black Eagle nest.*

1 The African Hawk Eagle builds a large nest for its size. ***2*** *A Wahlberg's Eagle nest, on the other hand, is relatively small, being built quickly by this intra-African migrant.* ***3*** *A Booted Eagle has placed its nest securely between the base of a small tree and the steep cliff face.* ***4*** *Martial Eagles normally nest in trees, but where these are not available they may breed on rocky outcrops.*

however, construct small flat ones. An experienced observer is usually able to recognise which eagle originally built a nest, although sometimes it may have been taken over later by another species. Most of the species build in trees, with the exception of Black and Booted Eagles which habitually use cliffs. A recent phenomenon, however, has been the use of man-made structures. Black Eagles have been recorded breeding on microwave towers and electricity pylons, and Tawny, African Hawk and Martial Eagles, as well as Brown and Black-breasted Snake Eagles, have all been found breeding on electricity pylons. The use of pylons may enable some species to expand into areas where there are no suitable trees or cliffs: in the case of the Martial Eagle an aerial survey located 18 nests along a 420-kilometre stretch of power lines in the Karoo. This population was stable with good breeding success. The appeal of pylons is illustrated by the observation of a pair of Martial Eagles moving from their unsuitable site in a tree on a low cliff to a pylon within a year of its erection nearby.

The Black Eagle, a species of hilly or mountainous country, is probably the most intensively studied eagle in the world, thanks to a 20-year survey of its breeding in the Matobo Hills in Zimbabwe. The combination of suitable nesting habitat and a prolific food source in the form of hyraxes (98 per cent of the 1892 prey items recorded) enabled a population of 60 pairs to breed in an area of 620 square kilometres. This average density of one pair per 10.3 square kilometres is remarkable, and probably the highest known concentration of large eagles anywhere in the world. Cliff ledges are the usual site for Black Eagle nests although sometimes trees may be used, in which case they will still be located in a hilly area. In one record the disused nest of an African Fish Eagle in a baobab was taken over. Nests are usually about 1 metre in diameter, with a leaf-lined bowl that is 30 to 40 centimetres across. Their height varies depending on their site and how long they have been in use, but not infrequently it reaches 2 metres. The sticks they contain may be large, and one measured was found to be 6 centimetres in diameter and 1 metre long.

The characteristic nest of the Tawny Eagle is almost always situated on the crown of a tree, preferably a thorny acacia, and is open to the sky. When the tree is in leaf the nest is difficult to detect, but in a bare tree it can be seen from a considerable distance away. It is usually sited at a height of between 6 and 15 metres above the ground, but in some cases may be as high as 30 metres up. The structure is fairly flat and about 1 metre in diameter, but becomes bulky if used over a period of years. The bowl is 25 to 45 centimetres across and is lined predominantly with miscellaneous dry material such as grass, seed pods and maize leaves, although green leaves are also used and pieces of plastic, paper and sacking have been found in nests from time to time. Tawny Eagles may use the same nest for several years in succession, but sometimes move to a new site quite often. Frequent moves are probably caused by changes in the conformation of the tree's crown which make securing the nest difficult or inhibit easy access to it when branches grow upwards.

Wahlberg's Eagle is an intra-African migrant that arrives in southern Africa in the latter part of August and departs in March or early April. Because these eagles have a limited time in which to complete the breeding cycle, they begin courtship and nest-building soon after arriving and lay their eggs in late September and early October. Their small, basket-shaped nests are sturdily constructed and measure 60 to 70 centimetres in diameter and 30 to 40 centimetres deep, with a leaf-lined cup 20 to 25 centimetres across. The birds may repair existing nests by building up the perimeter, but when necessary they rapidly construct a new nest, and in one observation an egg was laid two weeks after building had begun. They prefer to breed in river valleys, probably because this is where the large trees they favour may be found, and the nests are situated beneath the canopy, usually between 8 and 12 metres above the ground. A pair and its successors may occupy a territory year after year, and in one case a particular territory was known to have been in use for at least 24 years.

Until fairly recently the Booted Eagle was thought to be just a migrant from the Palaearctic, but in 1973 it was discovered to be nesting in the Cape Province. By 1990, and by means of systematic research, 118 confirmed nesting sites had been located and it was estimated that at least 400 pairs were breeding in the Cape Province. It seems that the previously undiscovered Booted Eagle may be the most common breeding eagle in the Western Cape! In March, after the breeding cycle is complete, the Cape population appears to move northwards to the north-western Cape, Namibia and southern Angola. In 1983, however, nests which indicated a laying time of June were discovered on the Waterberg Plateau in northern Namibia. As Cape birds lay in the second half of September, this appeared to be a distinct breeding population. This discovery suggests that there are probably three Booted Eagle populations: Palaearctic migrants (present in southern Africa from November to March); birds which breed in the Western Cape from August to March and then migrate northwards; and an indeterminate breeding population in Namibia which lays in about June. The situation, especially with regard to the birds breeding in Namibia, requires further elucidation.

Like the migrant Wahlberg's Eagle, the Booted Eagle starts breeding with some urgency, and its six-month cycle lasts for most of the time that it is in southern Africa. Courtship and nest-building begin soon after its return in August, and in one nest eggs were laid 33 days after the pair had started building it. In this region most nests have been found at the base of a small tree or bush on a cliff, and only occasionally have they been located in trees. This practice appears to be in contrast to that of Palaearctic Booted Eagles, which build substantial nests in trees. Here the nest is a fairly small, flat structure, 45 to 60 centimetres across, and constructed of pencil-thick sticks. Its usually well-concealed position and the fact that the species breeds in remote areas undoubtedly contributed to it being overlooked for so long. Booted Eagles are remarkably tolerant of other members of the same species and sometimes locate their nests only 0.5 kilometres apart. In one valley four nests were found within a distance of 2.8 kilometres.

For its size, the African Hawk Eagle builds a very large nest, about a metre across with a cup 25 to 30 centimetres in diameter, constructing it with substantial sticks. Nests may be used for long periods and some grow to be 1 to 1.5 metres deep. They are normally placed within a tree's canopy in a main fork or on a lateral branch, usually at a height between 9 and 15 metres above the ground and, although some may be rather exposed, most receive shade for much of the day. There are records of this species nesting on cliffs in Kenya, but none as yet in southern Africa.

The biology of Ayres' Eagle has been little studied in southern Africa and it appears to be nomadic or perhaps an intra-African migrant. The few nests that have been found south of the Zambezi River have all been in Zimbabwe. They occurred in thickly wooded country, usually in a steep valley, and were not conspicuous from a distance as are those of the African Hawk Eagle. The Ayres' Eagle's nest is, however, similar to that of the African Hawk Eagle but smaller, being only about 70 to 90 centimetres in diameter.

Long-crested Eagles are nomadic and breed erratically in response to the availability of vlei rats, their main prey. They build their nests in tall trees, well concealed beneath the canopy, and favour eucalypts where these are available. The nest is rather small, usually about 60 centimetres across, and is similar to that of Wahlberg's Eagle. The Long-crested Eagle has a tendency to build new nests quite frequently, or to repair old ones and fail to use them, behaviour which is undoubtedly linked to a fluctuating food supply.

As befits a bird of its size, the Martial Eagle builds a large, impressive nest that may be used for long periods. In one record a nest was occupied for at least 21 years. On the other hand, some individuals may have several nests within their large territories and move from one to another at random. These eagles use sturdy sticks in the nest's construction and a large one may be 2 metres across and up to 2 metres deep if it is built in a substantial fork. Even an abandoned nest may last for many years. A nest may be either beneath the canopy or, if it is built on top of a Sociable Weaver's massive structure for example, it may be open to the sky. A tree is the preferred site

1

2

__1__ Tawny Eagles almost always place their nests in the crown of a tree.
__2__ The young Crowned Eagle chick is cared for by the female until it is well feathered and can be left alone on the massive nest.

for a nest, but if a suitable one is not available the Martial Eagle will also build on a pylon or a rocky outcrop.

The Crowned Eagle is the only eagle in southern Africa that regularly inhabits forest, but it is by no means confined to this habitat and may occur and breed in mature woodland too. It even nests in large trees in peri-urban areas, for example in Pietermaritzburg where its prey includes domestic dogs and cats. This eagle builds an enormous nest that is placed in the major fork of a large tree and is rarely accessible without ropes and ladders. Eucalypts and baobabs are utilised in addition to forest trees such as yellowwoods, and very occasionally the nest may be situated on a cliff. Nests used for long periods, sometimes for at least 50 years, may measure 2 to 2.5 metres across and 2.5 to 3 metres deep, and are large enough to accommodate a double-bed mattress. Some of the largest sticks used in construction have measured 8 centimetres in diameter and 1.2 metres long.

In sharp contrast to the massive nest of the Crowned Eagle, those of the four species of snake eagles are typically small and flat, built with pencil-thick sticks. The nests of the Brown and Black-breasted Snake Eagles are similar and measure 60 to 70 centimetres in diameter with a leafy bowl about 25 centimetres across. They are often placed on the crown of a flat-topped thorn tree, or sometimes on a euphorbia, and are open to the sky. Because of their situation and small size, they are very difficult to detect, especially as the brooding adult crouches flat and cannot be seen from below. Both these species have been recorded utilising the disused nests of other birds of prey, but they built their own typically flimsy nest on top of the existing structure. Very few nests of the Southern Banded and Western Banded Snake Eagles have ever been found. Not only are these birds secretive when breeding, but they conceal their small nests among creepers within the canopy of a tree.

Although the Bateleur is often considered to be closely related to the snake eagles, its nest is a sturdy, compact structure that measures about 60 centimetres across and 30 centimetres deep, with a neat, leafy cup 25 centimetres in diameter. It is usually situated 10 to 15 metres above the ground in the tallest tree in the area and is built within the canopy in a main fork, or occasionally on a lateral branch. A study in the Kruger National Park revealed that knob-thorns, whose thorns give protection against mammalian predators, were preferred, followed by jackal-berry trees, which are shady and conceal nests. If humans approach, a Bateleur either leaves its nest unobtrusively long before they are near, or remains on it, crouching flat. The foraging method of Bateleurs, which involves long absences by both parents, renders their broods very vulnerable, and one third of all breeding failures in the Kruger Park study were due to depredation of nestlings. Careful selection of the nest site appears to minimise this risk.

A variety of trees, including dead ones, and occasionally cliff ledges or rocky outcrops are the sites of African Fish Eagle nests, and they are quite often fully exposed to the elements. Although these eagles usually breed near water, if they cannot find a suitable site they may build their nests several kilometres away. The nests vary considerably in size, depending on the site and for how long they have been in use, but usually they are 1.2 to 1.5 metres across and 30 to 60 centimetres deep. The cup may be lined with greenery, as well as miscellaneous dry material, including the occasional weaver's nest. Like many other eagles, African Fish Eagles may use their nests for long periods – 21 years in one case – or a pair may have several sites in their territory. Nesting density may be high in prime habitat such as along the Chobe River in Botswana, where in 1971 there were 38 pairs of African Fish Eagles evenly spaced over a distance of 55 kilometres. In the Okavango Delta in Botswana some pairs were only 400 to 500 metres apart.

The breeding cycle and behaviour of most eagles conform to a pattern, although individual species show some variation. In courtship displays eagles soar together or perform undulating flights during which they may briefly touch talons. Encounters in which birds interlock talons and tumble downwards in whirling flight have traditionally been regarded as courtship. However, this belief has recently been convincingly challenged, and it appears that in most instances these whirling flights involve aggressive interactions with intruders into a territory.

Some species known for particularly impressive or dashing aerial displays are the Black, Booted, Crowned and Ayres' Eagles. Often displays are accompanied by loud calling, and in the case of the African Fish Eagle duetting in itself undoubtedly has

1

2

__1, 2__ African Fish Eagles usually nest in trees – sometimes dead ones standing in water – but also build on rock outcrops. In either situation the nests are often exposed to the elements.

a courtship function as well as maintaining the pair bond. The Bateleur, the most aerial of all eagles, appears to perform little in the way of courtship, and many of its spectacular aerobatics probably involve aggressive or social interactions. The pair bond of the Bateleur seems to be maintained by bouts of mutual preening, which is unusual behaviour for an eagle. In addition to an undulating courtship flight over its forest territory, the Crowned Eagle performs a non-aerial courtship display during which the male runs around the female on the nest with his wings raised to display his colourful chestnut underwing coverts. An important aspect of the formation and maintenance of the pair bond is the provision of prey by the male for the female. This not only enables her to build up her fat reserves prior to egg-laying, but also demonstrates the male's 'fitness' as a suitable mate.

How a pair shares nest-building is poorly documented for most eagles. It appears that both sexes usually participate, although the male is more likely to collect material which he brings for the female to arrange. He gathers sticks from the ground or breaks them off trees by hanging on them and bouncing up and down until they detach. The length of time it takes to repair a nest or construct a new one is very variable; it may take weeks or months. Once the nest is complete the cup is lined with greenery, and this lining is maintained throughout incubation and quite often well into the nestling period. Eagles are particularly noted for using green leaves to line their nests, the obvious explanation being that they provide a soft bed for the eggs which would otherwise be in contact with the rough sticks of the nest. There may, however, be other less obvious reasons for their use.

Some eagles regularly lay two-egg clutches and some lay only one egg. The one-egg species are Wahlberg's, Ayres' and the Martial Eagles, the four snake eagles and the Bateleur. Sometimes 'two-egg' eagles lay a single egg, and very rarely a 'one-egg' species, such as Wahlberg's Eagle, lays two. In two-egg clutches the first egg is laid several days before the second, and as incubation starts with the first egg, the first-hatched chick (Cain) has a considerable size advantage over its sibling (Abel). In several species sibling aggression, or Cainism, is regular and results in the death of the second chick which is inevitably weaker (see page 19 for details). Eagle eggs are either plain white, becoming nest-stained as incubation progresses, or they are variably marked with rust-red. Some are very handsomely marked, and there is considerable variation even within the same species.

The female does the major share of the incubation and is fed by the male, who may brood for short periods while she is away from the nest feeding on the prey that he has brought. The specialised foraging requirements of the Bateleur result in its behaviour being different, and more akin to that of the vultures. The male Bateleur undertakes a considerable share of duties at the nest, thus enabling the female to forage for herself. When on the nest many eagles sit exceptionally tight and will only flush at the very last minute. The incubation period varies from one species to another: the shortest, lasting 40 days, is that of the Booted Eagle and the longest is that of the Bateleur at 55 days. Most other species incubate for approximately 44 to 48 days.

From the moment it hatches the helpless eaglet is closely brooded and tended by the female while the male supplies prey, often dramatically increasing the rate he maintained in the incubation period when he was feeding only the inactive female. Once the nestling is well feathered and can be left unattended without suffering discomfort, the female may assist the male in hunting, although he still plays the dominant role as food provider. Like the incubation period, the nestling period varies from species to species. It is as short as 50 to 54 days for the Booted Eagle and as long as 114 days in the case of the female Crowned Eagle (the young male leaves earlier, at about 106 days).

Once the eaglet leaves the nest it remains dependent on its parents for a considerable time – usually several months – and initially the nest is still the focal point where prey is delivered. The post-nestling dependence period of the Crowned Eagle is of particular interest. It may last for up to a year, thus producing a young eagle that is better able to fend for itself and giving it an improved chance of surviving the high mortality common among young raptors. However, this means that the parents can breed only every second year, and this biennial cycle is a well-known feature of Crowned Eagle breeding biology. Interestingly, it is by no means invariable, and there are many records of Crowned Eagles in southern Africa breeding regularly every year, even when their previous year's offspring has survived. The reason for this variation is uncertain, but it may be linked to the food supply available in the home range of a pair. For example, hyraxes may be more plentiful and more easily caught in a particular area than monkeys and other prey.

GREEN LEAVES FOR A TABLE-CLOTH?

Eagles are well known for lining their nests, often copiously, with green leaves. What function does this serve? An obvious explanation is that the leaves provide a comfortable bed for the eggs and insulation against heat loss through the nest sticks. Another suggestion is that the leaves may have repellant properties to provide protection against ectoparasites. This is not such a far-fetched idea in view of the fact that European Starlings are known to line their nests with particular leaves which repel parasites. In some areas eagles are known to choose the leafy twigs of certain trees in preference to others, but it is not known whether the leaves have insect-repellant properties. The explanation may simply be that the twigs are more 'leafy' and thus better for use as lining.

Yet another idea put forward is that green leaves serve a hygienic function, acting as a 'table-cloth' that stops food items from falling between the nest sticks, decomposing and attracting flies. Certainly the regular addition of leaves tends to keep the nest fresh throughout the incubation period and during the first half of the nestling period.

The most unusual explanation may perhaps be the correct one: bringing green leaves serves an 'emotional' function. Many leaves are brought during the pre-laying period when the female is not hunting actively, so fetching greenery compensates for her inhibited hunting instinct. Some support for this theory was provided at the nest of a Booted Eagle during extremely inclement weather when the male was unable to catch prey for the small chicks during the course of a whole day. The brooding female became increasingly frustrated, emitting her food-begging call frequently, and she left the chicks four times to fetch green sprays. She even tore off leaves and placed them beneath her, a clear indication of displaced feeding behaviour. Next day the weather cleared, normal feeding resumed, and the female brought no more greenery ●

Typically of eagles, the Bateleur lines its nest with green leaves.

A Black-breasted Snake Eagle chick on its typically small nest which is situated in the crown of a thorn tree.

Buzzards

Four buzzard species breed in southern Africa, three of which – the Forest, Jackal and Augur Buzzards – are considered to be 'true' buzzards of the genus *Buteo*. The fourth, the Lizard Buzzard, is a taxonomic enigma belonging to the genus *Kaupifalco*. It is more closely related to sparrowhawks and goshawks, and perhaps the chanting goshawks in particular, as it resembles them both in its appearance and in its calls.

The Forest Buzzard, previously inappropriately known as the Mountain Buzzard, is distributed in a narrow coastal band from Cape Town to Kwazulu-Natal and then northwards into the Northern Province. Although it was originally thought to be sedentary, recent ringing recoveries have shown that birds from Knysna in the Western Cape have been recovered as far afield as Tzaneen in the Northern Province 1302 kilometres to the north and in Kwazulu-Natal 806 kilometres away. It appears that the species breeds in the Western Cape from September to January and that some individuals, possibly immature birds, make long-distance movements. As yet, there is no more than circumstantial evidence that the Forest Buzzard may breed in the Northern Province and Mpumalanga.

True to its name, the Forest Buzzard inhabits forest and, in more recent times, pine plantations too. It is usually seen in clearings or along the edge of these areas, but it also hunts within them. The first nest, found in 1939 at Humansdorp, was in a yellowwood tree, but from then until the 1980s very few nests were recorded. One found on the Cape Peninsula in 1960 was considered dubious as Forest Buzzards were not known to occur so far south but, in the light of more accurate distributional information, this breeding record has subsequently been vindicated.

Nests are built in indigenous trees or in pines; six out of eight in one study were in pines, but all of these were in plantations adjacent to indigenous forest. They are usually 14 to 20 metres above the ground, and those in pines are normally placed against the trunk and supported by the tree's sturdier lower branches. The disadvantage of nesting in pines is that a tree containing a nest may be felled, and in fact two of eight nests monitored failed for this reason. On the other hand, the increase in pine plantations has extended the potential habitat of Forest Buzzards. The nest is built of small sticks 5 centimetres thick and measures 60 to 70 centimetres in diameter and 30 to 35 centimetres deep. The cup is 20 centimetres across and is lined with greenery, including pine needles.

The Jackal Buzzard could more appropriately be called the Mountain Buzzard because of its preference for hilly or mountainous country. It is widely distributed in southern Africa, including the southern half of Namibia where its range overlaps that of the Augur Buzzard, but does not occur in Zimbabwe. Its nest is generally located on a cliff ledge, often at the base of a small tree or bush, or it may be built in a tree, usually a pine, in which case it is placed in a sturdy fork. The nest proportions are similar to those of the Forest Buzzard and the cup is lined with greenery.

The Augur Buzzard, now considered a true species although previously it was

1 *Jackal Buzzards usually nest on cliff ledges but they may also place the nest in the sturdy fork of a tree.* ***2*** *The Augur Buzzard, a mountain species, also nests mainly on cliffs, siting its stick nest on a suitable ledge.* ***3*** *A Lizard Buzzard's compact nest is lined with 'old man's beard' lichen.* ***4*** *Forest Buzzards have taken to nesting in pine plantations in addition to indigenous forest and have thus been able to increase their range.*

regarded as a form of Jackal Buzzard, occurs mainly in Zimbabwe and northern Namibia. Thus, except for some overlap in Namibia, the southern African ranges of the two species are mutually exclusive. Although predominantly a species of mountains and rocky koppies, the Augur Buzzard may also be encountered in the arid Namib far from the nearest hills. It is particularly common in the Matobo Hills in Zimbabwe, where breeding territories may be only 3 square kilometres in extent, and it is here that most of the details of this species' breeding biology have been obtained. Its nest and habits are very similar to those of the Jackal Buzzard, and it too usually locates its nest on a cliff ledge, although trees, including pine trees, in hilly terrain are also used.

The Lizard Buzzard breeds in trees in woodland, making use of eucalypts and pines as well as indigenous species. It places its nest within the canopy in a main fork against the trunk or on a lateral branch, usually at a height of between 6 and 10 metres. A small, compact structure of sticks, the nest measures about 40 centimetres in diameter. The birds line the cup, which is 15 centimetres across, with 'old man's beard' lichen where this is available and with dry grass or, occasionally, green leaves.

The courtship of the Forest, Jackal and Augur Buzzards includes a number of aerobatic displays, some of which are spectacular and may involve, for example, the male diving down towards the soaring female. Loud calling is a feature of these displays. The Lizard Buzzard, however, performs few aerial flights, indulging instead in prolonged periods of calling from a perch.

All four species lay two eggs, although three-egg clutches are sometimes recorded. The ground colour of the eggs is white, or pale green in the case of the Forest Buzzard when they are fresh. Most eggs are marked to a varying degree with red-brown blotches, although some may be plain and others very sparingly marked. It is not known for how long the Forest Buzzard incubates, the Jackal and Augur Buzzards do so for close on 40 days, and the Lizard Buzzard for 33 to 34 days. The nestling period of the three *Buteo* species is about 50 days, and that of the Lizard Buzzard 40 days.

In other respects the buzzards' breeding behaviour is similar to that of eagles: the male provides prey and the female tends and guards the young. There is evidence of Cainism (see page 19) in the Forest Buzzard and the phenomenon is regularly recorded in the Jackal and Augur Buzzards, although both species may occasionally rear two chicks. There is no evidence of Cainism in the Lizard Buzzard.

Sparrowhawks and Goshawks

The nine species of sparrowhawks and goshawks that breed in southern Africa can be divided among three genera: those of the *Accipiter* genus which are the Ovambo Sparrowhawk, the Red-breasted Sparrowhawk, the comparatively small Little Sparrowhawk and the considerably larger Black Sparrowhawk, the Little Banded Goshawk and the African Goshawk; the Pale Chanting Goshawk and the Dark Chanting Goshawk of the genus *Melierax*; and the controversial Gabar Goshawk which belongs to the genus *Micronisus*. Some cogent arguments have been put forward for placing the Gabar Goshawk with the two chanting goshawks in the genus *Melierax*, and it is interesting to note that all three apparently incorporate the nests of colonial spiders into their own.

All sparrowhawks and goshawks build stick platforms in trees, although the Pale Chanting Goshawk has also been recorded breeding on power pylons and telegraph poles. Sometimes a pair may make use of the nest of another species, merely adding some lining material before the female lays. There are records, for example, of Black Sparrowhawks utilising a disused Hamerkop's nest and a Crowned Eagle's nest; of an African Goshawk using an old Hadeda nest; and of Little Banded Goshawks having taken over nests that belonged to Little Sparrowhawks and to a Lizard Buzzard. Most species, however – the exceptions being the Black Sparrowhawk and African Goshawk – usually build a new nest each season. Although they may breed in different nests each year they remain in the same area, seldom moving from a particular clump of trees or plantation.

A feature of most of these raptors is their use of non-indigenous trees for nesting, and it is one that has important ecological implications. In the Karoo Pale Chanting Goshawks readily build in casuarinas and other introduced species, and in the same region the Red-breasted Sparrowhawk has extended its range by breeding in poplar groves. It has been argued that the presence of poplars in the Karoo is an indicator

1 *Sparrowhawks and goshawks have readily adapted to nesting in exotic trees, and Red-breasted Sparrowhawks may regularly be found in pine plantations.*
2 *A female Little Sparrowhawk at her nest. In this species usually only a single chick survives from a two-egg clutch.*

1 *A female Black Sparrowhawk with her well-grown chicks. The nests of this species may be re-used over a number of years, as indicated by their bulkiness.* **2** *Black Sparrowhawks line their nests with green leaves.* **3** *African Goshawks may also re-use their nests year after year. The chicks stay in the nest for up to 35 days.* **4** *A Little Banded Goshawk brings a lizard, its main prey, to its chicks.*

of habitat degeneration and that, in areas such as national parks, these groves should be removed to restore the pristine situation. Not surprisingly, the suggestion has had a mixed response from raptor enthusiasts. Other alien tree species regularly used are pines and eucalypts, and the nests are usually situated in a plantation. In the Northern Province and Mpumalanga it has been estimated that the introduction of alien trees into the previously treeless grassveld, combined with a good food supply, has resulted in an eight-fold increase in the Black Sparrowhawk population in recent times; in one breeding survey only nine out of 162 nests recorded were in indigenous trees.

Most sparrowhawk and goshawk nests are not easily accessible, although much depends on the type of tree in which they are built. Except in the case of the Pale Chanting Goshawk, which inhabits arid areas where the only trees may be low thorn trees, they are rarely lower than 6 metres above the ground. The Black Sparrowhawk has been recorded nesting as high as 36 metres although, at the other extreme, it has also been found at ground level: in one remarkable case a nest was found in a hollow in leaves at the base of a poplar tree.

The dimensions of the nests vary from one species to another, and range from the Little Banded Goshawk's platform which is 20 to 30 centimetres across and 8 to 15 centimetres deep to the Black Sparrowhawk's structure which is 50 to 70 centimetres in diameter and 40 to 75 centimetres deep. Some Black Sparrowhawk nests may be used for long periods – 29 years in one observation – and inevitably they become bulky.

The type of nest lining used also varies from species to species. The Red-breasted Sparrowhawk merely uses finer twigs in the cup, as does the Ovambo Sparrowhawk, although it also adds occasional bark chips or green leaves. The Little Sparrowhawk may add a few green leaves and the nest of the Little Banded Goshawk is characterised by the inclusion of bark chips. Both the Black Sparrowhawk and African Goshawk line their nests generously with green leaves, and are the only two species in this group to do so. As they are also the only two that regularly re-use their nests each season, this practice may lend support to the suggestion that green leaves repel parasites (see page 61). The Gabar Goshawk is unusual in that it regularly incorporates the nests of social spiders of the genus *Stegodyphus* in its own nest. The Dark Chanting Goshawk also does this but not on a regular basis, and it is more likely to line the cup with miscellaneous material such as dung, grass, a sunbird's nest, a penduline tit's nest, fur and rags. Pale Chanting Goshawks use a variety of materials to line their nests too, depending on what is available, and rags and pieces of paper, rope and plastic have been found in them. In one study in the Karoo dry dung was the main component, followed by Angora goat hair and sheep's wool. An interesting observation was that these goshawks may re-use the nest lining, often transporting material from an old site to a new one. This undoubtedly indicates a shortage of suitable material in the arid areas in which they breed.

Sparrowhawks and goshawks are generally thought to be monogamous, but unpublished research on the Pale Chanting Goshawk has revealed that about 20 per cent of the females studied were polyandrous. Each mated with two males, and the latter assisted with the various duties at the nest. When this research material is published it will provide a fascinating addition to the few known examples of polygamous behaviour in birds of prey.

The breeding cycles of the sparrowhawks and goshawks are similar in outline, although there is some individual variation. Their courtship involves aerial displays during which the birds soar and sometimes indulge in undulating dives and chases. A fluttering flight with slow, exaggerated wing-beats is performed by several species. They all become more vocal at the onset of the breeding season, and chanting goshawks call a great deal from a perch. The African Goshawk, the only species found regularly in indigenous forest, performs an aerial display flight at a considerable height above the canopy, emitting a sharp *whit....whit....whit* call which carries over a considerable distance. However, as this display continues year-round it probably serves more for territorial advertisement than for courtship.

In all species for which observations are available, both birds build the nest, collecting the material for it in a number of ways: they break off sticks with their feet as they fly past, they perch and snap off twigs with their bill, or they collect them from the ground. Most species lay in September and October (spring and early summer), although some, such as the Black Sparrowhawk, may lay earlier. The Pale Chanting Goshawk sometimes responds to favourable local conditions in its arid environment by raising consecutive broods. Conversely, during a period of drought lasting almost four years in Namibia, no breeding at all was recorded in one study area.

The clutch size in general is 2-4 eggs, but the Little Sparrowhawk and the two chanting goshawks regularly lay two eggs and sometimes only one. The eggs of the Little Sparrowhawk, Black Sparrowhawk, African Goshawk and the two chanting

goshawks are plain white, and those of the remaining four species are marked with red-brown blotches. In the case of the Little Sparrowhawk the incubation period lasts 31 or 32 days whereas for the larger species such as the Black Sparrowhawk and the chanting goshawks it is about 38 days. The female does most of the brooding, although she leaves the nest to feed on prey the male brings to her. The male may assist her, and in some species he has been known to sit for quite long spells. Some females sit extremely tight, and in one case an African Goshawk tolerated an observer in full view a metre from the nest. Bold defence of the nest is a feature of several species, especially the Red-breasted and Black Sparrowhawks which may attack and strike anyone climbing to it.

Newly-hatched chicks are covered in white or buffy down, depending on which species they belong to. In the case of the Gabar Goshawk which has a melanistic form – 6 to 9 per cent of the population in some counts – the chicks hatch with either white or grey down and will develop into either grey or melanistic individuals. The young of chanting goshawks have characteristic long down plumes on top of the head which make them look as if they sport a punk-rock hairstyle. Although researchers believe that Cainism is not a feature of the breeding biology of sparrowhawks and goshawks, there is circumstantial evidence that it may occur in the Little Sparrowhawk. Usually only a single chick survives from a two-egg clutch, and in one case a chick was found with abrasions on its back consistent with it having been pecked by its older sibling. When the parents feed the chicks, species that prey on birds always thoroughly pluck them away from the nest before delivering them to their offspring.

The nestling period varies in length from 25 to 27 days for the Little Sparrowhawk to about 50 days for the Black Sparrowhawk and chanting goshawks. Interestingly, it may vary even within the same species, as was observed in the case of the Red-breasted Sparrowhawk: a brood in the south-western Cape left the nest after four weeks whereas the chicks at a nest in Kwazulu-Natal took a week longer to leave. It would appear that the quality of the meals was the reason for the difference, as the nestlings in the Cape received more substantial prey items. Usually the young move out onto the branches of the nest tree before making their first flight, and at this stage are called 'branchers'. During the post-nestling dependence period, which lasts several weeks, they develop their flying skills and some of the more agile species practise impressive kamikaze-style aerobatics through the branches, building up speed and dexterity that will help them to catch prey.

GOSHAWKS AND SPIDERS – A TANGLED WEB

The Gabar Goshawk usually builds its nest in one of the topmost forks of a tree, preferably an acacia. A remarkable feature is that spider web always forms part of the structure, and in some cases it festoons the nest and even extends onto the surrounding branches. The goshawks have been observed carrying the nests of colonial spiders of the genus *Stegodyphus* to their own and incorporating them during construction. Sometimes this is done at an early stage, so it is unlikely that the web has been brought as lining. The spiders, unwillingly transported from low bushes at a height usually less than 2 metres above the ground, find themselves in the topmost branches of a tree. Here they have no choice but to spin their web until, in time, it may cover the goshawk's nest.

What is the meaning of this extraordinary association between the Gabar Goshawk and colonial spiders? From the spiders' viewpoint there would seem to be little if any benefit from the association. The fact that their nests are always low down indicates that this is what suits them best; otherwise they would be found naturally high up in trees. What, then, are the possible benefits for the goshawk?

Various suggestions have been advanced, but none is easy to test. The first is that the web serves as lining or padding: but why is it incorporated into the bird's nest early in construction in some cases? Another thought is that the web serves to camouflage the nest. Certainly, when viewed from below, the nest would not normally be recognised as that of a goshawk, and several observers have remarked on its disused appearance. The unanswered question is: why should Gabar Goshawks require this camouflage whereas other small hawks of similar habitat apparently don't? A third idea is that the web strengthens the nest by binding it, but again the question arises: why should the Gabar Goshawk need to have a strengthened nest when other similar species don't? Another suggestion is that the spiders may control parasites in the nest. This is not as far-fetched as it may seem, as some birds are known to use green leaves with insect-repellant properties for the same purpose. However, the question remains: why don't other species do this too?

It is interesting that Dark Chanting Goshawks also incorporate colonial spiders' nests into their own, even during construction, but this practice is not invariable as it is in the case of the Gabar Goshawk. Web has also been found on Pale Chanting Goshawk nests, but it is thought that this must have been derived indirectly from web already in the nests of penduline tits that the goshawk used for lining. The fact that chanting goshawks, and particularly the Dark Chanting Goshawk, use web in their nests has been put forward as an argument for linking the Gabar Goshawk with these species in the genus *Melierax*●

1

2

1 *A Gabar Goshawk's nest festooned with the webs of colonial spiders.* ***2*** *Viewed from below, it may easily be regarded as disused.*

1

2

1 *A thick lining of sheep's wool provides a soft bed for Pale Chanting Goshawk eggs. In the arid Karoo lining material is sometimes carried to a new nest and re-used.* ***2*** *In the naturally tree-poor Karoo, Pale Chanting Goshawks have taken to nesting in casuarinas and other exotic trees.*

Harriers

Five harriers occur in southern Africa, of which three – the European Marsh Harrier, Montagu's Harrier and the Pallid Harrier – are Palaearctic migrants. The two southern African breeding species are the African Marsh Harrier and the Black Harrier. The former is widely distributed except in the arid west, while the Black Harrier is endemic to southern Africa and has the smallest distribution range of any harrier species in the world. It breeds mainly south of 31°S and moves northward when not breeding.

At the onset of breeding both sexes of the African Marsh and Black Harriers perform high aerial displays known as 'sky-dancing' which consist of a series of U-shaped undulations, sometimes as many as 50 at a time, on a horizontal plane. The African Marsh Harrier also performs 'sky-spiralling' displays during which the birds descend in a series of twisting, U-shaped undulations on a vertical axis. Although these displays have been interpreted as being mainly territorial in function – and sometimes up to eight birds may perform within sight of each other – they appear to also serve the purpose of attracting a mate.

A characteristic feature of harrier behaviour used at all stages of breeding is the aerial food pass, whereby food is transferred from the male to the female or to a juvenile. The male flies in a horizontal plane with the food held in a lowered foot as the female ascends towards him. As she approaches he flies upwards and releases the prey, which is moving upwards as it is released and thus almost stationary as she catches it. In 305 aerial food passes observed in the African Marsh Harrier, 93 per cent were caught; most of those missed were dropped by inexperienced juveniles.

The African Marsh Harrier usually breeds in marshy habitat, but occasionally it nests in a dry area such as a wheat-field, and once a nest was found on a low bushy tree 3 metres above the ground. In a dense reedbed nests may be 1.5 metres above the water, but in marshy vegetation they may be situated as low as 40 centimetres above water level. They measure between 40 and 60 centimetres across, depending on the site: those in reedbeds are more bulky on a solid base of sticks, while those elsewhere may be little more than a pad of dry grass and reeds. Both birds are involved in building, with the male sometimes making a substantial contribution, and they add lining of dry vegetation throughout the breeding cycle. The nests do not last and a new one is built each season, but a pair tends to remain in the same area unless the habitat has altered. In the south-western Cape the African Marsh Harrier has declined or disappeared from areas where it used to breed because of habitat loss or alteration.

The breeding season of the African Marsh Harrier varies from one region to another: south of the Limpopo River laying has been recorded between June and November with a distinct September/October peak in the south-western Cape; and in Zimbabwe eggs are laid between December and June. In contrast, both the breeding range and season of the Black Harrier are fairly clearly defined. Most nests have been found south of 31°S, and for some time this latitude was thought to mark the extent of its range. However, nests have recently been observed in Kwazulu-Natal and Free State. Laying has been recorded between July and November, with a distinct peak in August and September. Although Black Harriers may be seen in their breeding range throughout the year, they appear to be nomadic and move northward between January and July.

Unlike the African Marsh Harrier, the Black Harrier does not prefer marshy areas for breeding, although it will nest in such habitats, as well as in wheat-fields, renosterbos scrub and strandveld vegetation. It also nests in mountain fynbos, sometimes in remote places so that breeding pairs are easily overlooked. This has given the impression that Black Harriers are rare, when in fact in the south-western Cape they are probably more common than African Marsh Harriers. The nest, usually a pad of dry vegetation lined with finer material, is placed on the ground except in marshy situations where sticks may be used to form a base for it. It is generally smaller than that of the African Marsh Harrier, measuring 35 to 45 centimetres across. Both sexes build and add lining during the breeding cycle, but the female takes the major share.

Black Harriers tend to breed in the same area over a period of years. Some pairs may nest fairly close to one another, and in one observation three nests formed the corners of a triangle measuring 800 x 1000 x 1500 metres. There is also a report of apparent polygyny, according to which two nests were 200 metres apart in the same vlei although only one male was seen in the area.

The breeding cycles of both harriers are basically similar and conform with those of other raptors in which the male is the main provider of prey. Both species usually lay three or four eggs, but the African Marsh Harrier has been recorded laying up to six and the Black Harrier five. Most eggs are white, although some may have red blotches, and they become nest-stained as incubation progresses. Incubation for both species lasts about 35 days, and the nestling period is between 35 and 42 days. During the latter the nest is kept scrupulously clean and any uneaten prey items are removed.

The eggs of the African Marsh Harrier hatch at intervals, with the result that the last-hatched chick may be considerably smaller than its siblings. Although it may be pecked by the other chicks, it seems that when a smaller chick dies it is the result of starvation rather than sibling aggression. The nestlings in a Black Harrier's brood may also vary in size, but there are no observations recording the death of smaller chicks.

1 *An African Marsh Harrier's nest in a reedbed is supported by a solid base of sticks.* ***2*** *Black Harriers readily breed in dry as well as marshy environments, and here one is seen at its nest in strandveld vegetation.* ***3*** *A Black Harrier's nest of dry plant material is placed on the ground in mountain fynbos habitat.*

The Gymnogene and the Osprey

Each of these two species is remarkable in its own way. The Osprey occurs in almost all parts of the world and has attracted much conservation attention because of its decline in some countries as a result of direct persecution and the use of pesticides. Its return as a breeding species to Scotland was a much publicised event and from this original nucleus the species has now become well established.

The Gymnogene is a highly specialised raptor with a 'knee' joint that is able to flex backwards at 40°. This enables the bird to insert its leg into holes in trees, including ones in which species such as barbets and woodpeckers are nesting, to extract prey. Its light body and broad wings enable it to fly slowly as it searches for likely sources of food, and it often hangs upside down beneath weavers' nests or the enclosed nests of swallows to tear them open and rob them. It is always immediately recognised by small birds as a potential nest robber and is vigorously mobbed.

Despite its unusual foraging habits, the Gymnogene's breeding biology is similar to that of most raptors. It performs undulating display flights and the male brings prey for the female as part of courtship. An interesting feature is its tendency to 'blush': the bare yellow facial skin turns orange or even deep red. This appears to be an emotional response and usually occurs in the breeding cycle during copulation, nest-building and prey delivery. It is also possible that it is a method of heat loss, as females have been seen to 'blush' when alone on the nest.

Gymnogenes build a substantial nest of sticks that measures 60 to 70 centimetres across and 20 to 30 centimetres deep, with a central depression that is copiously lined with leaves. They place it in the fork of a tree within the canopy or at the base of a small tree or bush on a cliff. Sometimes they use the nest of another species such as the Black Sparrowhawk or Martial Eagle; in one observation a Gymnogene pair raised a chick in a Martial Eagle's nest before the eagles returned to breed themselves. They may nest in the same area for long periods, and in one case were recorded in the same kloof for 27 years.

The eggs, normally two in a clutch, are extremely beautiful and they are incubated mainly by the female over a period of 35 to 38 days. When two chicks hatch the smaller is aggressively attacked by its sibling and severe peck wounds are inflicted. The weaker chick usually succumbs, but occasionally two chicks may be raised together. The chick leaves the nest when it is about two months old and may remain with its parents for up to seven months thereafter.

GYMNOGENE EGGS – NOT JUST A PRETTY SHELL?

The eggs of the Gymnogene are exceptionally handsome and were once much prized by collectors, especially freshly laid ones for their attractive bloom. The creamy ground colour is usually completely obscured by a red wash, and this in turn is overlaid with blotches of darker red that coalesce over the entire surface. The final effect is that the eggs resemble dark mahogany. However, occasional eggs are poorly pigmented, and in some clutches a well-marked egg and an almost plain one may be found together.

The central depression of a Gymnogene's nest is copiously lined with leaves throughout the incubation period and these camouflage the dark eggs. The fact that the eggs are inconspicuous may have survival value for a species that is not aggressive, reducing the chance of predation when the female leaves the nest to hunt for herself during the incubation period. In terms of camouflage strategy it may be tempting to compare the Gymnogene with the Cuckoo Hawk, but in the latter case it is the leafy nest itself that probably has survival value ●

The attractive markings of the Gymnogene's eggs may camouflage them on their leafy background.

Until recently, the Osprey was considered a non-breeding Palaearctic migrant in southern Africa. In 1917 the respected naturalist Austin Roberts recorded a nest at the Berg River in the Cape Province, but his observation was considered dubious on account of his description of the eggs' colour and size. It was thought that there had been confusion with Yellow-billed Kites that breed in the area. However, in the early 1930s an oviduct egg was taken from an Osprey shot on the Limpopo River, and in October 1963 a nest with two young was reported from the Ndumu Game Reserve in northern Zululand. These last two records appear to be the only authentic ones to indicate that this species may occasionally breed in southern Africa. There have been many reports of Ospreys flying out of sight carrying fish to potentially suitable nesting areas, but none has been confirmed as a nesting record.

The Osprey builds a very large and bulky nest in a tree, on a cliff, on the ground or on a telegraph or power transmission pole. In some localities it breeds in loose colonies, but it also nests solitarily. Two or three eggs are laid and they are beautifully marked with red-brown blotches.

Like many other birds of prey, the Gymnogene lines its large nest with green leaves.

Falcons and Kestrels

Fifteen species of falcons and kestrels occur in southern Africa but five of these are migrant visitors. Of the ten breeding species, nine belong to the genus *Falco* and the tenth, the Pygmy Falcon, is in a genus of its own, *Polihierax*. Its breeding biology and behaviour are quite distinct.

The members of the genus *Falco* share a number of similarities in their breeding, such as aerial courtship displays which are usually accompanied by loud calling, courtship feeding of the female by the male, no nest construction in the accepted sense, and a normal clutch of three or four well-marked eggs (although several species may lay larger clutches on occasion). *Falco* species also protect their nest areas aggressively and, unlike most other birds of prey, the male undertakes a substantial share of the incubation. The incubation period of most species is about 28 days and the nestling period between 35 and 42 days. Although there are exceptions, most species lay their eggs from June to October. Several species also have the habit of hiding and storing surplus prey.

In southern Africa the Peregrine was traditionally described as a rare breeding species, but recent research shows that it is far more widespread and common than was previously thought, and in Zimbabwe alone the population is estimated to be 350 to 400 pairs. As may be expected from probably the most agile of all raptors in flight, its aerial courtship displays are spectacular, but it also performs bowing displays on or near the nest ledge. A high cliff near water is the preferred nest site, but neither height nor water is essential and nests may be found on quite low cliffs and also in arid areas such as the Namib Desert. The birds lay in a mere scrape in the soil of the ledge of a cliff, mine excavation or quarry, but sometimes take over the disused nest of another cliff-nesting species such as the Black Stork, Black Eagle or White-necked Raven. Unlike the Peregrine on other continents, in Africa this species rarely nests on man-made structures, although in recent years more instances of this have been reported.

The Lanner is far more versatile in its nesting situations than the Peregrine and regularly breeds in the nests of other birds of prey or crows. This enables it to be independent of cliffs, although a survey undertaken in the former Transvaal revealed that 62 out of 91 nests were on cliffs and, as in the case of the Peregrine, sometimes eggs were laid in the nest of another cliff-nesting species. The use of crow nests on pylons has undoubtedly enabled the Lanner to extend its range into areas where previously there had been no suitable nest sites. It also readily breeds on buildings in towns, where it preys on the abundant population of feral pigeons. One nest site in the window-box of a Harare office block in Zimbabwe was kept under observation for several years.

The African Hobby, a small but extremely swift falcon, is known to breed in southern Africa, although very few nests have been found. The three that have been located were all in Zimbabwe, and were all in the nests of other birds of prey: two in Wahlberg's Eagle nests and the third in a Yellow-billed Kite's nest.

Considered by many to be a smaller version of the Peregrine, the Taita Falcon is certainly spectacular in flight. It has a discontinuous distribution in Africa and is nowhere common, although the Zimbabwean population, concentrated mainly in the gorges of the Zambezi Valley, has been found to be more substantial than was previously estimated. Recently Taita Falcons have been found breeding in the eastern escarpment region of the Northern Province, a discovery which testifies to the fact that these birds can easily be overlooked since they spend long periods perched on cliffs. Their nests in the Zambezi Valley are typically situated on a basalt cliff at the back of an erosion hole where they are sheltered from direct sunlight. They are usually sited on high cliffs and are almost always near water.

The Red-necked Falcon inhabits the more arid regions of southern Africa and is found mainly in Botswana and Namibia, although recently peripherally in Zimbabwe too. It breeds either in the disused nest of another raptor, including species as large as the Lappet-faced Vulture, or in a natural hollow in a tree, usually an ilala or borassus palm where the base of a frond joins the tree.

One of the most ubiquitous of southern Africa's raptors, the Rock Kestrel nests on a ledge on a cliff, in the nest of a bird of prey or a crow, or on a man-made structure, such as the ledge on a building or in the girders of an aircraft hangar. Despite its small size it is particularly vocal and aggressive towards other birds, even eagles and buzzards, near its nest.

The Greater Kestrel utilises the stick nest of another species and does not seem to

1 *Peregrine chicks crouch on the narrow ledge that is their home.* **2** *A Lanner at its cliff-edge nest. Like most falcons, it defends its nest area aggressively.*

1

2

3

4

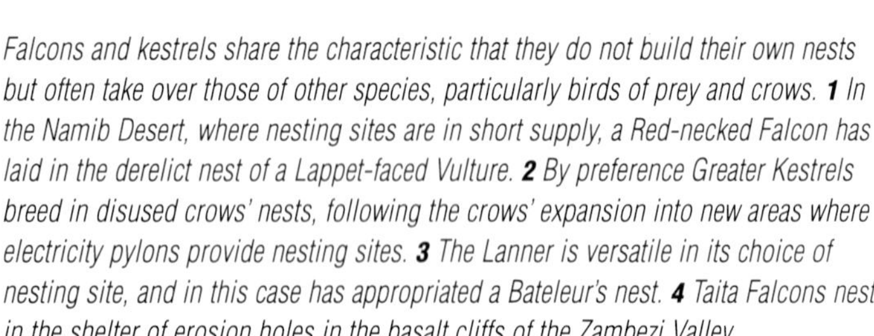

Falcons and kestrels share the characteristic that they do not build their own nests but often take over those of other species, particularly birds of prey and crows. ***1*** *In the Namib Desert, where nesting sites are in short supply, a Red-necked Falcon has laid in the derelict nest of a Lappet-faced Vulture.* ***2*** *By preference Greater Kestrels breed in disused crows' nests, following the crows' expansion into new areas where electricity pylons provide nesting sites.* ***3*** *The Lanner is versatile in its choice of nesting site, and in this case has appropriated a Bateleur's nest.* ***4*** *Taita Falcons nest in the shelter of erosion holes in the basalt cliffs of the Zambezi Valley.*

mind whether it is as large as that of a Secretary Bird or as small as that of a Black-shouldered Kite. The instance of a Greater Kestrel sharing a nest with a Lappet-faced Vulture has been mentioned (see page 52), and in another record Greater Kestrels were found breeding in an old crow's nest near a nest still occupied by Pied Crows. Indeed, they prefer to take over crows' nests, including those on power transmission pylons. By breeding on pylons the Greater Kestrel has been able to extend its range into areas where suitable nesting sites were not previously available. Interestingly, according to a study on the Transvaal Highveld, Greater Kestrels replaced Lanners when the latter inexplicably disappeared from their pylon breeding sites.

The Grey Kestrel occurs peripherally in northern Namibia where the only nest located was in a tree hole. From records elsewhere in Africa this seems to be a typical site, although the species may also breed inside a Hamerkop's nest. Dickinson's Kestrel also nests mainly in natural holes in trees, favouring the top of a dead palm where the crown has broken off, but also using a hole in a tree such as a baobab. Sometimes it moves into a Hamerkop's nest, and will even evict the rightful owners.

The smallest raptor in Africa is the Pygmy Falcon, whose breeding biology is quite different from that of kestrels and other falcons and, although its basic nesting habits are known, it is a species that would reward more detailed investigation.

In the arid western regions of southern Africa, where the Pygmy Falcon's distribution exactly coincides with that of the Sociable Weaver, the only known breeding sites of this tiny falcon are chambers in the weavers' communal nests. In east Africa, however, where Sociable Weavers do not occur, Pygmy Falcons utilise chambers in the nests of White-headed Buffalo Weavers. The only way to establish which chamber the falcons are occupying, other than by observing them fly in and out, is to find a nest in which chicks have hatched. A ring of dry white excreta at the chamber entrance pinpoints where the falcons are in residence.

Pygmy Falcons remain together as a pair throughout the year and have a courtship display that lacks any aerial element. Often after receiving prey delivered by the male, the female perches near her mate in a submissive posture to reveal her chestnut back and white rump to maximum advantage. She then wags her tail up and down, and the male may do likewise, so that the pattern of white spots on their tails is displayed. Copulation takes place on branches near the nest.

The breeding season is long, lasting from August to March with a peak in October and November, and its length is in part due to the fact that in seasons of good rain when food is plentiful two consecutive broods may be raised. The eggs are white and usually 2-4 in a clutch, although years of good rainfall may also result in larger clutches. During the incubation period, which lasts about 28 days, the male brings prey – mainly lizards – for the female and also does a share of the incubation himself. He continues to be the main food provider after the chicks have hatched. Both birds may be aggressive towards human intruders near the nest and sometimes attack them. The young leave the nest after about a month, but may remain with their parents for another two months before becoming independent.

PYGMY FALCONS AND SOCIABLE WEAVERS – GOOD NEIGHBOURS?

It is not easy to assess to what extent Pygmy Falcons prey on their hosts, the Sociable Weavers, but it appears that they may take nestlings while prospecting for a nest, and even catch adults – one was once observed seizing an adult in flight. However, the overall impact on the Sociable Weaver colony is probably minimal. In fact, it is more likely to be advantageous for the weavers to tolerate the falcons, as the aggressive little raptors probably deter potential predators such as snakes. Only detailed observations will establish whether the association between the two species is genuinely commensal ●

Chambers in a Sociable Weavers' nest which are occupied by Pygmy Falcons can be identified by a ring of white droppings.

1

Francolins, Quails, Guineafowls and Buttonquails

Three families – the Phasianidae, Numididae and Turnicidae – are represented in this section, and most of the species, generally known as gamebirds, are hunted on a commercial basis. All breed in a rudimentary grass-lined nest on the ground and sit extremely tight, relying on camouflage to avoid detection. The breeding season is variable but as a rule is associated with rainfall, and in periods of drought no breeding may take place at all.

Of the 20 breeding gamebirds in southern Africa, only the Chukar Partridge is not indigenous, having been introduced to various parts of southern Africa in order to build up a population for hunting. Fortunately the birds have died out except on Robben Island off Cape Town, where they are likely to remain isolated. Very few nests have been found, but this partridge's breeding habits are similar to those of francolins.

The nesting habits of all 12 indigenous francolin species in southern Africa are alike in many respects. As far as is known they are all monogamous, and they nest on the ground in thick grass or under a bush where their cryptic coloration conceals them from predators such as mongooses. Their camouflage, however, is not always sufficient and an incubating female is sometimes caught on the nest. Their courtship behaviour is poorly documented, but calling is believed to play an important role. Perhaps best known is the crowing call of the male Swainson's Francolin, emitted as he stands with beak pointed upwards and red throat inflated. When he sees a prospective mate he chases her with lowered head and, when she stops, he assumes an exaggerated upright stance with chest out and wings slightly extended and drooping.

Francolins normally lay 3-8 eggs which are thick-shelled and white, cream or

2

3

1 *Helmeted Guineafowls lay large clutches of distinctive pyriform eggs.* ***2, 3*** *An incubating Cape Francolin female sits tight on her nest, relying on her camouflage to conceal herself and her eggs. When she leaves the nest the eggs are conspicuous.*

yellowish buff in ground colour. They are mostly plain, but in some species may be speckled. It appears that only the female incubates, although in some species the male is known to be on guard in the area. The incubation period, where known, is 20 to 25 days, and when the cryptically patterned young emerge they are highly precocial, leaving the nest soon after hatching. They develop very rapidly and in several species have been recorded making short flights when only ten days old. The brood accompanies the female when she forages, scattering and lying flat if she gives the alarm. The chicks have a high-pitched, ventriloquial call and soon re-establish contact with the female when the danger has passed.

Three quail species – the Common, Harlequin and Blue Quails – breed in southern Africa. The Common Quail – which in contrast with the other two species shows only slight differences between the plumages of male and female – is migratory and occurs sporadically, taking advantage of local conditions such as rain in normally arid areas. The males arrive first, filling the air with their characteristic *wet-my-lips* call which continues throughout the day and night. When the females arrive they are courted by the males and, in the process, also fed by them. Having attracted a mate, the male walks round her and they copulate frequently. Pairs appear to be monogamous, but bigamy or successive polygamy may occur.

The Common Quail's nest is a simple scrape lined with a few dry grasses and it is well concealed in thick cover. Its usual clutch comprises 5-7 eggs which are laid at daily intervals, although larger clutches, of 14 eggs for example, are sometimes found and they are undoubtedly produced by two females. The eggs are creamy buff in ground colour with a pattern of brown blotches over the whole surface. Incubation, which is undertaken only by the female, lasts 17 to 20 days and when the eggs hatch they do so at the same time. Interestingly, research on other quails has shown that the chicks in the eggs 'talk' to each other and thus stimulate simultaneous emergence.

The nests of the Harlequin and Blue Quails are similar to that of the Common Quail, but in the case of the Harlequin Quail many nests may be concentrated in a small area. The incubation periods of the Harlequin and Common Quails are similar, but the duration of the Blue Quail's incubation is not known.

The breeding habits of the monogamous Helmeted and Crested Guineafowls are largely similar, but because those of the former are better known they are described here. Although male guineafowls chase other males aggressively throughout the year, their chases appear to intensify at the onset of the breeding season, and it becomes difficult to distinguish between an aggressive chase between males and the courtship chase of a female. Eventually male and female form a pair bond and move away from the flock to select a breeding site. Their nest, a grass-lined hollow measuring about 20 centimetres across and 7 centimetres deep, is very well concealed. Clutches number 6-19 eggs and larger ones – up to 50 eggs have been recorded – are undoubtedly contributed to by two or more females dumping their eggs in the nest of another female. The pyriform eggs are distinctive, being plain, creamy yellow in colour and very thick-shelled, with visible pores in the shell.

The incubation period lasts on average 26 days, and while the female sits on the eggs the male may keep a watch nearby for predators. The cryptic and attractively marked chicks, known as 'keets', leave the nest soon after hatching to forage with their parents. Sometimes other adults accompany the brood and chicks from different parents may join up to form a single mixed group. They are first able to fly from about 14 days old but remain with their parents for several months. The adults are remarkably bold in defence of their young and attack predators fearlessly.

Although the buttonquails resemble quails they belong to a different family, the Turnicidae. The two species that occur in southern Africa are the Kurrichane Buttonquail, which is resident or may make local movements, and the Black-rumped Buttonquail, which is partially migratory, and the nesting habits of both differ from those of the other species in this group. The nest, a small, sparse pad of grass 5 centimetres in diameter, is placed in a tuft of grass or under vegetation, some of which is sometimes bent over to conceal it. The Kurrichane Buttonquail lays 2-4 eggs which are thickly spotted and blotched, whereas the Black-rumped Buttonquail lays 2-5 eggs, but usually three, and these are finely speckled.

The most intriguing aspect of the buttonquails' breeding biology is the dominant part played by the female in courtship. In observations of captive birds the female Kurrichane Buttonquail made an *oowing* call to attract a male and then fed him as part of courtship. Once they had paired she laid a clutch of eggs which he alone incubated; she then courted another male and laid a clutch for him. This polyandrous behaviour extended to three males in the captive study. After 14 days the tiny chicks hatched and were cared for by the male and, once they were independent, he sometimes mated again with the female.

These captive observations may not necessarily pertain to birds in the wild. Indeed their validity in this regard has been disputed, and in the case of the Black-rumped Buttonquail polyandry has not been confirmed. Nevertheless, there can be no doubt that the breeding biology of buttonquails is unusual and it merits study in the wild, difficult as this may be.

1

2

3

1 *A female Grey-wing Francolin sitting on her eggs under this large grass tuft would be effectively hidden.* ***2*** *A clutch of 14 eggs in a Common Quail's nest is almost certainly the product of two females.* ***3*** *The Kurrichane Buttonquail's heavily marked eggs are less conspicuous than those of the francolins. They are apparently incubated only by the male.*

GAMEBIRDS AND ECONOMICS

Gamebird shooting is big business overseas and, it has belatedly been realised, could be in southern Africa too. In Zimbabwe gamebirds are already being exploited commercially, and there the two most commonly hunted species are the Helmeted Guineafowl and Swainson's Francolin. On the Stormberg Plateau in South Africa's Eastern Cape the commercial value of gamebirds has been strikingly illustrated by the hunting of Greywing Francolins on a sustainable yield basis. On farms in the region in 1992 hunting groups active on 24 days generated an income of R140 000, the equivalent of R110 per bird (compared with R120 for a sheep).

This income derives from a gamebird that requires virtually no economic input. Its environment merely has to be managed wisely, and it has been shown this can be achieved by maintaining the habitat diversity of an area. Hunters enjoy their sport and, as long as the bag limit is regulated on sound principles, the gamebird population remains stable. Indeed, hunting is beneficial because the immigration of birds from peripheral areas stimulates genetic interchange between populations.

Gamebird shooting on a sustainable yield basis has important implications for the economic viability of the birds' habitat.

As no two breeding seasons are alike, and they are dependent on rainfall to a considerable extent, it would be essential to monitor the shooting season on a regular basis. However, with this crucial factor in mind, there is no reason why gamebird shooting in southern Africa should not become a lucrative industry, and perhaps even attract hunters away from the grouse moors of Scotland! ●

Cranes

Three species of these large and spectacular birds occur in southern Africa, and although each belongs to a different genus, they share various aspects of breeding behaviour. Perhaps best known of these is a bounding courtship dance during which the cranes usually throw tufts of grass or other material into the air. Sometimes when they are gathered in a flock the dancing behaviour of one pair stimulates others to perform too. Calling varies from one species to another: the Wattled Crane is rather silent; the Blue Crane emits a harsh *kraaaaarrk*; and the Crowned Crane, the most vocal of the three, produces a haunting *mahem*, which has been taken over as the bird's Afrikaans name.

Cranes are long-lived and remain paired for life, and in all three species both sexes participate in defence of the territory, distraction displays, nest-building, incubation and the care of the young. Newly-hatched chicks are precocial and have a cryptic down pattern. The young remain under parental care for long periods – from ten to 21 months in the case of the Wattled Crane – and they may join a flock with others of their species. When breeding, cranes are alert and wary, and will creep away surreptitiously from the nest. Thus too much disturbance in an area can cause a breeding attempt to fail.

Wattled Cranes depend on a marshy habitat surrounded by grassland for breeding and feeding, and ideally require a home range that has a minimum of 20 – but preferably 40 – hectares of marsh surrounded by at least 150 hectares of grassland. Not surprisingly, encroachment by farming and forestry as well as the draining of vleis have had a serious impact on Wattled Crane populations. The breeding area is occupied year-round and if the birds are undisturbed they will use the same site for each breeding attempt. The nest, a large mound of plant material, is built in a shallow pond or on an island and generally measures 1 to 1.8 metres across and about 20 centimetres high, although its size varies depending on the site. There is no clearly defined breeding season and eggs may be found throughout the year, although usually they are laid in the dry winter months, especially in Zimbabwe.

The clutch comprises one or two eggs whose salmon-pink ground colour is overlaid with mauve and rust-red markings. When the first chick of a two-egg clutch hatches, after an incubation period lasting 35 to 40 days, the parents leave the nest with the chick and neglect the second egg, although they may return to the nest at night with the small chick to brood it there. The young crane is tended attentively by its parents and flies for the first time when it is 15 to 18 weeks old, although it continues to stay with its parents for a long while thereafter.

Blue Cranes breed in open country where the visibility is good all around and they may be found in the same general area year after year if the habitat is unaltered. The nest, quite often situated near water, is a scrape on bare ground, sometimes with a few small stones, animal droppings or plant material placed round the perimeter. The normal clutch is two eggs, which are similar in colour and markings to those of the Wattled Crane, and they hatch after about 30 days. In one observation the first-hatched chick was seen to leave the nest six hours after hatching and move 30 metres away while its sibling, which hatched shortly after it, remained on the nest until it was also strong enough to move off. The parents communicated with the chicks

1

2

1, 2 *Occasionally Crowned Crane nests have been found in trees, although usually this species nests on the ground in a marshy habitat.* ***3*** *Wattled Cranes breed in wetlands and build up a mound of plant material, often in shallow water, to serve as a nest.*

3

1 *Blue Crane chicks are precocial and leave the nest within hours of hatching. Their white 'egg tooth' disappears after about a week.* **2** *A Blue Crane nest is little more than a scrape.* **3** *The increased use of land for agriculture in the south-western Cape has allowed the Blue Crane population to expand into the region.*

using soft *grok-grok* calls. The young first fly when they are 12 to 13 weeks old but remain with their parents for up to ten months after hatching.

Like the Wattled Crane, the Crowned Crane prefers a marshy habitat, but it breeds during the rainy summer months when the cover is tall. The nest, constructed by both birds with plant material available nearby, is sited in or near water after the surrounding vegetation has been trampled flat. There are, however, occasional records of these cranes nesting in trees, some 5 to 6 metres above the ground. In two records the old nest of a Secretary Bird was used and in the third case the birds built at the top of a leadwood tree. A small chick was found unharmed at the foot of the tree and another was dead, possibly as the result of falling from the nest.

The clutch of the Crowned Crane comprises 2-4 eggs which are plain white but soon become nest-stained. They hatch after an incubation period of about a month and leave the nest shortly afterwards. At several nests it was found that the eggshells had been covered over with nest material, possibly to avoid attracting potential predators. As small chicks may return to the nest to be brooded, this precaution would seem to have survival value. Two or three young may be successfully reared, having remained with their parents for up to ten months. Although all three crane species may distract intruders from the nest, either by dancing or by feigning a broken wing, the Crowned Crane's display is probably the most intense. In one observation a man and his dogs were led 300 metres away from small chicks by means of an injury-feigning display.

CRANES AND CONSERVATION

Cranes can afford to reproduce slowly because they are long-lived, but once the adult breeding population declines it has difficulty in maintaining its numbers as its potential productivity has been lost. Thus no account of the breeding biology of cranes is complete without reference to the conservation problems faced by these magnificent birds. At a symposium hosted by the Southern African Crane Foundation in December 1989 an alarming picture of the decline of cranes in South Africa emerged.

At that time Wattled Crane numbers were low but stable at about 100 pairs, most of which were in what was then Natal, but some were in the eastern Transvaal. In the eastern Cape only one pair was left. The Crowned Crane population in Natal had dropped by 45 per cent between 1982 and 1989, and in the same period and region the Blue Cranes had declined by 90 per cent. In the eastern Cape the Blue Cranes had also decreased dramatically to a tenth of their former numbers, while in Namibia the small isolated population of about 80 birds in the grassland surrounding Etosha Pan had remained stable. The Karoo population had remained constant but, against the trend, Blue Cranes have increased in the south-western Cape and have expanded their range. The reason for their hundredfold increase is that formerly unsuitable habitat such as fynbos has been replaced by a mosaic of ploughed and fallow lands which is ideal Blue Crane habitat. Awareness programmes by farmers and conservationists in the south-western Cape have also helped to protect the cranes.

What are the main reasons for the decline of cranes? Loss of habitat is one, especially when vleis are drained, or forestry encroaches into grasslands. In some areas high-intensity rotational grazing – the 'wagon-wheel' system – causes the trampling of Blue Crane habitat and inhibits breeding. In Zimbabwe Wattled Cranes are affected by cattle moving into the vleis where they breed during the dry season. Poisoning, either indirectly from seed dressings or by deliberate persecution, has caused serious mortality and in one notorious incident in the eastern Cape 400 Blue Cranes were reported to have been killed. The reason for such a deliberate act of poisoning is that the cranes are seen as a threat to sprouting crops. Birds are also killed when they fly into power lines, but once high-risk areas are identified markers on the wires can solve this problem.

The Southern African Crane Foundation is doing excellent work in monitoring crane numbers, countering adverse attitudes towards cranes and preserving habitats. It is a sobering thought that were it not for the healthy Blue Crane population in the south-western Cape, South Africa's national bird could have been one of our rarest species ●

Rails, Crakes, Flufftails, Gallinules, Moorhens and Coots

The 18 southern African species in this group all belong to the family Rallidae. Two, the Corncrake and Spotted Crake, are non-breeding migrants from the Palaearctic. A third, the American Purple Gallinule, has been recorded in the south-western Cape on more than 20 occasions. Most of the birds found have been immatures and, having been blown across the Atlantic by westerly winds, they were often in an exhausted state. Usually they died, but it is not improbable that in time sufficient numbers may survive to establish a viable breeding population, in keeping with the well-known ability of the Rallidae to colonise new areas.

For convenience of discussion the Rallidae may be broken up into three groups: the rails and crakes; the flufftails; and the gallinules, moorhens and coot. Most of these species occur in an aquatic habitat, and generally their nests are simple bowl-shaped structures of reeds, grass or other plant material which are situated on the ground, in reedbeds or in clumps of grass. They are mostly skulking and secretive birds, and their nests are difficult to find.

Although the eggs of the Rallidae are extremely variable, the natal down of the chicks of most species is uniform black. Another characteristic of the group is that many of the species have loud calls and several of them duet. Vocal communication is important for a skulking species, not only for courtship and maintaining the pair bond, but also as a means of territorial advertisement in thick cover.

The rails and crakes include five breeding species, three of which – the African Rail, Black Crake and Baillon's Crake – are resident and two – the African and Striped Crakes – are intra-African migrants. All five species place their nests in reeds or rank grass in an aquatic habitat or in seasonally inundated grasslands, and some may pull down surrounding vegetation to form a canopy that conceals the nest. They all lay white or creamy eggs which are marked to some degree: the African Crake's eggs are blotched with red; those of the Striped Crake are handsomely marked with red over the entire surface; and those of the remaining species have spots or speckles of red or brown. The clutch usually comprises 2-6 eggs and the incubation period, where known, lasts 14 to 20 days. The incubating bird is reluctant to leave the nest and some species have been known to threaten humans who have approached the nest too closely. The young remain under parental care for some weeks and in the case of the African Rail they may sleep on rudimentary nests built by the adults.

The breeding biology of two species in this group is of particular interest. The Black Crake may raise consecutive broods, with immatures from earlier broods assisting with nest building and feeding the downy young. The habits of the Striped Crake involve a rare example of sequential polyandry, in which the female initiates courtship and lays a clutch for a second male when the eggs of her first mate are about to hatch. Only the male incubates and, although observations have been based on captive studies of these crakes over a period of 12 years, there is little reason to doubt that this is not normal behaviour, especially as only males have been known to incubate in the wild.

Of the five flufftails that occur in southern Africa, the Red-chested, Striped and Buff-spotted Flufftails are all resident and the Streaky-breasted Flufftail is an intra-African breeding migrant. The nest of the fifth species, the White-winged Flufftail, has yet to be discovered. Flufftails differ from other Rallidae firstly in having sexually dimorphic plumage – the distinctive reddish foreparts of the males contrasting with the uniformly drab plumage of the females – and secondly in that they

1

THE ADVANTAGE OF BROOD DIVISION BY THE RED-KNOBBED COOT

Red-knobbed Coots breed in open situations and are aggressively territorial. They are easily observed and their small chicks, although they have the dark down characteristic of the Rallidae, are rather bizarre creatures with a bald head, a reddish bill and a fringe of ginger down around the neck. Later in their development their down becomes uniformly smoky grey.

One researcher has observed that on a number of occasions coot parents divided their brood between them, each tending only their own half. In some divided broods helpers from a previous brood also tended the chicks, and in one case where there were more helpers than young each parent took care of a single chick and four helpers tended the remaining two.

What are the possible advantages of a divided brood? It would seem that such a strategy would give one or two dominant chicks less opportunity to monopolise the attentions of both parents and food would thus be distributed more fairly among them. The resultant higher survival rate is further enhanced if helpers are involved ●

2

3

***1** A Black Crake returns to its well-hidden nest in dense bulrushes.*
***2** A canopy of rank grass helps to conceal the Striped Crake's handsome eggs.*
***3** Red blotches, as opposed to speckles, distinguish the African Crake's eggs from those of other crakes.*

2

3

1 *A Purple Gallinule at its bowl-shaped nest in a reedbed. Both sexes incubate the eggs.* ***2*** *A male Red-chested Flufftail with its chick. The chick's black down is characteristic of the Rallidae.* ***3*** *The Buff-spotted Flufftail's nest in thick undergrowth is difficult to find.* ***4*** *The male Striped Flufftail shares incubation duties with the female.* ***5*** *Their eggs, like other flufftail eggs, are white and can be distinguished from the marked eggs of other Rallidae species.* ***6*** *The Streaky-breasted Flufftail breeds in the rainy season when there is grass cover for its nest.*

lay white eggs. All species have distinctive calls which are often the only means of recording their presence. The weird, foghorn-like banshee wail of the Buff-spotted Flufftail is remarkable even in a group known for its extraordinary vocalisations.

Only a few flufftail nests have ever been found, partly because the birds themselves are so elusive, but also because the nests are well hidden, often with surrounding vegetation pulled down over them for additional concealment. Some species, for example the migratory Streaky-breasted Flufftail, breed during the rainy season when there is maximum grass cover. The forest-dwelling Buff-spotted Flufftail has a nest that, like its call, is unusual and it is interesting that this species does not make its wailing call while breeding. It builds a ball-shaped structure with a side entrance on the ground amongst forest debris, using an assortment of materials such as skeletonised leaves and roots, and camouflaging the top with lichens and moss.

Observations at the nest of a Buff-spotted Flufftail have provided some of the most detailed information we have for any flufftail species. Once the clutch of four eggs was complete it was incubated by both sexes until the chicks hatched after 15 to 16 days. Eggshells were placed outside the nest entrance by the female where they were eaten by the male. The chicks were tended by both adults and remained in or near the nest for two days after hatching. After this they moved off into the forest with their parents and were not seen again. If a human approached the nest too closely the adults would emit a cat-like spitting noise and run around with drooping wings. In other observations adults with chicks have been known to attack the legs of humans near the nest.

From the few additional records that are available it appears that flufftails share incubation and care of the chicks, that the clutch comprises 2-5 eggs, and that the incubation period is not less than 14 days. Notes on a presumed single pair of Red-chested Flufftails in a territory watched over eight months revealed that five consecutive broods of two chicks each were reared. Both parents tended the chicks and fed them by offering them insects, and the female was seen to brood a chick and carry it under her wing in the manner of an African Jacana. The productivity of this pair suggests that immatures may have assisted in tending later broods, but no direct evidence of this was available. However, one published source states that such co-operative behaviour does occur.

The elusiveness of flufftails may be gauged by the fact that the Striped Flufftail is a fairly common but localised resident in the south-western Cape, but it was not until 1992 that the first nest in the region was discovered. The few observations made confirmed that both sexes incubated, apparently alternating at the nest according to a daily pattern (as has also been witnessed at a Red-chested Flufftail's nest), and the eggshells were eaten by the female when the eggs hatched. As soon as the four chicks were mobile they were led away from the nest by the female.

Excluding the American Purple Gallinule, there are five species in the final group of Rallidae: two gallinules, two moorhens and the Red-knobbed Coot. Although they are more easily observed than the preceding species, the breeding

4

5

6

1

1 Moorhens often build their nests in shallow water in the open, away from the reedbeds and rank grass favoured by other members of the Rallidae.
2 The Red-knobbed Coot also prefers an open site, but its nest is supported by aquatic vegetation in deep water.

biology of several of them is poorly known. Only the Moorhen and Red-knobbed Coot have been studied to any extent in southern Africa. The Purple Gallinule, Moorhen and Red-knobbed Coot are resident and the Lesser Gallinule and Lesser Moorhen are intra-African migrants which arrive to breed in seasonally inundated areas in the rainy season.

The Purple and Lesser Gallinules and Lesser Moorhen conceal their nests in reedbeds or rank grass over water, usually pulling down surrounding cover to conceal them, but the nests of the Moorhen and Red-knobbed Coot are often out in the open. The coot, in fact, builds a bulky floating nest that is anchored to vegetation or submerged sticks. The clutches of all five species comprise 3-9 eggs, and larger clutches of 11 in the case of the Moorhen may contain the eggs of two females. The eggs are buff or pinkish in ground colour with extensive brown or reddish spotting which is often concentrated at the larger end. The incubation period, where known, is between 19 and 25 days, depending on the species, and both sexes incubate and care for the young.

A study of the Moorhen in the south-western Cape revealed that because a good artificial food supply was always available, two colour-marked pairs were able to breed continuously over a period of 48 months, raising 33 and 32 broods to independence. Detailed notes taken during the incubation period established that the eggs were brooded continuously, by the male for 72 per cent of the time. Only the male incubated overnight, for an unbroken 14½ hours. It appears from this study that Moorhens are able to breed opportunistically and raise multiple broods in response to favourable conditions as long as they last. The chicks are tended by both parents who, on occasion, are helped by immatures from a previous brood. Sometimes the brood is divided between male and female, as has also been observed in the case of the Red-knobbed Coot. Generally, however, the male takes the greater share, allowing the female more time to feed herself and thus build up her reserves to lay another clutch.

The African Finfoot

The African Finfoot is an elusive species that frequents the margins of perennial waters that are well wooded with overhanging cover. Although it looks somewhat like a darter, it is apparently most closely related to rails. When it swims it spreads its tail on the water just as a beaver does, and it can submerge its body so that only its head and neck protrude above the water. Its red lobed feet have strong claws, probably for clambering up branches on the riverbank. The birds are shy and rarely seen, and as relatively few nests have been found the breeding biology of the species is poorly known.

The breeding season falls mainly between September and March, and as a rule laying coincides with a period of relatively high water levels in the various parts of the species' southern African range, after flood debris has been deposited. Monogamous and territorial, the African Finfoot does not indulge much in displays. In the only courtship behaviour described, one of a pair repeatedly swam out into open water and raised its wings alternately; the second bird made a clapping noise, probably by snapping its bill, and each time emerged from the bank to escort its mate back into cover.

The nest, an untidy structure, is placed up to 4 metres, usually less, above the water on an overhanging limb or amongst flood debris caught up in branches. According to one record, it measures 30 centimetres in diameter and 12 to 14 centimetres deep, with a cup 16 centimetres across and 8 centimetres deep. It is constructed with a miscellany of dry materials such as grasses, reeds and thin pliable twigs, and a finer lining in the cup comprises mostly dry leaves. Because of its situation and resemblance to debris the nest is very easily overlooked.

The African Finfoot usually lays two, or occasionally three, eggs which are somewhat rounded and resemble those of a korhaan. Their creamy buff ground colour is heavily overlaid with reddish blotches that coalesce to form a cap at the large end. Only the female has been observed incubating, but it is possible that the male may also do a share and sometimes he is perched nearby while the female is on the nest. She sits tight but if disturbed off the nest tries to slip away as unobtrusively as possible. The incubation period has yet to be recorded but it is not less than 12 days.

According to one observation, the downy chicks hatched on the same day and were alert and co-ordinated, remaining in the nest for at least two days where they were brooded by the female. They might have remained longer but took fright at the observer's presence on the third day and dropped 2.5 metres into the water below. There they emitted peeping calls like ducklings and swam off strongly with their parents.

So little is known about the breeding biology of the African Finfoot that even a few sustained observations would break new ground.

The African Finfoot often builds its debris-like nest on a branch overhanging a quiet backwater.

Bustards and Korhaans

Three bustard and six korhaan species breed in southern Africa, of which Ludwig's Bustard and the Blue, Karoo, Rüppell's and Black Korhaans are endemic to the region. The number of korhaan species would increase to seven if recent research that splits the Black Korhaan into two species, the White-winged and Black-winged Korhaans, is generally accepted, but in this text the Black Korhaan will be treated as a single species. Outside southern Africa the species in this group are all referred to as bustards.

Bustards and korhaans live in open arid country, in grassland or in bushveld, depending on the species. Some, for example Ludwig's Bustard, are nomadic, forming flocks of up to 40 birds and following seasonal flushes of food. Others, such as the Karoo Korhaan, are found in the same area year round. The three bustards are all vulnerable in southern Africa, mainly because their habitat has altered, but also because hunting pressure in the past had an impact on birds with a slow reproductive output.

Normally unobtrusive to avoid detection by predators, these ground-dwelling birds become conspicuous when they call and when they execute their territorial and courtship displays. In the latter the males' performances are often impressive, successfully advertising their presence to potential mates. The bustards inflate their gular pouches, puff out their white neck feathers, droop their wings, raise their tails to the vertical and fluff out their undertail coverts, sometimes enhancing their performance with an exaggerated bouncing gait. The Stanley's Bustard male becomes so puffed up that he resembles a white balloon from a distance, and may be visible to the naked eye from 2 kilometres away. Several males dispersed over an area are visited by females who presumably select the males with the most impressive displays. These males are polygynous and play no part in assisting the females once mating has taken place.

Amongst the korhaans, the male White-bellied and Blue Korhaans display by stretching forward their necks horizontally and puffing up their head and throat feathers. The Karoo and Rüppell's Korhaans are mainly vocal and perform their characteristic croaking duet mostly in the early morning and evening. The male Black-bellied Korhaan has a remarkable call which he makes from a favourite prominence. He sounds the first note of the *quick*-(pause)-*burp* sequence with his neck fully extended, retracts his neck for a few seconds, then raises it again to emit the popping burp note. He also flies around over several hundred metres with slow wing-beats that reveal his white wing patches, then parachutes slowly to the ground with legs extended.

The colourful male Black Korhaan is the most consistently noisy and aerial of all the korhaans. He emits his raucous *krracker, krracker, krracker* call as he flies round with slow wing-beats, finally dropping down with legs extended. Several males within a small area may be stimulated to perform their aerial display if a female is nearby.

Of all the korhaans' calls and displays, some of the most remarkable are those of the Red-crested Korhaan. Its call, a series of tongue-clicking notes followed by loud piping tones, is one of the most characteristic and evocative sounds of the bushveld and may be heard throughout the year. In the breeding season, however, when the male sees a female he uses deeper croaking calls to court her, making them from a regular calling spot. When a female is attracted he pursues her, puffing out his red nape patch and thus revealing how this species derived its name. The third element of his exhibitionist advertising display is the most impressive of all: he runs a short distance, flies vertically upward for 10 to 30 metres, turns upside down with his legs

1

2

4

3

__1__ Kori Bustard eggs laid in a shallow scrape. When the incubating bird is on the nest the rocks and grass clumps nearby help to disrupt its outline. __2__ The small stones surrounding a Ludwig's Bustard's nest have been accumulated by the sitting female. __3__ The Red-crested Korhaan's crest is usually seen only when raised as part of the male's display in pursuit of a female. __4__ The Karoo Korhaan lays single-egg clutches in the unpredictable conditions of its arid habitat.

A newly-hatched Red-crested Korhaan chick is effectively camouflaged by its down pattern.

thrown skyward and plummets back to earth, opening his wings at the last moment. The whole remarkable performance gives the impression that he has been shot at the zenith of his flight. A number of females may be impressed by his displays, and he mates with them on a polygynous basis.

Although the calls and displays of this fascinating group are conspicuous and generally well known, many of the basic facts relating to the different species' breeding biology have yet to be recorded. It is known that they all lay their eggs on the ground, usually in a shallow scrape. Sometimes the nest is surrounded by small stones, a phenomenon resulting from the incubating female's habit of collecting the stones from around the nest and stretching over her shoulder to place them on her back; they roll off and accumulate to form a ring. The nest is usually situated near a small bush, clump of grass or rocks which disrupt the sitting bird's outline.

The breeding season falls between August and February for most species, but the laying time varies, especially in arid areas, and depends on prevailing local conditions. The clutch of most species is one or two eggs and only the Blue Korhaan has been recorded to lay three. Single-egg clutches may indicate an adaptation to uncertain conditions. Both the Karoo and Rüppell's Korhaans, denizens of the drier parts of southern Africa, lay single eggs, although there are as yet unconfirmed reports that the former sometimes lays two. The eggs are oval in shape and cryptically coloured; most are olive green, dark khaki or buff in ground colour, with darker streaks or blotches of brown or reddish brown.

As far as is known only the female incubates and in most species her plumage is more drab than that of the male. All species are extremely wary at the nest, approaching it cautiously and leaving surreptitiously at a crouched walk. Sometimes the sitting bird will flatten itself, relying on its camouflage to avoid detection. The incubation period, recorded for only a few species and sometimes in captivity only, is usually 21 to 28 days, but in the case of the Kori Bustard it is 28 to 30 days.

The downy chicks are extremely cryptic and have characteristic markings on the head and neck which effectively camouflage them. They are precocial, leaving the nest with the female soon after hatching. They stay with her and in some species immatures and adults form small family groups. The young are first able to fly when they are about 35 days old. Observation of these birds is difficult in the wild, and accurate details about the individual species' breeding behaviour have yet to be collected.

Jacanas

There is a strong contrast between the behaviour of the African Jacana and that of the Lesser Jacana, the only two jacana species that breed in southern Africa. The former, a widespread and conspicuous species, is colourful, raucous and aggressive, whereas its smaller relative is comparatively silent and unobtrusive, and localised in its distribution. And, although in many respects their breeding biologies are similar, there are also certain important differences. Perhaps the most obvious of these is the fact that the African Jacana is polyandrous, whereas the Lesser Jacana is monogamous.

Both species make a rudimentary nest which is usually little more than a small sodden platform of aquatic vegetation, often near a clump of reeds or grasses. Sometimes the eggs are laid on a lily leaf or even on a block of floating peat. If flooding occurs the nest may be built up or the eggs may even be moved to a new site a few metres away. It seems probable that only well-incubated eggs that can float and be pushed to a new site by the adult are transported in this way.

Courtship in the African Jacana involves the male calling the female to one of several partial platforms that he has built. When she arrives they walk round each other in a crouched position before the male mounts the female and copulation takes

1

2

1 *A Lesser Jacana returns to its nest. The highly glossed eggs of both jacana species blend well into their watery environment.* ***2*** *Beautifully scrolled African Jacana eggs lie on a typically rudimentary nest amidst aquatic vegetation.*

place. Jacanas may breed at any time of the year, depending on local conditions. On the Nyl River floodplain in the Northern Province a researcher conducting a detailed eight-year study of the African Jacana observed that in 85 per cent of 256 records the eggs were laid between November and March.

The exceptionally beautiful eggs of jacanas are tan-yellow in ground colour, heavily overlaid with scrolls of black, and highly glossed. It was originally thought that the gloss contributed to water repellency, but in fact its function is to give the eggs a 'wet look' which makes them less conspicuous in their aquatic environment. In the Nyl River study it was found that several African Jacana females laid eggs that were unique in their dimensions, colour or markings and these eggs could readily be recognised as belonging to a particular female. The normal clutch of this species is four eggs and they are laid on successive mornings, with incubation beginning once the third egg has been laid. The Lesser Jacana usually lays three eggs.

The incubating jacana broods the eggs in a characteristic way. On arrival at the nest, usually after feeding, it adopts a knock-kneed stance and then settles so that its large feet are splayed out on either side of its body. It tucks its wings in tightly at a sharp angle beneath it, so that the eggs are pushed in against the brood patch and probably also lifted slightly off the damp nest. The belief that the eggs are held and incubated *under* the wings is incorrect. The incubation period of the African Jacana is 23 to 26 days; that of the Lesser Jacana has yet to be established.

It is in their incubating behaviour that one of the main differences in the breeding biology of the two species is found. In the African Jacana simultaneous polyandry is regular – although monogamous situations do occur – and only the male incubates. In contrast, the Lesser Jacana is monogamous and the sexes share almost equally incubation duties and care of the chicks. Another significant difference between the species is in the amount of time each attends the eggs. Lesser Jacanas spend 82 per cent of the daylight hours on the nest and incubate for alternating shifts averaging 39 minutes. The male African Jacana, on the other hand, broods intermittently, leaving the nest for short periods 35 times a day on average, and he sits for spells of between 15 and 30 minutes, averaging 20 minutes. The eggs are attended for 53 per cent of the daylight hours. If birds of the same or other species come too close to the nest they are aggressively chased away.

Newly-hatched jacanas are remarkable little creatures, appearing to be all toes attached to an incongrously small, downy body. They are precocial and become mobile within hours of hatching. As the African Jacana chicks hatch the male has been observed removing the eggshells from the nest. Once all the jacana chicks have emerged they leave the nest and are able to feed themselves. If danger threatens they dive under water and hide in vegetation or run to the nearest cover. In the African Jacana the male gives his alarm call and the chicks run to him to shelter under his wings. He then carries them to safety with just their long toes protruding. The bow-shaped radius or inner 'wrist' of the African Jacana's wings is an adaptation for carrying chicks which the Lesser Jacana lacks; it broods chicks beneath its wings but has not been observed to carry them.

The chicks of the African Jacana are usually first able to fly when 39 to 44 days old, but in some broods 59 to 75 days passed before the first flight was made. The reason for this marked difference appears to relate to the food supply, with chicks taking longer to develop in lean years. The young remain in the area in which they were raised for about three months before they disperse.

Although the nests of African Jacanas are highly vulnerable to predation, particularly by water monitors and Purple Gallinules, the survival rate of the chicks (at about 80 per cent) is high because of the tactics used to escape danger. Sometimes the adult may slap the water with its wings or use a weak fluttering flight to distract predators from the chicks.

A male African Jacana carries his chicks under his wing, with only their long toes protruding.

WHY IS THE AFRICAN JACANA POLYANDROUS?

The polyandrous breeding strategy of the African Jacana, which is found in a number of other jacana species, raises a number of questions. At the same time, the monogamous behaviour of the Lesser Jacana also requires explanation.

Let us first consider the facts of the African Jacana's polyandrous behaviour. According to the detailed study of African Jacanas undertaken on the Nyl River floodplain, there were fewer females than males; females were dominant and 68 per cent heavier than males; they aggressively defended a territory against other females; they usually had four mates but sometimes as many as seven in a season; and they laid as many as eight or nine clutches over a three-month period, an ability for producing multiple clutches unrivalled amongst the waders of the order Charadriiformes.

During the incubation period clutch losses were high, varying between 67 and 89 per cent. The extent of the losses depended on the conditions prevailing, and in years of drought they were higher. Predators such as water monitors and Purple Gallinules accounted for 83 per cent of them, and flooding for another 8 per cent. The explanation for the polyandrous mating system of the African Jacana is linked to this high degree of predation. If the female were to invest her time in incubation it would prove wasteful, especially if the clutch were lost towards the end of the incubation period. It is a far better strategy for her to become emancipated from nest duties and lay multiple clutches for several males in order to offset the high rate of nest failure.

To facilitate the laying of multiple clutches she lays small eggs, and a clutch of four is equivalent to only 15 per cent of her body weight (compared with 75 per cent of the body weight of other female waders of the order Charadriiformes that regularly lay four eggs). Smaller eggs usually require a longer incubation period, but those of the African Jacana are an exception so in this instance there is no disadvantage. Additionally, the male's frequent absences from the nest to feed are compensated by a good food supply and the relatively high ambient temperature at the time of breeding. On cool days it has been observed that the eggs require more attention.

The size of the egg seems to be the clue to the Lesser Jacana's monogamous mating system and shared incubation. Its egg is half the volume of an African Jacana's, and because of its very small size it requires more actual incubation. This can only be achieved if the male and female Lesser Jacana share these duties.

Finally, the fact that there are more male African Jacanas than females has an explanation. Although the female is heavier, she does not have proportionately longer toes than the male and is not able to exploit floating foraging areas that are accessible to the lighter male. Because the wetlands inhabited by jacanas are ephemeral, the birds have to move to new foraging areas from time to time, and it is the females who are under pressure to do so before males. This in turn exposes them to a higher risk of predation. The polyandrous strategy of the African Jacana is a system where the disadvantages are outweighed by the advantages in a remarkable way ●

Only the male African Jacana incubates, sitting with his wings tucked in tightly beneath his body and his large feet splayed on either side of him.

Snipes

The Painted Snipe and the Ethiopian Snipe belong to two quite distinct families – the Painted Snipe to the Rostratulidae and the Ethiopian Snipe to the Scolopacidae – but are placed together here purely for convenience. The former is widely distributed in Africa and beyond, its range extending from Madagascar into Asia and Australia. The Ethiopian Snipe is confined to Africa and in southern Africa it is the only breeding representative of some 35 species of Scolopacidae that occur in the region.

It is fitting that the account of the Painted Snipe should follow that of the jacanas because, like the African Jacana, it has a polyandrous mating system. However, there is as yet no explanation for the Painted Snipe's behaviour. The suggestion has even been made that the snipe may be related to jacanas for several reasons, including the similar down patterns of the chicks. Although the female Painted Snipe is like her African Jacana counterpart in that she is larger than the male, the two species are dissimilar in that the Painted Snipe, with her rich chestnut foreparts, is also more colourful than her partner. Once she has completed a clutch for one male she will mate with another, sometimes laying four consecutive clutches for different males. However, as in the case of the African Jacana, there are occasions when monogamous situations may occur.

The female defends a territory against other females and initiates courtship by preening the male; then she faces him with her wings outspread and walks round him emitting a mellow *boooo* call, similar to the sound produced by blowing across the top of a bottle. This is but one of a wide repertoire of calls, some of which carry long distances, and their resonance is facilitated by a specially adapted trachea.

The Painted Snipe's breeding season extends from August to November in the Western Cape, but north of the Vaal River and in Zimbabwe it lasts from July to March. The nest is situated in marshy locations and may be a shallow scrape lined with dry vegetation or it may be built up into a substantial pad of material. It is always well concealed by surrounding cover and when the male leaves and returns he does so stealthily, often crouching low to avoid detection.

The clutch is normally four eggs, but 2-5 have been recorded. They are laid at 24-hour intervals and incubation begins once the clutch is complete, at which stage the female leaves the male to seek out another mate. The slightly glossy eggs are pyriform and their buffy yellow ground colour is heavily overlaid with dark brown blotches. One incubation period in southern African has been recorded at 19 days. Once all the chicks have hatched the male leads them away and they do not return to the nest. The male guards and feeds the small chicks initially and they remain with him for one to two months. It is not known when they are first able to fly. Immature birds resemble the males and observations that there are more males than females in a population could be erroneous because of this factor.

The Ethiopian Snipe is best known for a display known as 'drumming', in which the male flies to a height of up to 40 metres, dives steeply with his tail feathers fanned, then zooms skyward to gain height and repeats the process. During this performance the outer tail feathers produce a humming sound which can be imitated by attaching a snipe's two outer tail feathers to a cork and swinging it round rapidly. The display usually takes place in the evening, but also on moonlit nights and sometimes on cool days. On the ground the male courts the female by approaching her with his tail cocked and fanned.

The breeding season is timed to coincide with wet conditions, such as those in winter and early spring in the south-western Cape. The nest is always located in a marshy area and forms a saucer of dry vegetation that is well concealed by surrounding cover. The clutch normally comprises two eggs, rarely one or three, and they are olive-green or khaki in ground colour with blotches of brown that are often concentrated at the large end. The newly-hatched chick has dark brown down flecked with chestnut.

When incubating, the Ethiopian Snipe sits extremely tight, flattening itself with its bill tucked down flat in front of it. It flies off only at the last moment, emitting as it does so a sucking call reminiscent of a boot being pulled out of the mud. Little more is known about the breeding biology of this species, although at one nest observed the adult removed the eggshells after the chicks had hatched and dropped them a few metres away, and when approached it flapped away from the chick, feigning injury.

1

2

1 *The male Painted Snipe incubates the eggs and rears the chicks while the female goes in search of another mate.* ***2*** *The Ethiopian Snipe lays its two eggs on a pad of dry vegetation in a marshy environment.*

The African Black Oystercatcher

The endemic African Black Oystercatcher, the only breeding oystercatcher in southern Africa, is a striking bird with uniform black plumage set off by its red eye, orange eye ring and red bill and legs. Its call, a strident, piping *klee-eep... klee-eep... klee-eep* series, is as unmistakable as its appearance and often attracts attention to the bird as it flies along. This is a typical seashore bird, found on sandy beaches, rocky shores and on offshore islands. The islands, in fact, are breeding havens for the species as there are no mammalian predators on them, but both the number of suitable ones and the space on them are limited. The lack of space for nesting and feeding forces some birds to breed inland on the islands and commute to the mainland to feed, carrying back one food item at a time to their chicks. However, the energy expended in such a laborious exercise is rewarded by security.

The African Black Oystercatcher is monogamous and a pair will breed in the same locality year after year, staying in the same general area throughout the year. Because a regular pair bond is maintained, there appear to be no elaborate courtship displays apart from aerial chases and exaggerated 'butterfly' flights with fluttering wing-beats. These are performed at the onset of the breeding season, which extends from October to March but peaks between December and February. The nest, a simple scrape 21 centimetres across and 4 centimetres deep, is usually situated on a sandy beach or a rocky shoreline, but one has been found on an island 9 metres above sea-level. Several scrapes may be made before one is finally selected for laying in. Nests on beaches are frequently near or amongst dry, blackened kelp which camouflages the incubating bird very effectively. Although they may be located either on small raised dunes or on flat areas above the high tide mark, some are still swamped by spring high tides.

The normal clutch comprises two eggs, but sometimes one, or very rarely three, are laid. They are oval and buffy-stone in ground colour, with scrolls and spots of dark brown. Incubation lasts 32 days on average, but ranges from 27 to 39 days, and both adults share the duty. At one nest the female was observed to leave the nest every 12 minutes, run down to the water and soak her belly feathers, then return to the nest, turn the eggs and settle on them once more. The male relieved her twice for spells of 16 minutes each, but he was not seen to leave the nest to wet his feathers.

If humans are seen approaching, the incubating bird either flies directly off the nest, or walks or runs away from it. It sometimes adopts a crouched position to lure intruders away or flies ahead of them, emitting its piping calls, and intensifies this behaviour when the chicks hatch. The chicks are semi-precocial and do not remain long in the nest after hatching, but when they crouch motionless next to seaweed or similar cover they are well camouflaged by their slate-grey down with dark lines on the back. Both parents tend the young, bringing shellfish to them whole and then extracting the fleshy interior for them. It is unusual for semi-precocial young to be fed by their parents until they are fully grown, but in this case it is necessary as they are unable to open shellfish until their bills have fully developed. The young accompany their parents to the shore and are capable swimmers if the need arises.

At 87 per cent, the mortality of chicks is highest during the first week after hatching when they are most likely to be preyed on by Kelp Gulls or mongooses, trampled by humans or run over by vehicles; after that they have a fair chance of survival. The fledging period is 35 to 40 days, and after another two to six months the young become independent, dispersing from their natal sites.

BREEDING OYSTERCATCHERS AND HUMAN DISTURBANCE

Unfortunately human pressure for recreational space on the coast has had a very harmful effect on the breeding success of oystercatchers. To make matters worse, the summer holiday season when humans are most active on the seashore coincides with the oystercatchers' peak breeding season.

Despite the fact that the birds replace lost clutches, the breeding productivity of African Black Oystercatchers is low. The species faces several threats: birds are poisoned by contaminated shellfish after a 'red tide' along the west coast of South Africa; nests are flooded by spring high tides; eggs and chicks are preyed on by Kelp Gulls and mammals such as feral cats and mongooses; and breeding attempts fail because of human disturbance. Of all the threats, the last-mentioned is particularly serious.

The situation was monitored over seven breeding seasons from 1978/79 to 1984/85 along a 12-kilometre stretch of coast west of Cape Agulhas. Sixteen or 17 pairs of oystercatchers were breeding in the area during the period covered, and the number of fledged chicks they produced each year dropped steadily from 13 in the first season to one in the 1982/83 season. No young were reared at all in the last two seasons, despite the fact that 24 replacement clutches were laid and sometimes three attempts were made in a season. During the period monitored the number of first breeding attempts showed little variation, but the rate of their failure increased progressively each season. While observing the oystercatchers the researcher noted a marked increase in the number of off-road vehicles along that stretch of coastline, and the extent to which they disturbed the beach-nesting birds, causing their breeding attempts to fail ●

1

2

3

1 *African Black Oystercatchers nest on beaches where the pressure of human disturbance often causes breeding attempts to fail.* ***2*** *An African Black Oystercatcher nest among rocks has been built up with stone chips.* ***3*** *When this species nests on a beach it often deliberately selects a site near blackened kelp which helps to conceal the incubating bird.*

Plovers

This fascinating group is represented in southern Africa by 11 breeding species, four of which – the White-fronted, Chestnut-banded, Kittlitz's, and Three-banded Plovers – are small and were originally called sandplovers. The other seven species – the Crowned, Black-winged, Lesser Black-winged, Blacksmith, White-crowned, Wattled and Long-toed Plovers – are all much larger and, as a rule, more strikingly marked; they also reveal a bold pattern of white in the wings in flight. Because of their terrestrial habits most species are drab above to make them less visible to aerial predators, especially when they are on the nest. Even the boldly patterned plumage of the Blacksmith Plover serves as disruptive camouflage when the bird is incubating.

Some species are nomadic, often flocking to areas where conditions are favourable. The White-fronted Plover, however, maintains a pair bond and defends a territory throughout the year, apparently doing so firstly to ensure that it has an adequate area in which to forage even when food is scarce, and secondly to enable the plovers to breed opportunistically when conditions are favourable.

The breeding behaviour of plovers is well documented and the territorial, courtship and anti-predator displays of the different species show certain similarities. They are monogamous (although occasional cases of polygamy have been recorded) and a pair will defend a territory even when breeding in loose colonies, chasing intruders away by running at them in a crouched position with lowered head.

Plovers' courtship behaviour is mostly terrestrial, although it may also involve 'butterfly' displays when the bird flies with slow, exaggerated wing-beats. In the 'scrape' ceremony, an important element of courtship that may be performed by the male alone or by both sexes, the plover leans forward at a sharp angle, tail in the air, until it rests on its breast. Rotating on its own axis with the breast as a fulcrum, it kicks sand backwards with its feet, forming a neat scrape with a slight rim. Several such scrapes may be distributed over a small area, and one is selected for laying in. Copulation often follows a 'scrape' ceremony.

The nests of plovers, being on the ground, are highly vulnerable to predation, and the birds have developed a number of behavioural patterns to counteract this. They are always very alert and the larger species usually dive-bomb a predator, calling vociferously to distract it. If other plovers are nesting in the area they join in to confuse the intruder even further. Often an adult runs away from the nest dragging one or both wings to feign injury and sometimes flapping weakly to enhance the effect. Once the predator has been lured away to a safe distance the plover flies off normally. Another effective deception is 'false brooding', whereby the bird runs some distance from the nest or chicks and pretends to be brooding elsewhere. It may repeat the process several times, luring the predator further away. Sometimes the brooding plover stands still and raises its wings to reveal a striking pattern, thus causing oncoming herds of animals to avoid the nest. The success of this ruse is demonstrated by nests that have been found with hoofprints all round them. A species such as the Crowned Plover, which nests where human disturbance is regular, is extremely bold in defence of its nest, advancing towards the intruder with raised wings.

1

In some cases the breeding season is variable, depending on local conditions, and in arid regions certain species breed opportunistically after rain. In other cases species breed regularly at a particular time of year. The White-crowned Plover, for example, lays mainly between August and October, when water levels are at their lowest and the sandbanks on which they breed are exposed.

All plover species make a simple scrape or lay their eggs in a hollow such as a hoofprint. At first the nest is usually not lined, but as incubation progresses the brooding bird flicks small pieces of earth, eggshell, animal droppings or dry vegetation towards and into it. The incubating bird may also gather surrounding material and pull it in towards the nest. In some nests accumulated lining may partially cover eggs, but this does not result from deliberate egg-covering behaviour like that of Kittlitz's Plover or, to a lesser extent, the White-fronted Plover.

Some species have particular or preferred requirements for their nesting habitat: White-fronted Plovers breed on beaches or on exposed sandbanks of inland rivers; Chestnut-banded Plovers choose the perimeter of brackish lagoons and saltworks; the Three-banded Plover prefers a gravelly or stony substrate near

2

3

4

5

6

***1** Rocks near a White-fronted Plover's nest on a beach make the incubating bird less conspicuous. **2** Similarly Three-banded Plovers, which nest on stony ground, choose a site where large stones disrupt the outline of the sitting bird. **3** The Black-winged Plover's eggs – khaki with dark brown blotches – are well camouflaged in a recently burnt environment. **4** White-crowned Plover eggs are laid on exposed sandbanks along rivers, where the heat is often extreme. The adult bird may cool the eggs by soaking its belly feathers before settling down to incubate. **5** Chestnut-banded Plovers prefer to nest at the edge of a salt pan. **6** Very soft mud surrounding a Long-toed Plover's rudimentary nest may help to protect it from predators.*

water; Black-winged and Lesser Black-winged Plovers show a preference for recently burnt areas; and Blacksmith and Wattled Plovers usually nest near water. Although the Long-toed Plover breeds on mudbanks, making the usual plover scrape, it also nests on floating matted weeds and constructs a rudimentary nest of aquatic vegetation like a jacana's.

A feature of most plover nests is that they are situated near some object that enhances the camouflage of the eggs. The White-fronted Plover, for example, lays its pale eggs with relatively sparse markings near driftwood, dry kelp or beach plants. In contrast, the eggs of the Three-banded Plover, heavily overlaid with fine blackish lines which virtually obscure the ground colour, are inconspicuous among the small stones and pebbles of the substrate on which they are laid. All the larger plovers lay eggs that are varying shades of khaki in ground colour with spots and blotches of dark brown. Many species lay near animal droppings, and the eggs of those that lay on recently burnt ground, such as the Black-winged and Lesser Black-winged Plovers, are especially well camouflaged.

Clutch sizes vary, although the four small species regularly lay two eggs. The exception is the White-fronted Plover which sometimes lays three eggs and very occasionally four. The seven larger species lay 2-4 eggs, most often three or four, but the Blacksmith Plover has been recorded to lay six and, in one exceptional clutch, seven. Plover eggs are characteristically pyriform, and in clutches of three or four the pointed ends incline inwards towards the centre of the nest, so that the area the incubating bird has to cover is reduced.

The incubation period is usually between 21 and 28 days, but there is considerable variation, even within the same species. The White-fronted Plover, for example, may incubate for any period between 26 and 33 days, which is little different from the 28- to 32-day term of the much larger Crowned Plover. The reason for such a range in the duration of incubation within the same species is not clear, but it may be linked to the ambient temperature at the time of breeding.

Both sexes share incubation, and if a nest is watched for a long period regular change-overs will be seen. The sitting bird remains alert and is often warned of approaching danger by its mate's alarm calls. In hot weather the attending parent may stand over the eggs to shade them, at the same time raising its back feathers to cool itself. Several species have been recorded to run down to water, soak their belly feathers, and then return to the nest and settle on the eggs to cool them. The White-crowned Plover does this regularly, probably because it breeds on exposed river sandbanks where the heat is intense, and one was even seen to place vegetation over the eggs, presumably to cool them. However, White-fronted Plovers nesting on beaches have never been seen to wet their eggs.

When the eggs hatch the pieces of shell are removed so that their white interiors do not attract predators. The chicks are covered with cryptically patterned down which makes them even more difficult to detect than the eggs. They are precocial, leaving the nest soon after hatching, and before long are able to feed themselves. They are still attended by both parents and respond instinctively to their alarm calls, flattening themselves and lying motionless, usually near some cover. The young plover is able to fly when about 35 days old and it remains with its parents for two to three months from the time of hatching.

A Wattled Plover settles to brood its eggs. This species usually nests near water.

THE CASE OF THE DISAPPEARING EGGS

Kittlitz's Plover has the remarkable habit of covering its eggs. Detailed observations, both in the wild and in captivity, have shown how this is done, but why it is done has yet to be explained.

When a human intruder approaches, the plover stands up, straddles the nest in a half crouched position, and kicks sand over the eggs with rapid inward kicks, first with one foot and then the other. As it does this rapid shuffle – for so it appears to the observer – it rotates, ensuring even coverage of the eggs. The operation is so deft and rapid that the eggs may be completely buried in as little as three seconds. The plover then pauses momentarily before running away. The eggs are so well concealed that, unless the spot is marked precisely, they would be almost impossible to find. Only when a strong wind is blowing or when the ground is wet are the eggs not completely covered. On returning to the nest once danger has passed, the plover flicks away any small stones with its bill and settles on the nest. Then it assumes a squatting position, rotating and kicking vigorously backwards until the eggs are uncovered.

Both sexes practise this egg-covering behaviour, although they do not do so when changing over at the nest or if disturbed at night. The evidence available indicates that Kittlitz's Plovers cover their eggs in an instinctive and invariable reaction to the approach of humans; they do not do so when approached by cattle or by mammalian or avian predators.

The case of the White-fronted Plover is somewhat different. It was originally thought that sand blowing over the nest covered the eggs of this species. However, observations have revealed that the White-fronted Plover also covers its eggs with a kicking action, but rarely buries them completely. It does not rotate around the eggs as it kicks, nor does it excavate them when it settles on them after returning to the nest. It does not always cover the eggs, but seems to do so only in the early morning.

Why should these species cover their eggs? There are two possible explanations: to maintain the temperature of the eggs, or to conceal them – or perhaps both. Because there appears to be no correlation between the egg-covering behaviour and prevailing weather conditions, it would seem that the second explanation is more likely. Yet we still don't know why these two plovers in particular should want to conceal their eggs, as they are as cryptic as the eggs of other plover species ●

1

2

3

__1, 2__ A Kittlitz's Plover's nest, with eggs covered and uncovered.
__3__ Having covered the eggs with a few deft kicks, a Kittlitz's Plover pauses over its nest before running off.

The Avocet and the Black-winged Stilt

Belonging to the same family but different genera, the Avocet and the Black-winged Stilt are similar in many respects, particularly in their breeding biology. Indeed, when they breed together it is easy to confuse their nests. Both have a widespread distribution outside Africa, and occur in an aquatic environment.

Both species are monogamous, and observations of nesting pairs have been made easier by the fact that the sexes can be distinguished. In the case of the Avocet, the male has a bright red eye whereas the female's eye is brown or hazel. The female Black-winged Stilt differs from the male in having brownish instead of black scapulars.

Avocets breed in colonies which may number 50 or more pairs, or may comprise just a few pairs in a recently inundated area. Their nests may be close to each other and are sometimes intermingled with those of Black-winged Stilts. The breeding season is mainly from August to November, but in arid areas the birds will breed opportunistically at any time of year. Their nomadic habits enable them to take advantage of locally favourable conditions, and in Bushmanland, for example, they were observed breeding in April after good rains.

At the onset of the breeding season Avocets perform a group ceremony in which several pairs assemble in a circle and, facing inwards, make bowing movements. Pairs form when a female accompanies a feeding male and he accepts her presence. He then initiates a 'scrape' ceremony similar to that of the plovers, and subsequently both birds work on the nest scrape, sometimes excavating several before one is selected. Copulation usually takes place nearby in shallow water.

Nests vary considerably: on hard ground a mere scrape with a few pieces of lining may suffice, whereas in moist situations the scrape is likely to be lined with more substantial pads of dry plant material. If flooding occurs, the nest may be built up to keep the eggs above the rising water. Usually four eggs are laid, but three or five have been recorded. They are greenish grey in ground colour with small, dark brown spots over the entire surface, and are laid at intervals of one or two days. Both adults incubate, changing over about once every hour. Although incubation may begin before the clutch is complete, the young all hatch at the same time, emerging after between 22 and 27 days, but on average after 24 days.

At one nest watched during the hatching period, the first egg showed signs of cracking at midday. By 08h00 the next day all four eggs were pipping and three hours later the first chick hatched. The adults removed the pieces of eggshell, eating some of the smaller fragments. At 13h00 the second chick hatched. At 15h30, while the female remained on the nest, the first-hatched chick and the male went down to the water a few metres away and in the shallows the chick was brooded under the male's wing, its feet protruding like those of a young jacana.

The newly-hatched chicks, covered in grey down with black markings on the back, are delightful creatures and even at this early stage the slight upward curve of their bills is discernible. Tended by both parents, they are precocial and able to swim well, and after about a month they can fly weakly.

Black-winged Stilts also breed colonially, in small groups with several nests scattered over an area or, in suitable conditions, in large colonies numbering hundreds of birds. In the 1950s about 700 birds were counted on the flooded fields of the Athlone sewage works near Cape Town, and three observers were easily able to measure 100 eggs in two hours. The area is now under housing.

In the south-western Cape breeding takes place mainly in October and November, but elsewhere the season is longer and, like the Avocet, the Black-winged Stilt breeds opportunistically in arid regions. Pair formation is similar to that of the Avocet, whereby the female feeds alongside the male until she is accepted by him, but in courtship the stilts perform 'butterfly' flights and dancing ceremonies in shallow water, where copulation takes place. Both sexes build the nest which, in general, is similar to that of the Avocet. However, the Black-winged Stilt also breeds in shallow water and there it builds up a substantial volcano-shaped nest of plant material, adding to it if the water level rises. Occasionally the nest may be situated on a thick floating mat of weeds.

The normal clutch comprises four eggs, but three or five have also been recorded. In one exceptional case a clutch of seven eggs was observed, but this was possibly produced by two females. The eggs, dark khaki in ground colour and heavily blotched with dark brown, are incubated by both parents, although the female does the major share. They hatch after 24 to 27 days. Both parents tend the precocial chicks, which are able to fly when about a month old.

As a general rule Avocets are very confiding at the nest and readily accept a photographic hide nearby. In contrast, Black-winged Stilts are extremely wary, and disturbance should be kept to a minimum while they are breeding. When they see a predator the adult birds hysterically emit their high-pitched piping call and jump up and down flapping their wings feebly. They sometimes also perform an injury-feigning display to lure a predator away.

1 *The bills of Avocet chicks, which are precocial, show an upward curve from an early age.* ***2*** *Black-winged Stilts sometimes nest in shallow water, where they build a substantial mound on which the eggs are kept dry.* ***3*** *Avocets line their scrape nests with varying amounts of plant material, using more if the substrate is moist.*

Dikkops

The two dikkops in southern Africa – the Spotted and Water Dikkops – are very alike in their habits and, apart from the habitats in which they nest, their breeding biologies are very similar. Both species are crepuscular and nocturnal, have loud, musical calls and form flocks outside the breeding season. The Spotted Dikkop is widely distributed, whereas the Water Dikkop is confined to rivers, lakes and other aquatic habitats. The former, in particular, would not be difficult to study, and it is surprising that most of our knowledge of this species – such as the incubation and fledging periods – is based on captive studies. Courtship behaviour has been described for neither species, probably because it would take place mainly at night.

The Spotted Dikkop, like the Crowned Plover, has adapted to an urban environment and may nest on the perimeter of school playing fields, in parks and in suburban gardens. It lays its eggs on the ground, usually without making a scrape but invariably in a position where its cryptic markings blend into a background of dry leaves, bark chips or animal droppings. Often some feature nearby, such as a piece of wood or a rock, helps to break up the outline of the sitting adult. Not infrequently the eggs are situated so that they are in shade for much of the day.

The Water Dikkop has not adapted to human proximity and breeds in more remote locations on the sandy banks of rivers, on the shores of lakes and lagoons, occasionally on a beach, and sometimes on a rock outcrop in the middle of a river. Its nest, like that of the Spotted Dikkop, may be situated near a feature that will break up the sitting bird's outline, but usually the eggs are laid in a scrape or a hoofprint.

Dikkops are monogamous and defend a breeding territory against other members of the same species. Both species breed from about August to December but their breeding season is by no means confined to these months. One pair of Spotted Dikkops nested in a suburban garden in Cape Town over a period of 12 years and laid eggs in every month except May and June. In urban areas in general Spotted Dikkops may lay multiple clutches, and there are records of one pair laying four times a year and another laying five clutches between 16 October and 3 March. The high number of consecutive clutches in suburbia may be explained by pairs losing their eggs or chicks to cats and dogs.

The clutches of both species usually comprise two eggs – very occasionally only one – and their creamy buff ground colour is thickly blotched with dark brown. Incubation lasts 24 days and is shared by both sexes. The cryptically patterned chicks leave the nest soon after hatching and are tended by both parents, taking food directly from the adult or picking it up when it is placed on the ground in front of them. In captivity a young Spotted Dikkop was first able to fly at 56 days old and presumably the fledging period of the Water Dikkop would be similar. The young may remain with the adults in a family party for some time after they fly, but it is not known when they become independent.

Dikkops threaten predators by extending their spread wings forwards, cocking and fanning their tails, and emitting a growling call, sometimes rushing forwards in this position. This behaviour is more commonly observed in Spotted Dikkops since they regularly encounter people, but in one record a Water Dikkop was seen to threaten and successfully ward off a boomslang that was heading towards its nest.

1

2

3

***1** The finely speckled down of Spotted Dikkop chicks provides effective camouflage, whatever the background. **2** A Water Dikkop has laid its eggs in a scrape, taking advantage of whatever cover that may be available for the sitting bird. **3** A Spotted Dikkop stands over its eggs, which are incubated by both sexes.*

Coursers

The five southern African coursers are birds of arid environments, including dry savanna, and even in dry woodland they all favour habitats where the ground is bare and they can run about freely. Burchell's, Temminck's and Bronze-winged Coursers are particularly attracted to recently burnt areas in which they feed and breed. Although the different species prefer to nest in different habitats, their breeding habits are relatively similar. The exception is the Three-banded Courser which buries its eggs, and in this respect it is unique among coursers.

The breeding season usually occurs before the rains begin, but nesting may be opportunistic. The Double-banded Courser, for example, lays at any time of the year, an adaptation to the arid environment in which it lives. All coursers are monogamous and nest solitarily. Little is known about the courtship behaviour of most southern African species except the Double-banded Courser. In its courtship ritual a pair walks together with a stiff-legged gait, then the male dances around the female with short hopping steps. As she watches him she emits high-pitched calls and finally solicits copulation by crouching.

Coursers lay their eggs on bare ground, making no nest scrape although a slight depression may form as incubation progresses. The rounded eggs sometimes move some distance from their original position as they are turned during the course of incubation. Only the buried eggs of the Three-banded Courser are immobile. A feature of most courser nests is that the eggs are usually surrounded by small objects such as animal droppings which help to disrupt their outline and, to a lesser extent, that of the sitting adult. The Double-banded Courser, like plovers, 'throws' this material towards the nest site.

Burchell's, Temminck's and Three-banded Coursers normally lay two eggs and the Bronze-winged Courser lays two or three. The Double-banded Courser invariably lays a single egg, and this is undoubtedly an adaptation for breeding in marginal conditions where food may be scarce and breeding is opportunistic. To make up for laying only one egg, a pair may produce several clutches during the course of a year.

The eggs of most species are rounded, except those of the Three-banded Courser which are somewhat elongated. The ground colour is usually almost obscured by fine black and grey scribbles, sometimes to such an extent that the eggs appear blackish, an effect that enhances their camouflage, especially on burnt ground. The Double-banded Courser incubates for 26 to 27 days, and the Three-banded and Bronze-winged Coursers for 25 to 27 days. The incubation period for Burchell's and Temminck's Coursers is unrecorded.

In general, both sexes incubate the clutch and by alternating on the nest are able to share the load of heat stress. The off-duty parent often stands in the shade of a small tree or bush, and both birds try to keep cool by raising their dorsal feathers and by panting. Rather than brooding the eggs as such, the parent bird often squats over them with its back to the sun so that they are shaded. The adults do not leave the eggs unattended when it is hot as the embryos are likely to die, as happened to Double-banded Courser embryos which were left uncovered for more than 15 minutes.

1

2

3

4

5

***1** A Temminck's Courser squats over its eggs to shade them. **2** Its chick, like the chicks of many ground-nesting species, has a cryptic down pattern. **3** A Double-banded Courser's single rounded egg is barely discernible among stones on a gravel plain. **4** Numerous fine scribbles almost obscure the ground colour, enhancing its camouflage. **5** If a Double-banded Courser is detected at its nest it may threaten an intruder boldly.*

When it comes to incubating the eggs, the Three-banded Courser again appears to be an exception. An observer watching a nest for periods of two and four hours during the heat of the day on two consecutive days noted that no change-overs took place. The incubating bird's partner simply stood motionless in the shade of a tree 5 metres away.

Coursers are very alert and wary when breeding. Because their nests have good all-round visibility they often perceive an intruder first and run off unobtrusively long before being seen. The Double-banded Courser, which nests on gravel plains surrounded by low scrub, may have concealed itself amongst bushes before an observer is anywhere near the nest, or it may sit tight like the Three-banded Courser and rely on its camouflage to avoid detection. If it is detected and approached too closely it may threaten an outstretched hand by gaping and spreading its wings and tail. Some birds even advance towards the intruder with outstretched wings touching the ground. When a predator approaches the nest coursers usually run away, sometimes false brooding as plovers do to lure it away from their eggs or chicks.

As soon as the eggs hatch the shells are removed and dropped away from the nest. The down pattern of the chicks is cryptic, the dark colouring of a Temminck's Courser chick, for example, rendering it virtually invisible on burnt ground, and the disruptive pattern of a Double-banded Courser chick providing extremely effective camouflage on stony ground. A chick of the latter species remained where it hatched for 24 hours and did not wander far during the first few days. Three-banded Courser chicks have been observed leaving the nest less than 24 hours after hatching.

In all species both parents tend the chicks. In the case of the Double-banded Courser they feed their offspring at frequent intervals, offering it insects in their bills. Although the chick is probably able to feed itself at the age of about a week, it may still occasionally be fed until it is able to fly at about six weeks old; this is the only fledging period recorded for a southern African courser.

BURIED EGGS

The Three-banded Courser is found in dry savanna and mopane woodland, and like several other coursers it is active mainly at dusk and at night. In contrast to other coursers, however, it lays its eggs in a deep scrape, usually covering them to two-thirds of their depth by throwing soil sideways into the scrape with its bill – it does not use its feet as Kittlitz's and White-fronted Plovers do. Sometimes, as incubation progresses, the eggs are almost completely buried, becoming so firmly embedded as the soil compacts that they cannot be moved by the courser. Indeed, the sitting adult merely covers the exposed part of the eggs, and when chased off the nest it makes no attempt to conceal them, or unearth them on its return. However, when an observer uncovered a nest – no easy task as the eggs were firmly buried – the incubating parent covered them again by flicking soil into the nest at intervals over a period of $3^{1}/_{2}$ hours.

Why should the Three-banded Courser bury its eggs? The two likely explanations are to conceal them or to keep them cool, and a combination of both may apply. It is generally thought that Kittlitz's and White-fronted Plovers cover their eggs to conceal them, and this may also be the reason for the Three-banded Courser's behaviour. But the enigma remains – why should only these species cover their eggs and not other members of their respective groups? ●

The buried eggs of the Three-banded Courser are immovable.

Pratincoles

Pratincoles are classified with coursers in the family Glareolidae but differ from them in that their proportions are tern-like and their feeding habits are aerial. Of the three species that occur in southern Africa, two – the Red-winged and the Rock (formerly White-collared) Pratincoles – are breeding intra-African migrants, while the third – the Black-winged Pratincole – is a non-breeding Palaearctic migrant. Neither of the breeding species has an extensive distribution in southern Africa, the Red-winged Pratincole occurring in a band that runs northwards from Kwazulu-Natal to Mozambique and then westwards through northern Zimbabwe and Botswana to Caprivi. The Rock Pratincole's range is even more limited, as this riverine species is found mainly along the Zambezi River and along the Save River in Mozambique.

Although it is now rare in South Africa, the Red-winged Pratincole still breeds in fairly substantial numbers in Kwazulu-Natal, mainly in the Richard's Bay area. There its breeding season is November and December, but further north it lasts from August to November. The birds nest in open situations such as on floodplain mudflats, in ploughed fields or on overgrazed or recently burnt veld, and almost always near water. They are monogamous and breed in loose colonies of ten to 100 pairs, sometimes no more than a metre apart. The two birds perform fast undulating flights as part of their courtship behaviour, and on the ground assume various erect and bowing postures, displaying their throat patches and repeatedly raising and lowering their tails. Sometimes a scrape ceremony is performed, and before copulating the male may offer food to the female.

The eggs are laid directly on the bare ground or in a slight natural depression such as a hoofprint, and whenever possible on dark soil or in burnt areas. In situations such as these they are extremely difficult to detect, as their buff ground colour is heavily overlaid with irregular dark brown blotches. The usual clutch is two eggs, but sometimes one or three are laid.

Both male and female incubate the eggs, which hatch after 17 to 19 days. When the chicks emerge, dark down covering their backs makes them even more difficult to detect than the eggs, especially when they hide in a hoofprint. If predators do approach, the adult birds lure them away by feigning injury. The chicks, which are

1

2

1 *Red-winged Pratincoles lay on dark soil to enhance the camouflage of their eggs.*
2 *Once hatched, the chick crouches flat, relying on its down pattern to escape detection.*

__1__, __2__ Rock Pratincoles breed on exposed rocks along the Zambezi River, usually laying their eggs in a hollow on the bare rock. ***3*** *Both sexes incubate the eggs, changing over regularly.*

tended and fed by both parents, leave the nest after two to three days. It takes them 21 to 28 days to fly, and they are proficient on the wing at the age of 30 days. The parents continue to feed them until they can fly and hawk insects for themselves.

In southern Africa the Rock Pratincole has been studied mainly on the upper and middle sections of the Zambezi River in Zimbabwe, where the total population numbers only about 1300 birds. They are present there from July until early January, the timing of their arrival and their subsequent distribution on the river being linked to falling water levels when the rock outcrops on which they breed become exposed. The breeding season extends from August to December, but most clutches are laid between September and November, with a peak in October. At individual colonies egg-laying appears to be synchronised.

Rock Pratincoles are monogamous and arrive at their breeding sites in pairs. Aerial chasing flights form part of courtship and the birds also pursue each other on the ground, one displaying to its mate by bobbing its head and standing erect with its bill open and the white collar on its nape flared. Although this species breeds in loose colonies, the nest sites are never as close as those of Red-winged Pratincoles. They are located on rocks in midstream in fast-flowing water, usually no more than half a metre above water level. Some are exposed to full sun, whereas others are shaded under rock overhangs or by vegetation. Sometimes the birds use exactly the same spot as they did the previous breeding season. The eggs are laid directly in a hollow on the bare rock or occasionally on sand and caked mud deposited by flood waters.

The usual clutch is two eggs but single ones are also laid. They vary in ground colour from off-white to pale khaki and, being heavily overlaid with scrolls of grey, black and brown, are well camouflaged on the grey rocks on which they lie. They are laid at two-day intervals and hatch after 20 days. Both parents incubate, but they may leave the eggs unattended in the cool early morning and late evening when they are hawking insects. As the daytime temperature increases the birds change over at the nest every 30 to 40 minutes, and try to keep cool by raising their dorsal feathers and panting. Often the relieving bird returns with wet belly feathers to cool the eggs. Instead of seeking shade, the bird that has been relieved goes to the water, stands in it and drinks. As it only wets its belly feathers just before returning to the nest, it seems that this behaviour is to cool the eggs rather than the adult.

The newly-hatched chick has greyish down which matches the rocks, so it is well camouflaged when, on hearing its parent's alarm call, it freezes and lies flat. Once it is a few days old it is able to swim to safety, often hiding in a crack under an overhang. The young are fed by regurgitation and when a parent returns they run eagerly towards it and solicit vigorously. They are able to fly when they are 40 to 45 days old, but remain with their parents until they are about three months old.

Although the present population of Rock Pratincoles on the upper and middle Zambezi is considered stable, the birds require a specific breeding habitat and available sites are limited. Dam construction and the silting of rivers could adversely affect not only their habitat, but also that of other Zambezi species such as the African Skimmer and White-crowned Plover which nest on exposed sandbanks.

Gulls

Only three gull species breed in southern Africa: the large Kelp Gull, and the smaller Hartlaub's and Grey-headed Gulls. Kelp and Hartlaub's Gulls are considered resident, although some birds may make minor movements; and the Grey-headed Gull is either resident or it may disperse over distances averaging 300 kilometres, but sometimes as far as 2000 kilometres.

Kelp Gulls occur along the entire southern African coast from Namibia to Maputo in Mozambique whereas Hartlaub's Gull, also a coastal species, is confined to the west coast and the south-western Cape, with occasional vagrants recorded on the east coast. The Grey-headed Gull is the only species that is regularly found inland, where it breeds on lakes and dams, but it also occurs on the coast, mainly in the east. The populations of all three species are increasing, especially that of Hartlaub's Gull; as areas of human population expand, it and the Kelp Gull find rich pickings at rubbish tips. The Grey-headed Gull's increase in population is linked to the building of new dams which it colonises.

Gulls are monogamous and territorial, and even in the tightly packed colonies in which they breed they maintain a minimum distance between nests. Aggression and courtship, the two main requirements for breeding, involve a wide range of complex ritualised postures which, with minor variations, are common to all three species. The best known aggressive display is the 'long call', a series of loud screams which, in the case of the Kelp Gull, are emitted with the head thrown back vertically. Hartlaub's and Grey-headed Gulls emit similar screams holding their heads forward and pointed down. During pair formation and when a pair is selecting a nest site they stand with body lowered and tilted forwards, sometimes side by side. In another courtship display, known as 'head flagging', they turn their heads from side to side.

1 *The red spot at the tip of a Kelp Gull's bill stimulates the begging reaction of its small chicks.* ***2*** *Surrounding vegetation provides cover for a newly-hatched Kelp Gull chick.*

Soliciting food also involves a certain posture, adopted by young gulls and females begging food from males. They lower their body into a crouched position, drawing their head in and pointing their bill obliquely upwards.

The nests of all three species vary from a sparsely lined scrape to, more usually, a fairly substantial pad of material with a hollow on top. Anything available, such as dry grass, twigs, fishing line or pieces of dry kelp, are used for building, and on one island Kelp Gull nests were found made with the fur of Cape fur seals. The nest is sometimes located next to a rock, bush or piece of driftwood. Both sexes incubate,

***1** The massed mobbing flight of breeding Hartlaub's Gulls serves to confuse and deter predators.* ***2** Gulls' nests, such as this of a Hartlaub's Gull, are constructed of whatever material may be available.* ***3** The chicks of Hartlaub's Gulls, like those of other gulls, can leave the nest soon after hatching. They are tended by both parents.* ***4** Grey-headed Gulls, which regularly breed inland, often build substantial nests of dry grass.*

and in cases where hybridisation between Hartlaub's and Grey-headed Gulls has been recorded, both species have shared incubation.

The chicks of these gull species have greyish down with black spotting on their backs. They are able to leave the nest within a few hours of hatching and soon learn to crouch or run to cover and hide. They are also able to swim competently if the need arises. Both parents tend the chicks.

Kelp Gulls breed from September to March, mainly in November. They nest in colonies on the mainland, on offshore islands or, very rarely, on inland islands near the coast. The site they choose may vary from flat, open ground on an island to a wooden platform, a steep coastal slope or a beach, but it must be remote and relatively free from mammalian predators. It should also offer some cover such as rocks or vegetation for the chicks to hide in, as long as the overall situation is still a fairly open one. Nests may be densely packed, sometimes as close as 20 to 40 centimetres apart or even as many as four to a square metre. The colony is normally restricted to Kelp Gulls but sometimes includes other species such as Caspian Terns.

Kelp Gulls usually lay two eggs, sometimes one or three, and the first egg is generally larger than subsequent ones. Moreover, the eggs hatch at intervals, after an incubation period of 26 to 27 days. The older chick, being larger, has the advantage at feeding time, and if food is scarce its weaker sibling is likely to die. By this means the brood size is reduced to ensure the survival of the fittest. The chicks are fed by regurgitation, and experiments have shown that the position and colour of a red spot at the tip of the adult's bill release a begging instinct in the chick, which pecks at the spot to obtain its food. Young Kelp Gulls first fly at 45 to 50 days old.

Hartlaub's Gull, which breeds from March to September, nests on offshore islands, occasionally on buildings on the mainland, and on wetlands near the coast. On offshore islands the gulls sometimes occur in huge colonies, the best known of which is probably that on Robben Island off Cape Town. Here they invariably breed in association with Swift Terns, either completely out in the open or in clearings among alien acacias. This particular colony has grown from 400 pairs in 1953 to approximately 2500 pairs in 1994. However, its numbers fluctuate considerably, and as there are often two breeding cycles – one in March/April and another in July/August – it is difficult to know to what extent the same birds may be breeding again. On the mainland breeding occasionally occurs at sewage works and on islands in vleis near Cape Town, and, in the early 1970s, small numbers of these gulls began nesting on the roofs of buildings in Cape Town. Usually Hartlaub's Gulls lay two eggs, although the clutch may comprise 1-5 eggs, and the incubation period is 25 days. The chicks fledge 40 days after hatching.

Grey-headed Gulls breed either on the coast or well inland, between June and November. Generally they nest on islands, in lakes and dams inland or in lagoons or estuaries at the coast, but in some cases a colony may breed on an undisturbed stretch of shoreline. Sometimes nests may be built on matted floating vegetation or on the old nest of a waterbird such as a coot. The clutch of the Grey-headed Gull comprises two or three eggs. These vary considerably in their ground colour and markings, even in the same clutch, and may be greenish, olive or dark khaki with brown and greyish spots. Neither the incubation period nor the fledging period of this species has been recorded.

A visit to a gull colony is not only an unforgettable experience – the noise made by the birds as they fly up to mob an intruder is overwhelming – but one that should be undertaken with great caution, as disturbance often results in high mortality. Chaos reigns when an intruder, human or otherwise, enters a colony, and in the confusion a gull will readily eat its neighbours' eggs or kill their chicks, or species such as Sacred Ibises take the opportunity to prey on eggs and young. If they feel threatened Kelp Gull chicks in a nest with a drop below merely crouch and face inwards, but Hartlaub's Gull chicks on a rooftop attempt to run away, fall over the edge and are killed. It takes a long time for a threatened colony to return to normal, and chicks which have become displaced have a torrid time trying to relocate their parents; many are viciously pecked and even killed as they move through the colony.

Terns and the African Skimmer

Although 19 species of terns occur in southern Africa, most are regular non-breeding migrants present in large numbers, some are rare vagrants, and only five – the Caspian, Swift, Roseate, Damara and Whiskered Terns – breed in southern Africa. Of these only the Damara Tern is endemic to Africa. The African Skimmer is placed in its own family, the Rynchopidae, but it is tern-like in both its appearance and its breeding habits.

The world's largest tern, the Caspian Tern is found mainly on the coast in southern Africa, but sometimes also far inland. The current population of approximately 500 pairs is distributed in 28 breeding localities from Swakopmund in Namibia to Lake St Lucia in Kwazulu-Natal. Most breeding takes place on or near the coast on offshore islands, on beaches and on islands in lagoons and estuaries, but this species also nests inland in localities such as the Makgadikgadi Pan in Botswana and on dams in the Free State. One of its most important breeding areas is Lake St Lucia, where it nests in June and July so as to take advantage of optimal fishing conditions when water levels are dropping. Elsewhere in its range the breeding season is from November to March. Although Caspian Terns may breed singly, they usually nest in loose colonies, sometimes with Kelp Gulls. The terns locate their nests on the periphery of the gull colony but join forces with the larger birds to mob predators. They are also aggressive towards humans and dive-bomb them.

The nest, a shallow scrape, is either unlined or sparsely lined with a few pieces of shell or other material such as dry vegetation. The clutch normally comprises two eggs but sometimes one or three, and these vary in ground colour from buff to greenish with dark spots and blotches. They are laid at intervals of two or three days and incubation starts with the first egg. Both sexes brood, changing over at regular intervals, and the off-duty parent often stands beside its mate. The chicks hatch after about 21 days, then remain in the nest for between one and seven days, where they are tended by both parents. When they leave the nest they seek refuge in cover until they can fly, at the age of 28 to 35 days. They may still be fed by their parents for six to eight months after their first flight.

The Swift Tern is the most common breeding tern in southern Africa, with two-thirds of the 6000-pair population nesting in the south-western Cape – on offshore islands and occasionally on islands in wetlands and sewage works near the coast – and most of the remainder off the Namibian coast. At the beginning of the breeding season, which extends from February to July, the birds arrive paired and soon establish a nest site. Like other tern species, Swift Terns are monogamous and territorial. Their courtship includes aerial chases, offering fish to the mate, and a strutting posture in which the bird stretches its neck and raises its crest, allows its wings to droop almost to the ground and holds its tail up.

This is a gregarious species which nests colonially, often in association with Hartlaub's Gulls. The nest is a shallow scrape, sometimes with a little lining, in which a single egg is normally laid. Two-egg clutches are very rare and may be the product of two females laying in the same nest. The eggs are usually creamy buff in ground colour but can be very variable, and they are covered with a scattering of dark brown spots and blotches. These markings vary considerably and no two eggs look alike.

Both birds brood, but it is not known for how long. If disturbed they fly up with

1

2

3

4

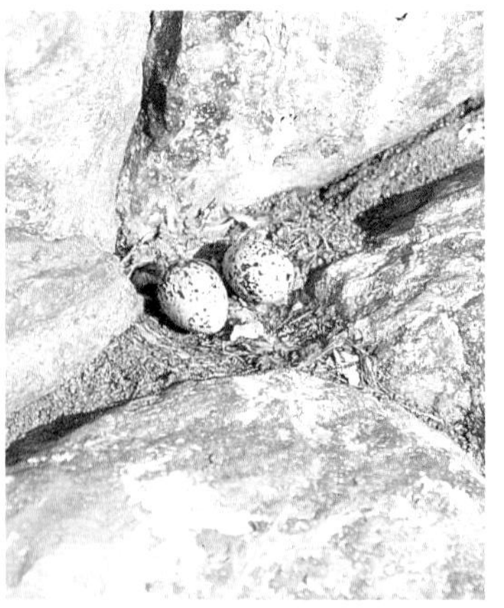

1 *Caspian Terns nest in a shallow scrape, sometimes on the edge of a colony of Kelp Gulls.* ***2*** *Nests are regularly spaced in a colony of Swift Terns, and no two eggs are alike.* ***3*** *Swift Terns usually lay a single egg, and both it and the chick are tended by both birds.* ***4*** *A Roseate Tern's nest on St Croix Island. A two-egg clutch is unusual, and when it occurs the volume of the first egg is greater than that of the second.*

1 *Damara Terns usually lay a single egg.* **2** *Both sexes incubate it, each one for about an hour at a time.* **3** *A Damara Tern's egg is easily overlooked among the stones on a stretch of beach near Cape Agulhas.*

harsh alarm calls, and in mixed colonies of Swift Terns and Hartlaub's Gulls the flock's combined clamour and mobbing deters avian predators such as Sacred Ibises and Kelp Gulls. Both birds also tend the chick, feeding it on whole fish until it fledges after 35 to 40 days. Once they have left the nest the young terns gather to form a crèche, although each chick is still recognised by its own parents. The juveniles remain dependent on the adults for at least four months after they fly.

The Roseate Tern has the unfortunate distinction of being the rarest breeding seabird in southern Africa, with a population that at present probably does not exceed 100 pairs, all of which are believed to breed in colonies on St Croix and Bird Islands in Algoa Bay. Various factors have contributed to the decline of this attractive species, notably human disturbance which drives birds away from their nests and allows Kelp Gulls to move in and prey on eggs and chicks. This problem is exacerbated by the terns' reluctance to return to their nests quickly. Other factors are chicks wandering into the territories of other pairs and being viciously pecked and sometimes killed, and the flooding of nests during the winter breeding season. In order to conserve the remaining population, strict measures are needed to protect the breeding colonies from human disturbance, especially from guano collection. Some form of control of the growing Kelp Gull population at breeding areas would also seem to be essential.

The Roseate Tern derives its name from the rosy wash on its ventral plumage when breeding. Its courtship behaviour is similar to that of the Swift Tern, and its nest, too, is a shallow scrape that is initially unlined, although material may be added as incubation progresses. Nests may be situated in hollows on sloping rock faces or on the ground where they are hidden amongst cover. Eggs are laid mainly in the midwinter month of July. The clutch usually comprises a single egg, but when two eggs are laid the volume of the first is greater than that of the second. Both birds incubate, although one takes a greater share, and when changing over the incoming bird may present a fish to its mate. The chick hatches after 25 days and is tended by both parents, remaining in the nest for seven to 12 days before leaving to hide in vegetation away from the main colony. It is able to fly at the age of between 23 and 30 days, but remains dependent on its parents for several weeks, possibly months.

The Damara Tern, the smallest breeding tern in southern Africa, ranges from northern Namibia, where about 98 per cent of its population breeds, as far east as Port Elizabeth. In Namibia it breeds from November to June, with a peak in January. The nests are situated on gravel plains, dry salt pans, stony areas and coastal marine terraces in the Namib Desert, usually 1 to 5 kilometres inland so as to avoid predation by black-backed jackals which regularly scavenge along beaches. In the south-western Cape, where the birds breed from November to January, they may be found in unvegetated troughs between dunes where there are areas of stones and shells.

Damara Terns show considerable fidelity to their nesting areas, and in one record a particular bird was recaptured on the same pan on which it had been ringed eight years previously. They breed in loose colonies – numbering 60 nests according to some observations in Namibia – and are tolerant of their neighbours, although chicks are attacked if they wander too near another pair. If the ground is not too hard, both birds form a shallow scrape by shuffling their feet and flicking small stones out of the hollow they create. In Namibia the rim of the nest may be ringed by small stones and pieces of lichen.

A single egg is usually laid, probably because the chick needs to be fed frequently as well as being brooded and shaded. In the harsh conditions in which these terns occur it would be more difficult to raise a larger brood, so the adults concentrate their energies on rearing a single chick successfully. It has been suggested that the occurrence of a number of two-egg clutches near Lüderitz may be linked to better food

In its inland aquatic habitat the Whiskered Tern builds a substantial floating nest of waterplant stems.

resources in that area. The egg of the Damara Tern varies in ground colour from creamy buff to khaki and is entirely covered with small spots. This cryptic colouring makes it difficult to detect, especially as it is laid next to a camouflaging feature such as stones, and it rarely falls prey to Kelp Gulls.

Both birds incubate, changing over regularly at intervals of about an hour. As soon as danger threatens the terns fly up and call overhead, but they return quickly to their nests once the source of their alarm has passed. The chick hatches after 17 to 20 days and is tended for two days on the nest before it wanders several metres away from it. If alarmed it hides in whatever cover may be available, relying on its cryptic down pattern to avoid detection. It is able to fly at 20 days old but remains dependent on its parents for a further two or three months.

Most Damara Terns nest in remote and often inhospitable places. However, some suffer from human disturbance, and in the south-western Cape in particular they are vulnerable to similar pressures as those suffered by the African Black Oystercatcher (see page 83). Another threat in the south-western Cape has been the loss of traditional breeding sites when dune areas were stabilised with vegetation.

The fifth breeding tern in southern Africa, the Whiskered Tern, is a widely distributed inland species of wetlands, vleis and lakes. Because of the ephemeral nature of its environment it is nomadic and breeds opportunistically in seasonally inundated pans. It breeds from September to April, when both sexes assume an attractive plumage, and forms loose colonies numbering from ten to 30 pairs. Each nest is a substantial floating mound made up of waterplant stems, with a hollow on top for the eggs. The male brings food for his mate during the building process, and they copulate on the nest.

Usually three eggs, but sometimes two, are laid, each on consecutive days, and the first may even be laid before the nest is complete. Both sexes brood, changing over every ten to 20 minutes. The non-incubating bird often stands on the nest beside its mate, and in hot weather it wets its belly feathers before settling on the eggs again. During the incubation period the nest gradually sinks and the terns build it up again, sometimes taking material from abandoned nests nearby. They may vigorously attack humans who approach the nest, sometimes striking them and even drawing blood with their bills. Their harsh squeaky alarm call sounds like a rusty nail being pulled from a plank.

The chicks hatch after an incubation period of about 21 days and can soon swim competently. When danger threatens they hide in vegetation, returning to the nest as soon as it has passed. While one parent broods and shades the chicks the other brings food, often passing it to its mate without alighting. The young first fly at the age of 21 days, but it is not known for how long they remain dependent on the adults.

The African Skimmer, an intra-African migrant that occurs in southern Africa from April to December, breeds mainly in the panhandle region of the Okavango Delta and along the Zambezi River. Its breeding season is timed to coincide with low water levels between the months of July and October, when the sandbanks on which it nests are exposed.

Skimmers are monogamous and breed in loose colonies. Their courtship involves aerial chases and displays at the nest site, where the male makes a cooing call, opens and closes his bill, and jerks his head from side to side. The nest is an unlined scrape, usually 4 to 8 centimetres deep and 15 to 25 centimetres in diameter. Often excavated in moist sand, it is formed by the bird pivoting on its breast in a hollow and kicking backwards with its feet. While making the nest and during incubation the bird rests its bill on the nest rim, and in doing so creates a series of grooves that resemble fork marks around a pie crust.

The female may sit in the scrape for some time before laying. When she does lay, she normally produces three eggs over a period of four to five days, usually at night. Sometimes, however, clutches comprise 1-4 eggs. Resembling tern eggs, the

1

2

3

__1__ African Skimmers breed when water levels along the Zambezi River are low, exposing the sandbanks on which they nest. __2__ At any sign of danger the incubating bird flattens itself in its deep nest. __3__ Three eggs is the usual clutch size of this species.

Skimmer's eggs are buff in ground colour with blotches of dark brown. Both birds incubate, changing over frequently especially when it is hot, and the non-incubating parent often stands nearby in shallow water. When the relieving bird returns to the nest it frequently wets the eggs, having moistened its belly feathers by splashing in the water with its feet or by flopping its belly briefly into the water. Both these actions are performed in flight. If the bird on the nest is alarmed it flattens itself and is difficult to detect in the deep nest hollow. Sometimes a Skimmer feigns injury to distract a predator by flopping along the ground with wings and tail spread.

The chicks hatch after 21 days and are kept constantly wetted and shaded, and are fed on whole fish which the adult carries crosswise in its bill. They leave the nest two days after hatching, and if alarmed they crouch in cover or on a dark patch of sand, blending into this background by means of their cryptic down pattern. One observer reported that a brooding adult transported a chick in its bill in flight when disturbed, but this is the only account of such unusual behaviour. Sometimes the chicks rest in a hollow they have made in moist sand. They can fly when they are 35 to 42 days old, but are still tended by their parents. It is not known at what stage they become independent.

The African Skimmer faces a number of threats, one of which is the reduction to its breeding habitat caused by the construction of dams and the silting of rivers. It is possibly significant that this species' first recorded breeding on the Mashonaland plateau in Zimbabwe occurred in 1992, after a prolonged drought had caused a dam's water level to drop, exposing a piece of land on which the birds nested. Other hazards faced by skimmers include the trampling of their nests by wild and domestic animals coming to the water to drink, and adults being snared for food and chicks being caught for fishing bait. It is unfortunate that they breed in areas used for recreation in the panhandle region of the Okavango Delta and above the Victoria Falls, where waves created by irresponsible speedboat drivers swamp their nests and where they are kept off them by people fishing and picnicking on sandbanks.

Sandgrouse

The sandgrouse of southern Africa are birds of dry savannas, deserts and semi-deserts, and are adapted for survival in inhospitable conditions. Some of their breeding habits reflect these adaptations, and thus are of particular interest. The Namaqua Sandgrouse, known for its musical *kelkiewyn* call, is the most common species and is found in true deserts as well as other arid regions. Burchell's Sandgrouse occurs predominantly in areas of red Kalahari sand which its plumage matches; the Yellow-throated Sandgrouse prefers an environment where short grass grows on dark alluvial soils; and the Double-banded Sandgrouse inhabits dry woodland such as mopane and arid stony areas. The ranges of the sandgrouse overlap to a certain extent and in some areas, such as the Nata region of the Makgadikgadi Pan in Botswana, all four species may be seen at the same waterhole. In their arid environment, sandgrouse depend on a reliable source of water and they sometimes fly up to 80 kilometres to reach a drinking place, gathering there in flocks of hundreds, sometimes thousands. They feed on seeds that are small but highly nutritious.

All four species are monogamous, sexually dimorphic and gregarious, especially when drinking. Although they usually lay in the cooler months between April and September, breeding is not confined to any particular season and may take place opportunistically after rain. Little is known of their courtship behaviour, except that the Yellow-throated Sandgrouse executes a bobbing dance, and the Namaqua Sandgrouse performs a strutting display in which the male, with his head drawn in and tail raised and fanned, pursues the female. Sometimes he chases her so vigorously that she is literally bowled over.

The nest is a rudimentary scrape or a natural hollow such as a hoofprint which is often located near a tuft of grass or between small bushes. Lining may accumulate in it when a pair changes over as they, like plovers, throw pieces of dung and vegetation sideways into it with their bills. Three oval, cryptically marked eggs are laid and are incubated for long unbroken periods by both sexes. The Namaqua and Yellow-throated Sandgrouse females sit on the nest during the day and the males take over at night, changing over in the early morning and late afternoon. In the case of the Namaqua Sandgrouse the male incubates for 14 hours at a stretch in summer and a mammoth 16 hours in winter. The Double-banded Sandgrouse's pattern is different, with the female incubating overnight and in the early part of the morning, while the male takes over during the rest of day. The breeding habits of Burchell's Sandgrouse are poorly known and there are no details of how the sexes share incubation.

The chicks of the Yellow-throated Sandgrouse hatch after 27 days and those of the other species after 21 days. They all emerge covered with down that is cryptically patterned with white markings on a dark background, and within a day are feeding on seeds indicated to them by their parents. Both adults care for them, but it is the male who is responsible for meeting their moisture requirements, transporting water to them in his specially adapted belly feathers.

In the case of the Namaqua Sandgrouse, whose breeding biology has been studied in greater detail than that of the other species, the chicks either hatch at the same time or over a period of two to three days. In the latter case, they stay in the nest until the whole brood has hatched. The eggshells are removed and are dropped away from the nest to avoid attracting predators. The chicks, like those of the other sandgrouse species, are highly precocial and feed themselves at a day old, keeping cool by remaining in the shadow of the adult bird. If only two of the brood survive, as quite often happens, each parent looks after one of them, although the family stays together as a unit. When danger threatens the chicks respond to their parents' alarm calls by flattening themselves, usually near some object or under a bush, and relying on their camouflage for concealment. The adult birds sometimes lure predators away with an injury-feigning display, spreading their tails and dragging their wings along the ground.

Namaqua Sandgrouse chicks can fly weakly at a month old and competently two weeks later, but they remain with their parents until they are about two months old. The fledging details of the other species are likely to be similar.

1

2

3

4

1 *A Double-banded Sandgrouse settles on the nest to brood the eggs and a newly-hatched chick. In sandgrouse both sexes share nest duties.* ***2*** *A well-camouflaged female Namaqua Sandgrouse faces into the sun when incubating so that less of her surface area is exposed to radiation.* ***3*** *A Yellow-throated Sandgrouse nest, like those of other sandgrouse, comprises no more than a scrape in which lining material accumulates.* ***4*** *A Namaqua Sandgrouse's nest is positioned so that surrounding bushes will help to conceal the incubating bird.*

ADAPT OR DIE

The breeding habits of the Namaqua Sandgrouse are well known mainly through the efforts of a single observer in the Kalahari Desert, who has recorded various strategies for survival in harsh conditions. How is this species adapted to cope with producing offspring in circumstances of extreme aridity?

In the first place, it usually breeds during the cooler months from autumn to spring. However, when it does nest opportunistically after rain the female may be exposed to intense heat during her ten-hour spell of incubation. How does she cope with this stress? One strategy is to face into the sun to reduce the surface area exposed to radiation. If there is a breeze she faces into it, lifting herself off the eggs, drooping her wings and raising her back feathers to allow maximum circulation of air past her body. If she is still uncomfortable she 'pants', fluttering her throat rapidly.

Because the eggs are laid on bare ground the temperature of the soil is critical if the embryos are to survive. It has been shown experimentally that when the surrounding soil temperature reaches 60 °C the incubating female is unable to keep the eggs sufficiently cool beneath her and the embryos die. The problem of overheated eggs also applies to the laying period. The female lays the three eggs at 24-hour intervals and incubates only when the clutch is complete. She needs to leave the nest to feed and build up resources for egg production, but the first egg she laid would overheat if it were left exposed. To solve this problem the male takes over, staying on the nest to keep the egg cool while the female feeds.

The most remarkable aspect of parental care exercised by sandgrouse is the means by which the chicks obtain their daily water requirements. The male flies to a waterhole to drink and, after satisfying his own thirst, stands in the water to allow his belly feathers to absorb moisture. Specially adapted for this purpose, the feathers have coiled and intertwined barbules which uncoil on contact with liquid. Hair-like filaments on the barbules soak up water like a sponge, absorbing about 22 millilitres by capillary action. By the time the sandgrouse has returned to his family between 10 and 18 millilitres of moisture remain on his feathers, and the chicks strip it off with their bills. Because the moisture evaporates during transport a nest cannot be located too far from a water source. It has been calculated that 32 kilometres is about the maximum distance the male can travel if sufficient water is to be carried back to the chicks.

Why should water be transported by this means, and why only by the male? The male sandgrouse requires all the water he can drink during his brief visit to water. If he were to regurgitate it to his chicks, he would diminish his own essential supply of liquid. The evolution of a second mode of water transport solves the problem. However, this second mode has not evolved to the same extent in the female: she lacks the specialised feather structure and, if her mate were to die, she could only soak up about 9 millilitres of water for her chicks. A possible reason for only the male soaking his feathers could be that the female needs to brood the chicks on cold desert mornings and for this purpose her feathers must be dry ●

A Namaqua Sandgrouse chick strips water from the male's specially adapted belly feathers.

Pigeons and Doves

This group includes some species that, being conspicuous, confiding and attracted to human habitation, are some of southern Africa's best known birds. Conversely, several forest species, such as Delegorgue's Pigeon and the Blue-spotted, Tambourine and Cinnamon Doves, are shy and elusive.

Of the 14 species in the group, five are pigeons and nine are doves. Apart from minor variations, the characteristics and habits of pigeons and doves are remarkably similar. In all species except Delegorgue's Pigeon and the Tambourine and Namaqua Doves, male and female are alike, and only in the Namaqua Dove is the difference between the sexes a striking one. Most species emit cooing calls, except the Green Pigeon which gives vent to a series of weird whistling and growling notes. Most are also gregarious and assemble in large flocks at a good food source or at a waterhole. Contrary to popular belief, they are not good symbols for peace and can be extremely aggressive towards other members of the same species, fighting with slapping blows of their wings.

Pigeons and doves are monogamous and may pair for life. Common courtship behaviour includes 'bowing', in which the male follows the female with his neck inflated and calling loudly; and 'towering', when the male flies above tree-top level with loud wing claps and then glides in a circle with tail spread and wings held stiffly out. This flight also serves as territorial advertisement. Other aspects of courtship include mutual preening and 'billing', in which the female grasps the male's bill to solicit feeding. This latter behaviour is often a prelude to copulation.

It is difficult to define the breeding seasons of most species, and nests may be found in all months. Moreover, a pair's breeding may extend over several months, as it often raises multiple broods. In a species such as the Rameron Pigeon the breeding season usually coincides with a seasonal abundance of fruit, and in the case of the Namaqua Dove, which inhabits arid regions, breeding may be in response to rainfall.

The nest, a simple platform of twigs, leaf-stalks and rootlets, may appear to be flimsy as the eggs can be seen from below, but the material is criss-crossed in such a way that a relatively strong structure is formed. Moreover, once the eggs have hatched the rim of the nest is strengthened, cemented together by the chicks' droppings. Sometimes the nest of another species, such as an egret or a thrush, is taken over. Nests range in size from the Rock Pigeon's, which is about 20 centimetres in diameter, to that of the Namaqua Dove, at approximately 7 centimetres across. However, when nests are used repeatedly – and in some species a pair has been known to use the same nest up to eight times – they grow considerably, especially in depth. The amount of material that goes into a nest is very variable, and may range from the 345 sticks and 97 stiff grass stalks that were found in a Rock Pigeon's nest to the mere 28 twigs and 26 leaf-stalks of a Green Pigeon's nest.

The Rock Pigeon is versatile in its choice of nest site and may breed on cliffs or mine dumps, in mine shafts, on buildings, in large trees or on discarded machinery. On offshore islands where there are no mammalian predators, its nest may be situated close to the ground. This ability to utilise a variety of sites has undoubtedly facilitated the Rock Pigeon's expansion into new areas. The Feral Pigeon is closely linked to human habitation and breeds almost exclusively on buildings in towns.

1 

 2

1 *A Green Pigeon broods its offspring on its flimsy nest. Both sexes share this duty.* **2** *Tambourine Dove nests are characterised by the use of black rootlets.* **3** *Rock Pigeons often nest on man-made structures.* **4** *Laughing Dove chicks hatch naked but are soon covered with yellow down.*

3 

 4

A male Namaqua Dove feeds his two chicks simultaneously by regurgitation. The chicks' droppings around the edge of the nest help to strengthen it.

All the other species nest mostly, but not exclusively, in trees or bushes. Green Pigeons favour leafy trees, which may explain why aggregations of their nests are found near human habitation such as farmhouses or national park rest camps where the trees tend to be well watered. The Red-eyed Dove often chooses to nest on a branch overhanging water, and Cape Turtle and Laughing Doves, which live in association with humans, sometimes use sites such as a flower pot on a verandah or a ledge on a pillar. The height of the nest varies, although in its arid habitat where there are few tall trees the Namaqua Dove tends to nest low down, usually not more than a metre off the ground, and sometimes even on it.

Both partners build the nest, the male bringing material to the female who does the construction. He may be remarkably fastidious in his choice of material and often picks up and drops several twigs before he finds one to his liking. The Rameron Pigeon, an arboreal species which rarely descends to the ground, breaks twigs off trees for the nest. All the species may add material to the nest during the incubation period.

Most pigeons and doves usually lay two eggs, although sometimes one or three are laid, and very rarely four. Only the Rameron Pigeon regularly lays a single egg. The eggs are white or creamy yellow and have no markings. Both sexes incubate, and in some species each takes a regular shift. Occasionally the male feeds the female on the nest. Should the incubating parent feel threatened it usually sits tight, but if it is flushed off the nest too suddenly it sometimes drags an egg out with it. During the incubation and nestling periods birds of several species have been observed dragging their wings along the ground in an injury-feigning display similar to that of ground-nesting birds such as plovers. The length of the incubation period is variable, ranging from about 13 days for the Green Pigeon and Laughing Dove to 15 to 18 days for the Rock Pigeon.

When the chicks hatch the adults drop the empty shells some distance from the nest so as not to draw attention to it. The chicks, or squabs as they are often called, emerge naked, blind and helpless, but soon develop a coat of yellowish or whitish down and they open their eyes when four to six days old. As they grow older they defend themselves by snapping their bills and slapping at an intruder with their wings. They are brooded and fed by both parents, emitting a squeaky call and quivering their wings excitedly when an adult returns to the nest. The adult feeds the chicks simultaneously by regurgitation, with one chick on each side inserting a bill into its crop. The substance they receive, known as 'pigeon's milk', is an exudation from the parent's crop which is rich in protein and fat, looks and smells like soft cheese, and has a nutritional value similar to that of mammalian milk. When the chicks are older they are given whatever food the parents have been feeding on, but it is probably still mixed with a certain amount of 'milk'.

Why should most pigeons and doves lay only two eggs? It is possible that the size of the clutch and the production of 'pigeon's milk' are related, as the parents can presumably produce only enough of this sustenance for two offspring. Another factor is that the parents feed their young simultaneously, and it would not be practical for them to cope with a larger brood. Instead of laying a larger clutch in response to optimum conditions for breeding, raising several consecutive broods would appear to be the best means of increasing productivity.

The young of some species, such as the Laughing Dove, may leave the nest when they are no more than 12 to 13 days old. At this stage they cannot yet fly properly, and return to the nest to be fed for a few more days until they are more proficient in the air. The length of the nestling period appears to bear no relation to the size of the species: Green Pigeon chicks may leave the nest after 12 to 13 days, whereas the smaller Namaqua Dove chicks leave after 16 days. The subsequent period of dependence on the adults varies from about a week to several weeks.

Parrots, the Rose-ringed Parakeet and Lovebirds

Members of this group are familiar to many people because they are kept in captivity. Indeed trapping, despite being illegal, is partly responsible for the decline of the Cape Parrot in South Africa. However, much of what is known about these species' breeding biology is gleaned from captive studies, as there has been very little research on most species in the wild. One reason for this is that their nests are in cavities which are often inaccessible, and the birds are alert and wary.

There are four parrots, one parakeet and three lovebirds in southern Africa. The Rose-ringed Parakeet, which is feral and has become established in Kwazulu-Natal and around Johannesburg, nests in holes, competing with indigenous hole-nesting species. Its breeding habits are similar to those of parrots. Lilian's and the Black-cheeked Lovebirds were originally considered to be a single species and, as may be expected, their breeding habits are similar.

Parrots, parakeets and lovebirds share a number of common features. They are gregarious, noisy and fly fast and strongly; the Cape Parrot, for example, may regularly travel 80 kilometres to feed. The sexes are similar and the birds are monogamous, sometimes pairing for life. Their courtship behaviour is quite complex, but it includes some simpler aspects such as the male feeding the female, head-bobbing, and a sidling walk that involves successive approaches and retreats by the male to and from the female. Mutual preening is a regular feature of their behaviour, and it is from this activity that lovebirds derive their name. The lovebirds are unusual because females are dominant over males and need to indicate their submissiveness before copulation takes place; this they do by fluffing out their feathers. The copulation of all species in this group is protracted, lasting five or six minutes.

Parrots and parakeets nest in holes in trees, laying the eggs on wood chips at the bottom of the cavity. The nest may be a natural hole or one that has been excavated by a barbet or woodpecker, and the birds often enlarge the entrance by gnawing at it. Lovebirds also breed in holes in trees, but they differ from most of the other members of the parrot family in that they build a nest in the cavity. The Rosy-faced Lovebird builds in a variety of situations, including in the nest chambers of Sociable Weavers, in crevices in cliffs, in holes under bridges and even under the eaves of buildings in large towns such as Windhoek in Namibia. Although it roosts in the nests of White-browed Sparrow-weavers, there has been no confirmation that it breeds in them. Its nest, a well-made cup fashioned from uniform strips of vegetation, is added to throughout the breeding cycle, so that when the chicks hatch the additional material serves as a means of sanitation, covering eggshells, faeces and even dead chicks. Lilian's and the Black-cheeked Lovebirds usually build in a hole in a tree, the female constructing a sturdy domed chamber with a tunnel leading to the nest entrance. The pieces of material used are not as uniform as those used by the Rosy-faced Lovebird, and the inclusion of twigs results in a more complex structure. It is suspected that these two lovebirds breed in the nests of Red-billed Buffalo Weavers, but there has been no confirmation of this.

A natural hole in a dead tree provides a nest for a pair of Meyer's Parrots.

Parrots lay 2-4 eggs and lovebirds 2-6 (and as many as eight in captivity). The pure white eggs are laid at intervals of about two days, with the result that the chicks in a brood may vary considerably in size. As far as is known only the female incubates, and while doing so she is fed by the male. She sits for long periods, and in the case of the Rosy-faced Lovebird leaves the nest only three or four times during the day. The three smaller parrot species – the Brown-headed, Meyer's and Rüppell's – incubate for about 21 days and the Cape Parrot for 28 days. The average incubation period for lovebirds in captivity is 23 days.

The chicks of parrots are covered in whitish or smoky grey down, and those of lovebirds in reddish down. All are weak and helpless, and are closely brooded by the female. The male brings food which he regurgitates to the female, and later she in turn regurgitates it to the chicks. Once the nestlings can be left unattended she also collects food for them. At the nest of a Meyer's Parrot an observer noted that both parents brought food to the nestlings at intervals of one to three hours, remaining at the nest for about half an hour and changing places at the entrance about five times during this period. The parent who was not feeding hung upside down, possibly to facilitate the regurgitation of more food.

Studies of captive birds show that Cape Parrots first fly at 63 to 77 days old and Meyer's Parrots at 56 to 63 days or longer. Young lovebirds leave the nest between 40 and 44 days after hatching. The comparatively long time that parrot chicks spend in the nest has given rise to a feature that is common to a number of hole-nesting species: their feathers develop slowly so that they are not fouled by the nestlings' excreta. Once the young birds have fledged little is known about their behaviour, but it is assumed that they accompany their parents to feeding areas and remain in a group. Young lovebirds form a pair bond while they are still in juvenile plumage and when they become independent of their parents at about two months old.

Because there are so few breeding records it is difficult to define the breeding seasons of most species. As a rule the parrots breed between March and August, but nests have also been observed in other months. Rosy-faced Lovebirds breed mainly from February to April or May. Little is known about the breeding season of the other two lovebird species.

REAR-END LOADING: AN UNUSUAL METHOD OF TRANSPORTATION

Not only are lovebirds unusual within the parrot family in that they construct a nest inside the cavity in which they breed, but in the case of the Rosy-faced Lovebird the means of carrying the building material is quite remarkable. The female cuts strips of material such as bark, leaves and grass to a uniform shape and size with her bill and tucks them into the contour feathers of her lower back and rump. These feathers have microscopic hooklets holding them together so that the material does not fall out during the flight to the nest.

The two other lovebird species in southern Africa, Lilian's and the Black-cheeked Lovebirds, carry nest material in their bills. However, lovebird species elsewhere in Africa, as well as the Asiatic hanging parakeets of the genus *Loriculus*, also tuck nest material into their feathers, and not only into the rump feathers as the Rosy-faced Lovebird does, but also into the body feathers. This behaviour, when linked to a variety of other features, indicates a common ancestry. We have yet to explain why lovebirds should have evolved such a remarkable means of carrying nest material, and why in some species this behaviour is absent ●

Specially adapted contour feathers enable Rosy-faced Lovebirds to carry nesting material on their rumps.

Louries

The louries are unique to Africa and in the southern part of the continent four species – the Knysna, the Purple-crested, Ross's and the Grey Louries – occur. Of these, Ross's Lourie is a rare vagrant in northern Botswana and is not known to nest there. Louries are characterised by their arboreal habits and their loud croaking calls. Indeed, the Grey Lourie's harsh *go-wheeeeh* has given rise to its alternative name of 'go-away bird'. These species are often gregarious, but when breeding they are solitary, monogamous and territorial. As in the case of parrots, much of what is known of their breeding biology is derived from studies of captive birds.

Little is known about their courtship behaviour, but it involves feeding of the female, displaying the crimson on their wings and, in the case of the Knysna Lourie, flashing their white cheek marks. The breeding season of the various species is not clearly defined and appears to depend on the availability of food. The Grey Lourie may nest in any month, although in Zimbabwe there is a marked October peak and in Namibia breeding is mainly in summer and autumn. The Purple-crested Lourie breeds mainly in summer, from October to February.

Like pigeons, louries build flimsy platforms of twigs through which the eggs can often be seen from below. The nest, measuring 20 to 30 centimetres across, is placed in a tree usually at a height of between 3 and 10 metres above the ground, and quite often on top of a creeper which conceals it. Grey Louries show a preference for the crown of a thorny tree and sometimes place their nest on top of a clump of mistletoe or similar parasitic plant. Both birds are involved in nest construction, one partner (presumably the male) supplying twigs broken off from surrounding trees while the other builds. This technique is also reminiscent of the behaviour of pigeons and doves. In one observation, it took a pair of Knysna Louries five days to finish building their nest.

Usually two or three eggs are laid, each on successive days, although rarely the clutch may comprise one or four eggs. They are round and white, but sometimes tinged with blue, and the yolk of the Knysna Lourie's egg is bright vermilion. The fact that chicks of different sizes have been reported suggests that incubation may begin before the clutch is complete. In captivity the incubation period of Knysna and Purple-crested Louries was about 22 days, and in the wild that of the Grey Lourie has twice been recorded as 26 days. Both sexes share incubation, and in the case of a captive pair of Purple-crested Louries both birds were observed to be incubating on the nest together.

When the eggs hatch the shells are sometimes eaten by the parents. The chicks emerge with their eyes open and with a thick coat of grey-brown down. Their first feathers appear when they are about seven days old and by the time they are 21 days old they are well feathered. Both parents brood them for the first week, feeding them by regurgitating items such as berries which are usually undigested. Since the chicks do not pass faecal sacs, the adult pecks at the anus and then eats the liquid excreta. An aspect of lourie behaviour that merits further investigation is co-operative breeding. In two instances helpers – in one case a single bird and in the other four birds – have been observed at Grey Lourie nests, and it is possible that these may have been the offspring of previous broods.

When the chicks leave the nest at about 21 days old they are not yet able to fly, but they clamber about the tree on their strong legs. A special call used by the parents guides them back to the nest, but they are adventurous and do not stay in it for long. Their premature departure from the nest makes it difficult to establish when they can first fly, but they are usually between 28 and 35 days old before their flight feathers are fully developed. The young birds remain dependent on their parents for several weeks after they can fly. On one occasion a Grey Lourie was seen to drink, then fly up to a young bird recently out of the nest and regurgitate water into its mouth.

The flimsy platform nests of Purple Louries are often concealed by vegetation in a tree's canopy.

Cuckoos

Cuckoos are well known for laying their eggs in the nests of other birds and taking no further part in the rearing of their offspring, a practice known as brood parasitism. Eleven species breed in southern Africa, although little is known about one, the Barred Cuckoo, which is a summer visitor to the area around the confluence of the Haroni and Lusitu rivers in Zimbabwe. Evidence of it breeding in the region is circumstantial: when a white egg with a zone of faint reddish brown speckles at the large end was found in the nest of an African Broadbill, it was attributed to the Barred Cuckoo partly by elimination and partly because it was similar to an egg laid by a captive bird of the same species in Tanzania.

The remaining ten cuckoo species are divided among four genera: the African, Red-chested and Black Cuckoos are in the genus *Cuculus*; the genus *Clamator* includes the Great Spotted, Striped and Jacobin Cuckoos; the Emerald, Klaas's and Diederik Cuckoos (often known as the glossy cuckoos) belong to the genus *Chrysococcyx*; and the Thick-billed Cuckoo is in the genus *Pachycoccyx*.

Most of the southern African cuckoos are intra-African migrants, although populations of certain species are resident in at least part of their range. The Diederik Cuckoo is probably the most common of these visitors, although its numbers fluctuate from year to year, and birds return to the same area each season. The status of the Thick-billed, Emerald and Klaas's Cuckoos is uncertain; further details are given in the breeding summary overleaf.

All cuckoos have loud, characteristic and often monotonous calls which they make from a regular song perch, sometimes even calling through the night. The strident triple-noted call of the Red-chested Cuckoo is typical, and has given rise to the bird's Afrikaans name, *Piet-my-vrou*. In the case of the male the calls are used to advertise his territory and to attract a mate.

Although cuckoos are highly vocal, most of them are extremely elusive and it is difficult to observe their courtship behaviour. Depending on the species, males may mate with several females, or females may mate with several males. Where promiscuous behaviour such as this occurs, as it does in the genera *Cuculus* and *Chrysococcyx*, the song perch is a focal point for attracting a mate. *Clamator* cuckoos are often found in pairs and, as they co-operate to distract the host away from its nest at laying time, they may well be monogamous. Males of this genus perform a slow 'butterfly flight' courtship display. The breeding behaviour of the Thick-billed Cuckoo also includes courtship display, in which the male makes erratic buoyant undulating flights above the woodland canopy to advertise his presence. However, he does not co-operate with the female when she lays. A feature that seems to be common to most of the African cuckoos is courtship feeding of the female by the male, and his offering is often a prelude to copulation. Frequently the food item is a hairy caterpillar, a regular component of the cuckoos' diet.

***1** A case of mistaken host identity: the dark egg of a Red-chested Cuckoo has been laid in a Cape Siskin's nest that has been built in a site normally used by a Dusky Flycatcher. **2** A Black Cuckoo chick fills the nest of a Crimson-breasted Shrike. It will have evicted the shrike's own eggs or chicks. **3** The spotted egg of an African Cuckoo was later rejected by the host Fork-tailed Drongo because it did not match its own plain eggs. **4** The Cape Robin is one of the main hosts of the Red-chested Cuckoo.*

The breeding seasons of the southern African cuckoos are linked to those of their host species, falling mainly during the spring and summer months and sometimes continuing into autumn. In the south-western Cape the Red-chested Cuckoo arrives in September and begins breeding in October, so one of its hosts, the Cape Robin, escapes parasitism during the early part of its own breeding season. In Namibia several cuckoos breed later than they do elsewhere in their range to coincide with the later rainy season.

As a general rule each cuckoo species tends to specialise in a particular host – usually the species which raised it – or group of hosts, and it is intriguing to consider the variety and number of hosts that are parasitised. They range from a single species in the cases of the African and Thick-billed Cuckoos to as many 24 in the case of the Diederik Cuckoo. Often there is regional variation: Klaas's Cuckoo frequently parasitises sunbirds, except in the south-western Cape where the Cape Batis is one of its main hosts. In some cases a host may be parasitised in error, as apparently happened when a Red-chested Cuckoo egg was found in the nest of a Cape Siskin which was at a site normally used by a Dusky Flycatcher, one of the known hosts of the cuckoo. Sometimes, when a female cannot find a suitable host and can no longer delay laying, she 'dumps' her egg in the nest of an unlikely host, or even in one that is deserted. This may explain why several cuckoo eggs may be found in a nest. A Starred Robin was once found to have been parasitised simultaneously by a Red-chested Cuckoo and an Emerald Cuckoo.

The pressure on certain hosts in an area may be considerable and it could be argued that the cuckoo is damaging its own long-term prospects by not allowing its host to breed. In one example a pair of Malachite Sunbirds was parasitised by Klaas's Cuckoo over a period of several years and during that time raised only a single chick of its own. In a study of Red-billed Helmet Shrikes and Thick-billed Cuckoos, 55 per cent of the nests were parasitised and some groups of this sociable shrike raised no young of their own in five years. It would seem, however, that things balance out in the long term, as the hosts survive and there are still cuckoos to parasitise them.

The eggs of cuckoos may be virtually identical in colour to those of the host, a partial match, or no match at all. A perfect match is achieved by the Thick-billed Cuckoo, a host specialist whose egg duplicates that of the Red-billed Helmet Shrike in colour and size so exactly that its presence can only be confirmed when it hatches. The blue eggs of the Striped Cuckoo closely resemble those of its Arrow-marked Babbler host and can be distinguished only by their slightly rounder shape, slight glossiness and pitting on their surface. Although the eggs of the Pied and Black-faced Babblers are similar to Arrow-marked Babbler eggs, they have a nodular surface, and perhaps for this reason these species are not parasitised. Great Spotted Cuckoo eggs match those of its main host, the Pied Crow, in colour, but they may be recognised because they are smaller. They differ from the eggs of the Black Crow, another host, and as there are no records of the Black Crow rearing chicks of this cuckoo, it seems likely that its eggs are rejected.

The Diederik Cuckoo is a good example of a species in which 'families' or 'tribes' have evolved in a long process of natural selection. Each 'tribe' within the species lays eggs that match those of its host, which is always the species that originally reared it. Thus one Diederik Cuckoo 'tribe' lays matching plain blue eggs in the nests of Red Bishops, another deposits perfectly matched blotched eggs in the nests of White-winged Widows, and yet another leaves white eggs with variable red and brown spotting in the nests of different weavers. Masked Weavers, however, lay extremely variable eggs, a strategy by this species to outwit the cuckoo. Interestingly, there are no records of the Diederik Cuckoo parasitising the Yellow-rumped Widow, even though its eggs are very similar to those of the White-winged Widow.

The Red-chested Cuckoo also lays different types of eggs. Some are plain chocolate- or olive-brown and glossy, and match the eggs of Natal, Heuglin's and Chorister Robins, whereas others are speckled or blotched on varying ground colours. Eggs that are laid in Boulder Chat nests, for example, are greenish white with red-brown speckles and are indistinguishable from those of the chat except that they are slightly smaller and rounder.

Other cuckoos' eggs rarely match those of the hosts. Because the Fork-tailed Drongo's eggs are extremely variable, those of the African Cuckoo seldom match and are frequently rejected. The Jacobin Cuckoo makes no attempt to emulate the eggs of the bulbuls, Fiscal Shrike and Fork-tailed Drongo which it parasitises, laying instead pure white eggs.

1 

 2

3 

 4

1 *A Red-chested Cuckoo's egg (front), although slightly smaller, closely matches those of its Boulder Chat host.* ***2*** *Two white Jacobin Cuckoo eggs have been laid in the nest of a Black-eyed Bulbul, which is not a discerning host.* ***3*** *The colour of a Striped Cuckoo's egg (bottom right) is a perfect match for the turquoise blue of the host Arrow-marked Babbler's eggs.* ***4*** *A deserted Bleating Warbler's nest has been opened to reveal an Emerald Cuckoo's egg alongside the warbler's own smaller egg.* ***5*** *A Red-chested Cuckoo chick evicts an egg of its Cape Robin host.*

It is generally accepted that some host species do not tolerate cuckoo eggs. The Fork-tailed Drongo, for example, may evict even those African Cuckoo eggs which, to the human eye at least, are extremely similar to their own. Masked Weavers, too, are very discerning and even in their enclosed nests they can recognise their own eggs, rejecting those of Diederik Cuckoos that do not match.

A cuckoo's main challenge is to deposit its egg in the nest of its host, and in order to achieve this it employs various strategies. *Cuculus* and *Chrysococcyx* females act on their own, patiently observing a nest for an opportunity to slip in and lay. Although this process takes only a matter of seconds, she is not always successful and cuckoo feathers lying beneath a nest indicate that she has been caught in the act and attacked by the host. A different problem faces the female Red-chested Cuckoo when she parasitises the Starred Robin. The latter species builds an enclosed domed nest with a small entrance which the much larger Red-chested Cuckoo must penetrate in order to lay her egg. How she does this without causing visible damage to the nest remains a mystery. A *Clamator* pair co-operates to get its egg into a host's nest and, while the male lures the parents away by being conspicuous nearby, the female slips in unobtrusively and lays. Such teamwork is particularly important in the case of the

 5

BREEDING SUMMARY OF CUCKOOS The authenticity of hosts requires meticulous verification for a number of reasons and the rule best followed is 'the proof of the cuckoo is in the hatching'. Thus, with very few exceptions, the hosts are listed in the following breeding summary on this basis.

	STATUS	BREEDING SEASON	HOST SPECIES
CUCULUS			
AFRICAN CUCKOO	Intra-African migrant; August-April.	September-December.	Fork-tailed Drongo.
RED-CHESTED CUCKOO	Intra-African migrant; September-April.	October-January.	Kurrichane Thrush, Olive Thrush, Cape Rock Thrush, Stonechat, Natal Robin, Heuglin's Robin, Chorister Robin, Cape Robin (main host), White-throated Robin, Starred Robin, Swynnerton's Robin, Boulder Chat, White-browed Robin, Bearded Robin, Dusky Flycatcher, Long-tailed Wagtail, Cape Wagtail, Cape Siskin (in error).
BLACK CUCKOO	Intra-African migrant; September-April.	November-March, mainly January-March in Namibia.	Southern Boubou, Tropical Boubou, Swamp Boubou, Crimson-breasted Shrike. Extralimitally African Golden Oriole, Puffback Shrike.
CLAMATOR			
GREAT SPOTTED CUCKOO	Intra-African migrant; September-April, but in Namibia arrives and departs later. Some birds overwinter.	August-March but variable: October-January in South Africa; December-March in Namibia; August (one record only) to January in Zimbabwe.	Pied Crow (main host), Black Crow, Indian Mynah, Pied Starling, Burchell's Starling, Long-tailed Starling, Glossy Starling, Greater Blue-eared Starling, Red-winged Starling, Pale-winged Starling. Hoopoe, Ground Woodpecker records considered doubtful.
STRIPED CUCKOO	Intra-African migrant; October-June.	October-May.	Arrow-marked Babbler (main host); Bare-cheeked Babbler in Namibia; White-rumped Babbler extralimitally.
JACOBIN CUCKOO	Intra-African migrant; September-April. Some birds overwinter.	October-April. Isolated records in May, July and September suggest opportunistic breeding by over-wintering birds.	Fork-tailed Drongo, Cape Bulbul, Red-eyed Bulbul, Black-eyed Bulbul, Sombre Bulbul, Fiscal Shrike. Other 'hosts' recorded but not confirmed by hatching; egg 'dumping' possibly involved .
PACHYCOCCYX			
THICK-BILLED CUCKOO	Uncertain, probably local (perhaps altitudinal) migrant.	September-January.	Red-billed Helmet Shrike.
CHRYSOCOCCYX			
EMERALD CUCKOO	Uncertain. Probably intra-African migrant; September-January. May be resident.	October-January.	Bleating Warbler (apparently main host), Starred Robin, Puffback Shrike. Blue-grey Flycatcher record not confirmed.
KLAAS'S CUCKOO	Uncertain; migratory in some areas and sedentary in others. November-April in Namibia.	September-July. Mainly October-December in South Africa; December-April in Namibia. Isolated records for May, June and July in South Africa suggest opportunistic breeding.	Bar-throated Apalis, Long-billed Crombec, Yellow-bellied Eremomela, Neddicky, Dusky Flycatcher, Cape Batis, Pririt Batis, Chinspot Batis, Collared Sunbird, Black Sunbird, Scarlet-chested Sunbird, White-bellied Sunbird, Dusky Sunbird, Greater Double-collared Sunbird, Marico Sunbird, Malachite Sunbird.
DIEDERIK CUCKOO	Intra-African migrant; October-April. Some birds may overwinter.	October-March.	Mainly Cape Sparrow, Spectacled Weaver, Cape Weaver, Yellow Weaver, Golden Weaver, Lesser Masked Weaver, Masked Weaver, Spotted-backed Weaver, Red-headed Weaver, Red Bishop; also Mountain Chat, Karoo Robin, Kalahari Robin, White-browed Robin, Tit-babbler, Rattling Cisticola, Karoo Prinia, Marico Flycatcher, Paradise Flycatcher, Cape Wagtail, Great Sparrow, Grey-headed Sparrow, White-winged Widow, Golden-breasted Bunting.

EGGS	INCUBATION PERIOD	NESTLING PERIOD	DEPENDENCE PERIOD
Pinkish with dark red-brown speckles concentrated at large end. Seldom match host's eggs and often rejected. Removes egg of host.	11-17 days, probably nearer the shorter time.	22 days. Evicts host's eggs/chicks.	Not known.
Vary, sometimes match host's eggs. Lays 1 per nest; usually removes egg of host.	12-14 days (shorter than Cape Robin's).	18-21 days. Evicts host's eggs/chicks until fourth day, then instinct lost.	20-30 days.
White or cream with red-brown blotches and speckles. Closely match host eggs, but noticeably larger. Lays 1-2 per nest; removes egg of host.	13-14 days.	20-21 days. Evicts host's eggs/chicks.	21-28 days; maximum recorded 42 days.
Greenish blue with brown and grey blotches, rounded. Match Pied Crow eggs but smaller. May lay 1-7 per nest (a record of 13 probably involved 2 females); does not remove host's eggs, but may damage them.	14 days (shorter than host's).	22, 26 days (2 records only). Host chicks usually disappear during nestling period, although some survive.	Not known.
Turquoise blue. Match host's eggs almost perfectly. Occasionally removes egg of host; may also damage eggs.	11 days (much shorter than host's). Sometimes embryonic development before egg is laid.	Not known. Does not evict host's eggs/chicks, which may be reared successfully.	Maximum 35 days recorded.
Immaculate white. Do not match host's eggs. May lay several eggs in same nest (possibly 2 females involved); does not remove host's eggs.	11 days (shorter than host's). Sometimes embryonic development before egg is laid.	16 days. Does not evict host's chicks, which may be reared successfully.	Maximum 33 days recorded.
Pale creamy green or blue-green with pale brown, grey and lilac spots forming a zone at the large end. Match host's eggs so perfectly that their presence confirmed only by hatching. Lays 1-2 per nest; removes egg of host.	About 13 days (4 days shorter than host's).	About 30 days. Evicts host's eggs/chicks.	Maximum 50 days recorded.
Plain white or white with a few brown speckles at large end. Larger than Bleating Warbler's eggs. Removes egg of host.	16 days (unpublished observation).	Not known. Evicts host's eggs/chicks.	About 14 days.
Variable, usually white with brown or reddish spots. Usually closely match host's eggs. Removes egg of host.	11-14 days.	19-21 days. Evicts host's eggs/chicks.	Maximum 28 days recorded.
Vary to match host's eggs, usually well. Removes, and sometimes eats, egg of host.	11-12 days.	20-21 days. Evicts host's eggs/chicks.	Maximum 38 days recorded.

***1** A Klaas's Cuckoo has laid its egg (with white ground colour) in the nest of a Cape Batis. **2** The chick, which will have evicted the host's egg or chick, demands food from the female batis.*

Striped Cuckoo, since its babbler hosts are communal breeders and it has to outwit not only the two parents, but the entire group.

The Thick-billed Cuckoo also parasitises a communal breeder, the Red-billed Helmet Shrike, but the female, unaided by the male, employs a different technique. When the shrike has completed its clutch and is incubating she flies directly at it, frightening it off the nest, and quickly deposits her egg. This aggressive behaviour works because, like many cuckoos, she looks like a hawk and uses the resemblance to scare off the sitting shrike. In one remarkable instance a female picked up two seven-day-old host nestlings and dropped them over the nest edge.

Cuculus and *Chrysococcyx* cuckoos usually remove an egg of the host when they deposit their own, carrying it away in the bill and sometimes even eating it. This has given rise to the misconception that a cuckoo carries its own egg in its bill and deposits it by this means. *Clamator* cuckoos rarely remove an egg of the host but they may damage eggs already in the nest, causing cracks or punctures.

Usually cuckoos lay while the host's clutch is being laid, although sometimes they may deposit their eggs before the host has done so, or after the clutch is complete. Certain cuckoos – and perhaps all of them – are able to retain their egg in the oviduct for about a day before they lay. This has been illustrated by newly laid eggs of Striped and Jacobin Cuckoos having been found with the embryos already about 20 hours developed. If advanced embryo development such as this is combined with an incubation period that is shorter than that of the host, the young cuckoo gains a considerable advantage at hatching and is able to evict the host's eggs or, if it lacks the instinct to evict, to outgrow its nest-mates.

Cuckoos lay many eggs at intervals of about 48 hours. It is thought that most lay 'clutches' of 3-5 eggs, rest for a while, and then start a new series. In this way some species, such as the Great Spotted Cuckoo, may lay up to 24 eggs in a season. The eggs are usually deposited in different nests, but sometimes there may be several in the same nest. It is not unusual to find 5-7 Great Spotted Cuckoo eggs in the nest of a Pied Crow, and in one record 13 cuckoo eggs were found, in addition to the crow's four. Some large clutches may be the product of more than one female cuckoo. By laying a series of eggs individual cuckoos insure their breeding productivity against the risk of rejection and predation.

Chicks of *Cuculus* and *Chrysococcyx* species, as well as the Thick-billed Cuckoo, evict the eggs or chicks of the host within the first few days of hatching, after which the instinct is lost. The behaviour of the Red-chested Cuckoo is typical: the naked and blind chick, stimulated when it feels eggs or chicks against the sensitive skin of its back, manipulates itself until it has an egg or chick on its slightly hollow back. It then uses its legs to brace itself against the side of the nest and levers the egg or chick over the rim where it lies ignored by the host or falls to the ground below. If the operation does not succceed the first time, the cuckoo repeats its attempts until it achieves its purpose. When there are chicks and eggs in the nest, it seems that the chicks are evicted first, perhaps because their movement stimulates the cuckoo. Sometimes, if the cuckoo has mistimed its parasitism, eviction is not possible. In a Boulder Chat's nest a newly-hatched Red-chested Cuckoo was unable to evict the host's week-old chicks and by the following day the cuckoo had disappeared.

Not only do *Cuculus* and *Chrysococcyx* chicks eliminate competition for food by evicting the host's eggs or young, but they also have brighter gapes than their host's offspring, and these enhance the foster parent's stimulus to feed them. Also in *Clamator* species, a brighter gape deceives the host into feeding the cuckoo chicks preferentially and they outgrow its own chicks, which may die if they are unable to compete successfully. The Great Spotted Cuckoo chick has a more heavily spurred palate than that of its crow nest-mates and this serves the same function as the brighter gape. Some chicks, such as those of the Great Spotted and Striped Cuckoos, also deceive their foster parents by mimicking the begging calls of crow and babbler chicks respectively. It is debatable whether such mimicry is genetically innate or learnt by imitation. The dorsal feathering on the young of *Clamator* cuckoos is very similar to that on the host chicks, and if only a cursory inspection of the nest's contents is made, it is often difficult to detect whether a cuckoo is present.

The bright colour of the gape has another function in that it helps the chicks to deter intruders. They lie quiescent in the nest, but if a predator approaches too closely they suddenly lunge at it with gape wide open. They may also defend themselves by exuding watery, foul-smelling excreta if handled.

Most young cuckoos continue to be fed by their foster parents for some time, usually a few weeks, after they have left the nest. There have been reports of species such as Diederik Cuckoos feeding their own young at this stage. However, as immatures resemble females, it is probable that such incidents involve instances of misplaced courtship feeding by a male which has mistaken the young cuckoo for a female. It has been observed that young cuckoos out of the nest often sit very still except when their foster parents are nearby with food. A possible explanation for this behaviour is the cuckoos' resemblance to hawks, especially when they fly. On several occasions the host species has been seen to attack the young cuckoo when it flew, although it continued to feed it when it was perched.

Coucals

Although classified in the same family as cuckoos, coucals do not display parasitic behaviour. However, some of their breeding habits, notably the role reversal of the sexes, are unusual. There are five species in southern Africa: the Green Coucal, which is in its own genus, and the Black, Coppery-tailed, Senegal and Burchell's Coucals which all belong in another.

Coucals are elusive birds with skulking habits, and in their preferred habitat of shrubby thickets and rank grassland their nests are not easy to find. Thus their breeding behaviour is poorly known, except for that of the Black Coucal which has been revealed through the efforts of a single observer and is markedly different from the other coucals' behaviour. Very few nests of the elusive Green Coucal have been found in southern Africa, and none at all of the Coppery-tailed Coucal, although this large coucal is relatively common in suitable habitat in northern Botswana. Its nests have been found elsewhere in its African range and they are typical of the genus.

In Zimbabwe Black Coucals breed in rank grassland during the rainy season from December to March. Most birds probably migrate when the grasslands dry up, but some may overwinter if the habitat remains sufficiently moist. Only one Black Coucal nest is recorded from Natal and that was for October. Senegal Coucals breed from October to March, Burchell's Coucals from August to February, and the few Green Coucals nests recorded were found in November and December.

Coucals are perhaps best known for their musical glugging calls which sound like liquid being poured from a long-necked bottle and, it is widely believed, presage rain. Some species have been observed calling in duet, and in these cases the calls may serve a courtship function. There has been little observation of other forms of courtship behaviour. As in the cuckoos, the male approaches the female with an offering of food, although he may wait until copulation is complete before passing it to her. Sometimes the female remains motionless at his approach, or she may quiver her wings and spread her tail. In some species, particularly the Black Coucal, the female is the dominant partner and advertises a territory by calling. When she emits a special call her authority is such that the male interrupts his normal activities, even feeding chicks, to bring her an offering of food. Tail-wagging behaviour by a pair of Green Coucals has been observed and was presumed to be part of courtship.

Coucals are considered to be monogamous, except for Black Coucals which are either monogamous or polyandrous. This species also differs from other coucals in that the female is much larger than the male and defends a territory, and both sexes assume a more colourful breeding plumage for the duration of the rainy season. In one study a female laid clutches for three different males, each of which had his own nesting area within her territory. She took no part in nest-building, incubation or care of the young, but during the course of the breeding season laid a total of six clutches for her three mates. The nest, which is probably built by the male in most species, is an untidy ball-shaped structure with a side entrance. It is made of dry grass and twigs and lined with green leaves which continue to be added during the incubation period. Nests are always well concealed in thick cover and are placed at varying heights. However, the Black Coucal invariably builds its nest low down, using growing vegetation to form a bower-like shell and then finishing off the interior with dry grass and sedge and lining it with greenery. The nest of the Green Coucal is completely different: an untidy platform of twigs with the leaves still attached, it resembles debris caught in the tree and is extremely difficult to find.

Coucals lay 2-5 eggs, rarely six, which are initially white but become nest-stained as incubation progresses. Sometimes the first egg is laid while the nest is still being built. Incubation begins with the first egg and subsequent ones are laid at intervals that vary between 24 and 72 hours. Burchell's Coucal incubates for between 14 and 17 days and the Senegal Coucal for not less than 17 days. The incubation periods of the other coucals are not known. From the scant information available it appears that only the males incubate and they sit tight; Black Coucals flush only when the nest is almost trodden upon.

A newly-hatched coucal is a remarkable creature. Its leathery black skin is covered on the dorsal surface by stiff, white, hair-like down which projects forward and hangs over the bill in a style that would be the envy of a punk rocker. An enormously distended belly is covered by a translucent skin through which the viscera can be clearly seen, adding to this extraordinary appearance. By the age of two weeks the nestling looks, if possible, even more bizarre, being covered by a mass of spiky quills that make it look like a hedgehog. Only when the feathers appear does the chick begin to resemble a bird, and by then the transverse black barring on its back effectively camouflages it when it leaves the nest.

Because incubation commences when the first egg has been laid, the young vary in size. Their eyes open from the age of five days, and from that stage they defend themselves by hissing like snakes and by voiding extremely nauseating, foul-smelling liquid faeces, quite different from their normal encapsulated faecal sac. In the latter regard they are similar to the chicks of certain cuckoos.

In most coucal species both sexes feed the nestlings, although the male appears to do the major share. In the Black Coucal, however, only the male tends the chicks, and broods them during the first week. If a predator threatens the nest the parent coucal emits clucking notes and may feign a broken wing to lure it away. In one remarkable instance in Uganda a Burchell's Coucal was reported to have transported its chicks in flight, one at time, from the path of an advancing fire by carrying them between its legs.

The nestling period lasts approximately 21 days, but it is difficult to establish this accurately because the young may leave the nest after about 14 days, before they can fly. If they are disturbed they may even move out when only 11 days old. They can move quickly through the undergrowth on their strong legs, hiding when threatened, and if they are caught they hiss and growl. They are able to fly when about a month old. The period of dependence on the adults is sometimes quite brief because a second brood may be started soon after the young leave the nest.

1 

2 

__1__ A Burchell's Coucal approaches its well-hidden nest to feed its chick.
__2__ With its covering of long, stiff hairs, a newly-hatched Senegal Coucal appears bizarre.

2

Owls

Difficult to observe because of their nocturnal habits, owls are nevertheless fascinating subjects of study. Some aspects of their breeding behaviour have been revealed by overnight monitoring of their nests, as well as by the technique of radio-telemetry which allows researchers to monitor their movements at night. There are 12 species in southern Africa, most of which have a wide distribution south of the Sahara. The Barn and Grass Owls belong to the family Tytonidae, and the remaining 'true' owls are in the family Strigidae.

Owls are monogamous and in most species a pair probably remains together for life, using the same nest site over a long period if it is left undisturbed. Because the pair bond is maintained year round and is reinforced by nightly duetting, courtship behaviour, where known, is not elaborate. However, feeding of the female by the male forms an integral part of courtship, and serves to build up the female's reserves before she lays. The breeding season varies from one species to another, and in the case of the Barn Owl nests may be found at any time of the year. Nesting sites also vary, although a number of species lay their eggs in holes in trees, or make use of the abandoned nests of other birds. Some species form a scrape, but generally owls do not make nests in the accepted sense.

The eggs of all species are round and white, and only the Barn, Grass and Marsh Owls regularly lay clutches comprising more than three. Intervals of usually two or three days pass between the laying of each egg, so a single nest will contain chicks of different sizes. It is difficult to establish when the eggs are laid, although a study at a Barn Owl's nest showed that in that instance the female laid the eggs at night.

It appears that the females do most of the incubation, certainly during the day, and although there are a few records of males sitting at night, the periods they spent on the eggs were brief. The male feeds the incubating female and she sits extremely tight, in several species relying on camouflage to avoid detection. Hole-nesting species such as African Scops, Pearl-spotted and Barred Owls often lie flat, face down, so that their plumage deceives predators into believing that there is only debris lying at the bottom of the dark nest hole.

3

4

5

__1__ Wood Owl chicks at the entrance to their nest. At this stage both parents make regular visits to feed them. __2__ Barn Owls nest in various sites, but often select one that is associated with man, such as the loft of a barn. __3__ The lack of pellets at a Barn Owl's nest indicates that the site is being used for the first time. __4__ Like other owls, Barn Owls lay their eggs two or three days apart, resulting in a brood containing chicks of various ages. This nest, judging by the large number of pellets around it, has been in use for several seasons. __5__ The Grass Owl's nest is similar to that of the Marsh Owl, except that a tunnel at the back of it leads to a 'funk hole' to which the chicks can disperse.

Considering the great disparity in the size of owls, their incubation periods do not vary greatly, and those of most species fall within the range 28 to 34 days. Several species have been seen eating the eggshells once the chicks have hatched, and this behaviour may be general. The chicks emerge blind and helpless, and they are sparsely covered in white or buffy down which later thickens as a second coat is acquired. During the first few weeks the female tends her brood closely, feeding each chick initially by dangling food against its bill, usually having positioned herself so that it faces forward between her legs. The male is the main food provider, but towards the end of the nestling period the female may also hunt. When large broods such as those of Barn Owls are reared, the survival of the smaller chicks depends on the food supply. If there is sufficient they catch up with their siblings, but if not they die of starvation. In the cases of the Giant Eagle and Pel's Fishing Owls, even though they lay no more than two eggs, only the larger chick survives.

Some owls are very protective of their offspring and they either perform injury-feigning displays to distract intruders or they actively attack them, even humans. The nestlings themselves are fierce in their own defence and, once they can stand, they spread their wings to appear larger and thus intimidate intruders. Most snap their tongues when alarmed, and some hiss at predators to deter them. It is difficult to establish the length of the nestling period as young birds often leave the nest before they can fly. However, for most small- and medium-sized owls it is between 28 and 35 days. The post-nestling dependence period is also very variable and may last from a few weeks to several months. During this time apparently abandoned young owls may be found and 'rescued' by well-meaning people. In fact the young birds maintain contact with their parents at night by means of extremely insistent begging calls, either wheezing sounds or high-pitched shrieks, that carry a long distance. If one is found it should be left alone or placed in a less vulnerable position nearby, as its parents are guaranteed to locate it when it calls at nightfall.

Barn Owls usually breed after the rains – from February to May in the summer-rainfall regions and from August to December in the winter-rainfall region of the south-western Cape – although they may take advantage of local conditions and breed in any month. In one remarkable study in Zimbabwe a pair raised consecutive

1

2

3

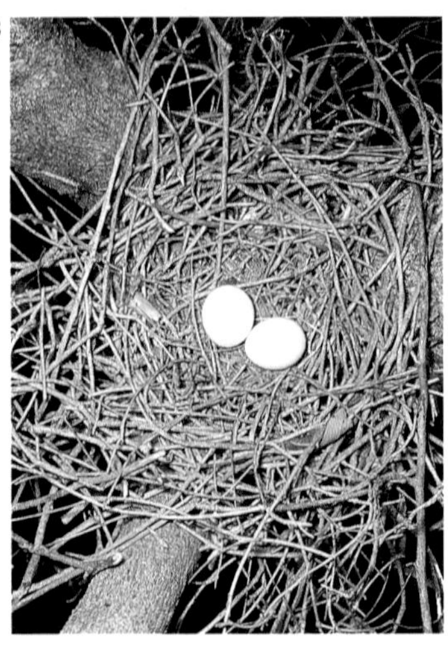

4

1 *A female Pearl-spotted Owl brooding her offspring waits for the male to arrive with food for herself and the chicks.* ***2*** *An African Scops Owl at its nest, a natural hole in a dead tree.* ***3*** *White-faced Owls often breed in the disused nests of other birds, in this case the nest of a Little Banded Goshawk.* ***4*** *A Pearl-spotted Owl emerges from its nest-hole, which in most cases has been excavated by a barbet or woodpecker.*

__1__ Cape Eagle Owls lay their eggs directly on the ground, often on cliff ledges. __2__ Spotted Eagle Owls nest in a wide variety of sites, including on the ground. The sitting female has closed her eyes and elongated her ear tufts to make herself as inconspicuous as possible. __3__ Protected by a thick coat of down, Spotted Eagle Owl chicks await food in their rot-hole nest.

1

2

3

broods of nine, seven, eight and eight chicks in 11 months during a population explosion of mice. Sometimes the first egg of the next clutch was laid while there were still chicks in the nest. Prey was so plentiful that the owls were observed to catch 24 mice in 17 minutes.

During courtship Barn Owls increase the intensity of their screeching calls and indulge in aerial chases. They often nest in association with man, utilising house roofs, farm outbuildings, church steeples, wells, mine shafts and, on one occasion, a compartment in the girders of a crane that was in daily use. Hamerkop nests are also favoured, even if it means dispossessing the owners. Pellets are regurgitated at a new nest so that by the time the eggs are laid there is usually a substantial bed of broken pellets for them.

A Barn Owl clutch usually comprises five or six eggs but sometimes as few as three or as many as 12, the larger clutches coinciding with periods of food abundance. The eggs are laid at intervals of two or three days so that by the time the last chick has hatched its oldest sibling may already be well feathered. It takes 50 to 55 days before the chicks can fly, although when there is an abundant food supply they may do so at only 45 days old. They may disperse considerable distances; a ringing recovery has shown that over two years a Barn Owl moved 400 kilometres from its birthplace.

In its grassland habitat the Grass Owl breeds mainly after the summer rains, from February to April. Like the Marsh Owl, it nests in rank grass, but its nest can be distinguished by the fact that tunnels connect it to chambers nearby. These serve as 'funk holes' into which the chicks, once mobile, can disperse to spread the risk of predation. The eggs, usually 3-5, are laid on a pad of grass which is probably scraped together from the immediate vicinity, and sometimes grass is pulled down over the nest to form a rudimentary bower. The nestlings of both Grass and Marsh Owls retain their down for much longer than other species as it camouflages them well in their grassy environment. If they are discovered, young Grass Owls hiss loudly like snakes to deter predators.

Wood Owls breed in spring and early summer from August to October, nesting mainly in holes in trees, often low down, or in the old nests of birds of prey such as the African Goshawk. Sometimes they lay the eggs on the ground at the base of a tree or under a log, or occasionally they use an artificial site such as a dovecote. If a good site is found the owls may use it for many years. Usually the clutch comprises two eggs, and when they hatch the parent birds are sometimes fierce in the defence of their young, even attacking human intruders.

In their courtship displays Marsh Owls fly with slow, exaggerated wing-beats, sometimes clapping their wings, and they croak like frogs. They share the same habitat as Grass Owls, and like them breed mainly between February and April, laying their clutches of 3-5 eggs on a pad of grass. In defence of its nestlings the Marsh Owl performs a remarkable injury-feigning display: it flies above the intruder, croaking harshly, then crashes into the ground as if shot and squeals as it flops about. The young birds leave the nest at about 18 days old and set up various temporary roosts in the surrounding grass. Although they do not fly until they are about 42 days old, by dispersing at an earlier age they have a better chance of survival.

Like several other southern African owl species, the African Scops Owl breeds in spring and early summer, from August to October. It nests in holes in trees, usually natural holes that are open to the sky but also woodpecker holes and even artificial nest-boxes. One pair even made use of a hollow between the sticks in the side of a disused Lappet-faced Vulture's nest. The two or three eggs that are laid are brooded almost constantly by the female. At about 22 days, the incubation period is shorter than for most other owls.

The male White-faced Owl courts the female by walking along a branch towards her, bobbing as he does so and to the accompaniment of his bubbling hoot. This species also breeds from August to October, laying its two or three eggs on the disused nests of birds of prey, crows, Grey Louries, Scaly-feathered Finches, Cape Sparrows and Wattled Starlings, among others. It also breeds in a natural hollow or fork in a tree.

The female Pearl-spotted Owl makes soft begging calls for long periods, during the day as well as at night, and the male feeds her as part of their courtship ritual. They breed in spring and early summer, nesting in holes in trees – mainly those of woodpeckers and barbets – and sometimes also using nest-boxes. Occasionally they line the holes with green leaves. Usually three eggs are laid.

1

2

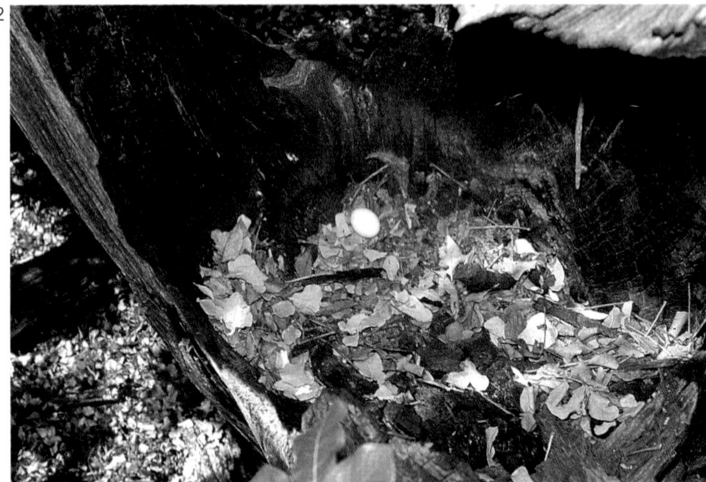

3

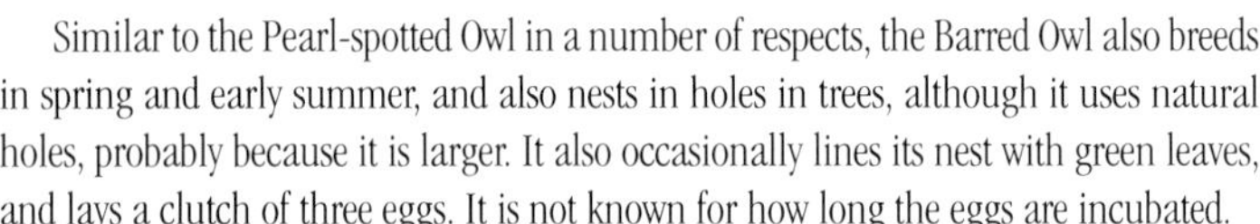

__1__ A Giant Eagle Owl chick crouches low on top of a Hamerkop's nest. __2__ Pel's Fishing Owl normally lays two eggs, and in this case the clutch is probably incomplete. __3__ Only one Giant Eagle Owl chick is reared, as the smaller one invariably dies of starvation.

Similar to the Pearl-spotted Owl in a number of respects, the Barred Owl also breeds in spring and early summer, and also nests in holes in trees, although it uses natural holes, probably because it is larger. It also occasionally lines its nest with green leaves, and lays a clutch of three eggs. It is not known for how long the eggs are incubated.

The Eagle Owls – the Cape, Spotted and Giant – form a natural group with similarities in their breeding behaviour. The male and female Giant Eagle Owl sit together and duet with deep dyspeptic grunts, jerking their bodies as they do so. They also reinforce the pair bond by preening each other around the face. This species breeds in winter, from June to August, laying mainly in the nest of another large species such as a bird of prey, but also on Hamerkop, Red-billed Buffalo Weaver and Sociable Weaver nests. Quite often it takes over the nest of another species, such as Wahlberg's Eagle, simply by squatting in it and preventing the rightful owner from using it. Hollows in trees may also be used. Normally two eggs are laid, but sometimes only one. In one record two chicks hatched seven days apart, presumably reflecting the laying interval. The chicks hatch after about 38 days, but only one is reared, as the smaller one is unable to compete and dies of starvation. In one nest observed the smaller chick had disappeared by the time it was 13 days old.

To draw attention away from its young the adult Giant Eagle Owl performs an injury-feigning display similar to that of the Marsh Owl, but it also alights on a branch and flaps its wings feebly, sometimes even hanging ludicrously upside down before falling into the undergrowth and flopping about. The young owl itself shams death if it feels threatened, lying flat until the danger has passed. It may remain with the adults until the next breeding season, and even then does not always move away.

Prolonged bouts of hooting, during which the white throat is puffed out as a visual signal, serve both courtship and territorial functions for the Cape and the Spotted Eagle Owls. The latter breeds from August to October, whereas the Cape Eagle Owl is a winter breeder, nesting from May to August. It lays its eggs in caves or on ledges, or on the ground among rocks or bushes. The nest is usually well concealed and may only be found when the owl is accidentally flushed. Spotted Eagle Owls lay in a wide variety of sites, even in urban areas, and eggs may be found under a bush or among rocks on the ground, in a hollow in a tree, in or on top of nests of species such as crows, ravens, Hamerkop or Sociable Weaver, and on buildings. In an analysis of 359 records 61 per cent of the nests were on the ground, 26 per cent in trees, 11 per cent were on buildings and the balance were in miscellaneous sites. Both species may use the same site for a number of years: Spotted Eagle Owls were found to be still breeding at a site first discovered 40 years previously, and a Cape Eagle Owl nest was still in use in a cave eight years after it had first been found.

The usual clutch is two or three eggs, although the Spotted Eagle Owl may lay four or, exceptionally, six. This species performs injury-feigning displays to distract predators and will also attack intruders at the nest, especially if it is breeding in an urban area. Cape Eagle Owl nestlings remain in the nest a little longer than other owls, leaving it only when they are about 42 days old, and they can only fly at about 70 days old.

Pel's Fishing Owl also uses its call as a form of courtship display, sitting on a dead branch over water and hooting mournfully through the night. It breeds from February to April in the Okavango Delta, timing its laying to coincide with the peak of the flood so that when the young are in the nest the water levels are dropping and fish are more easily caught. Occasional June records lower down the delta were timed to coincide with the later arrival of flood waters there. This owl nests in holes in large trees or in a fork where several branches converge, or occasionally on a collapsed Hamerkop's nest. It generally lays two eggs, but from time to time only one. Even when two eggs are laid, only one chick is ever raised, as the second to hatch dies of starvation. The adult birds feign injury to protect their chick.

Nightjars

Among the best camouflaged birds in the world, nightjars are virtually impossible to see when they are roosting in daylight. Their plumage resembles dead leaves so closely that their nests are usually found accidentally, when the sitting bird flies up at the last moment. Like owls, they are difficult to study because of their nocturnal habits. Only the Freckled and Fiery-necked Nightjars have been observed in any detail and although the breeding habits of other species, such as the Natal Nightjar, are poorly known, there is little doubt that they are basically similar to those of species that have been researched.

Six nightjar species breed in southern Africa, of which the Fiery-necked, Natal and Freckled Nightjars are resident, the Rufous-cheeked and Pennant-winged Nightjars are intra-African migrants, and the Mozambique Nightjar is resident in lowveld and migratory in highveld regions. All these species, with the exception of the male Pennant-winged Nightjar, are extremely difficult to identify by sight, but they may be distinguished by their calls. Perhaps the most familiar call is the clear, whistling *Good Lord, deliver us* of the Fiery-necked Nightjar. Both the Rufous-cheeked and Mozambique Nightjars make prolonged churring sounds, the Freckled Nightjar emits a *bow-wow-wow* like a small dog, and the Natal Nightjar makes a series of slow *chop- chop- chop-chop* ... calls. The male Pennant-winged Nightjar's call resembles the stridulations of a cricket and is often difficult to distinguish from insect noises. Perhaps his highly visual displays make a loud and distinctive call superfluous.

With the exception of the Pennant-winged Nightjar, little is known about nightjars' courtship apart from their calling, but possibly the white windows in the wings and white outer tail feathers serve as a visual display in flight. In the case of the Pennant-winged Nightjar, the male arrives in the breeding area first and sets up a territory in which he displays to attract arriving females. He shows off his 40 centimetre-long pennants by flying through the treetops followed by one or two females, and on a favourite raised perch he stands high on his legs with wings spread, revealing the broad white bands on his wings so that he resembles a giant swallow-tailed butterfly, and rotates to enhance the display. There is evidence that Pennant-winged Nightjars are polygamous and they are loosely colonial when breeding, sometimes siting their nests only 6 to 8 metres apart.

Most nightjars, however, are monogamous and maintain a pair bond, and Fiery-necked Nightjars are known to pair for life. They are also territorial and males use their calls both to proclaim a territory and to attract a mate. The breeding and feeding territory of a pair of Fiery-necked Nightjars is about 6 hectares in extent and they are faithful to their nesting area from one year to the next.

The breeding season for most species is relatively short, from September to November, although some species have been recorded breeding over a longer period, between August and January. Nightjars make no nest, depositing the eggs directly onto the substrate of the chosen site. The individual species differ in their choice of

1

2

3

4

1, 2 *A Freckled Nightjar is almost invisible on its granite nest site, especially when it flattens itself to reduce shadow. Once it leaves the eggs, however, they are conspicuous on the bare rock.* ***3*** *At dusk the male Rufous-cheeked Nightjar brings food to the chicks and relieves his mate at the nest.* ***4*** *Most nightjar eggs are easily seen, but those of the Pennant-winged Nightjar blend well with the leaf litter on which they are laid.*

__1__ As a male Fiery-necked Nightjar alights it reveals its white wing patterns which may be used in display. __2__ The position of an incubating Fiery-necked Nightjar is revealed only by the bird's eye slit.

location. Fiery-necked Nightjars always lay on leaf litter, preferably but not invariably beneath trees, and the incubating female is usually in dappled shade. Pennant-winged Nightjars nest in leaf litter too, usually beneath trees and often in miombo woodland, but they also lay in the open where there is no cover. Rufous-cheeked and Mozambique Nightjars deposit their eggs on bare, sometimes stony ground, and the presence of shade and leaf litter may be incidental. The latter species also chooses burnt areas or black clay soils. Natal Nightjars nest in the cover of grass or bushes in the swamp areas they inhabit. The Freckled Nightjar nests on bare rock, its plumage a perfect match for dark granite.

A nightjar clutch usually comprises two eggs but sometimes a single egg is laid. The different species' eggs vary not only in ground colour, from pale pink to deep salmon-pink, but also in the amount of underlying lilac and overlying brick-red blotches or marbling. The eggs of the Fiery-necked Nightjar have fewest markings, and sometimes none at all, and those of the Pennant-winged Nightjar are the most heavily and attractively marked. The Freckled Nightjar differs from the other species in that it lays white eggs with grey markings, and in the past these were often inaccurately described. Unlike the eggs of most ground-nesting species, those of nightjars are not cryptic, and only the well-marked eggs of the Pennant-winged Nightjar may be camouflaged if they are laid on leaf litter; in other situations they are easily seen.

There is usually an interval of a day, or sometimes two days, between the laying of each egg but, as the egg is so visible, incubation begins as soon as the first egg has been laid. The camouflage of the sitting female effectively hides the egg, and if danger threatens she closes her large eyes to mere slits and flattens herself to eliminate the shadow cast by her body. If the eggs are lost to a predator she may lay a new clutch. In the case of the Pennant-winged Nightjar only the female broods, and if she is flushed off the nest she is likely to desert it. In the other species the female sits by day and the male takes over at night, relieving the female shortly after dusk. In an exposed site the female sits with her back to the sun and pants with a rapid gular flutter to keep cool.

The Fiery-necked Nightjar incubates for 18 days, and the information available for other species indicates that their incubation lasts for a similar period or two or three days shorter. The chicks hatch at the same time or a day apart, and when they do so the eggshells are removed and dropped away from the nest. The newly-hatched chicks are precocial and are extremely mobile within a day, their cryptic down making them difficult to detect.

The pattern maintained during incubation is continued into the nestling period and the male assists with brooding at night. Shortly after dusk he visits the nest to feed the chicks and then the parents alternate on the nest, the one brooding while the other feeds. When the parent arrives the chicks run to it and are fed by regurgitation, the chick enfolding the bill of the adult in its own. It is also possible that the two chicks are fed simultaneously.

When the chicks hatch the adults may perform injury-feigning displays to lure away predators, spreading their wings and tail on the ground to reveal the maximum amount of white. As the nestlings become too large to be covered they sit tucked against the parent's breast facing forward, their camouflage so effective that they look like part of the adult's body. They quite often move away from the nest at night, but in the early stages of the nestling period they run back to it to be fed when a parent arrives. Large young respond to predators by lunging at them with wings spread and gape wide open, at the same time emitting a growling call.

Observations of the Fiery-necked Nightjar have shown that the young birds can fly weakly when they are about 20 days old and competently by the time they are 28 days old. They stay with the female during the day, roosting in front of her until the age of 42 days. There appears to be considerable variation in the dependence period, with young continuing to associate with their parents until 62 days old in one record and 145 days old in another. In a third instance, a female laid a second clutch when the young of the first brood were only 33 days old.

NIGHTJARS BY MOONLIGHT

Nightjars are crepuscular rather than nocturnal, and sing most vigorously on moonlit nights but are inactive when it is dark. A recent study of the Fiery-necked Nightjar revealed that most birds laid their eggs to coincide with the full moon so that incubation took place while the moon was waning. Thus when the chicks hatched the moon was waxing, and the parents had more time in which to catch food for them. Migratory Pennant-winged Nightjars which arrived after Fiery-necked Nightjars had started breeding did not lay until the following full moon.

It would be an interesting exercise if future breeding records of all the nightjars were to be analysed according to the phases of the moon – assuming that the laying date could be determined with reasonable accuracy – so that it could be established whether there is a link between the timing of the nightjars' breeding cycle and the moon's phases ●

Fiery-necked Nightjar chicks hatch when the moon is waxing, possibly to give the adult birds more time to catch food for them.

Swifts and Spinetails

Swifts are the most aerial birds in the world and their habits are remarkable. They eat, drink, bathe, gather nest material and, sometimes, mate in flight. The European Swift, which spends three quarters of its life on the wing, does not alight between one breeding season and the next. It has been calculated that one that lived to be 18 years old flew 6.4 million kilometres in its lifetime. Not unexpectedly, birds that have such an aerial existence are highly specialised in many ways, and particularly in their breeding habits.

Knowledge about the nesting behaviour of the 11 species of swifts and spinetails that breed in southern Africa varies considerably: some species, such as the White-rumped, Little and Palm Swifts, have been well studied, but we have little information about others, notably the larger species which nest in cracks in inaccessible cliffs. Indeed, the nest of the Scarce Swift has yet to be discovered, although the birds have been seen flying into fissures in high cliffs in eastern Zimbabwe.

Some of southern Africa's swifts are resident, others are regular intra-African migrants and others still are both. The Black Swift, for example, is a breeding migrant in South Africa, but in Zimbabwe is resident. Also in Zimbabwe, there are two distinct populations of Horus Swift, one of which breeds on the central plateau from October to April and then disappears until the next season, while the other is apparently resident in lowveld regions and breeds from April to September.

The breeding season of the different species varies depending on each one's status. Intra-African migrants such as the Black, Alpine and White-rumped Swifts breed mainly in the summer months. Bradfield's Swift, which inhabits arid regions, breeds later, mainly from February to April in Namibia. Nesting for resident Little and Palm Swifts has been recorded in all months, but with peaks at certain times of year, depending on the locality. In Northern Province the little-known Mottled Spinetail breeds in April and May, and the few recorded Bat-like Spinetail nests have been found between October and March. Sometimes breeding varies according to local conditions and after good rains may continue for longer than usual. Under particularly favourable conditions individual pairs may raise consecutive broods, and in the case of the White-rumped Swift have been known to produce up to three broods in a season.

'Circusing' flights round nest sites, during which the swifts wheel about excitedly uttering their shrill, screaming calls, probably fulfil a courtship function, but actual

1 

2 

3 

4 

***1** Mottled Swifts breed in cracks in caves such as these in the Matobo Hills, Zimbabwe. **2** A White-rumped Swift, a foodball in its throat, returns to the nest it has taken over from a Lesser Striped Swallow. **3** The nests in a colony of Little Swifts are closely packed under the platform of a water storage tower. **4** This species also takes over swallow nests, such as those of the Lesser Striped Swallow.*

1 *A Mottled Spinetail faces inwards on its nest in a hollow baobab, its long wings projecting over the nest's rim.*
2 *Originally excavated by White-fronted Bee-eaters, these holes in a mine dump were subsequently used by Horus Swifts.*

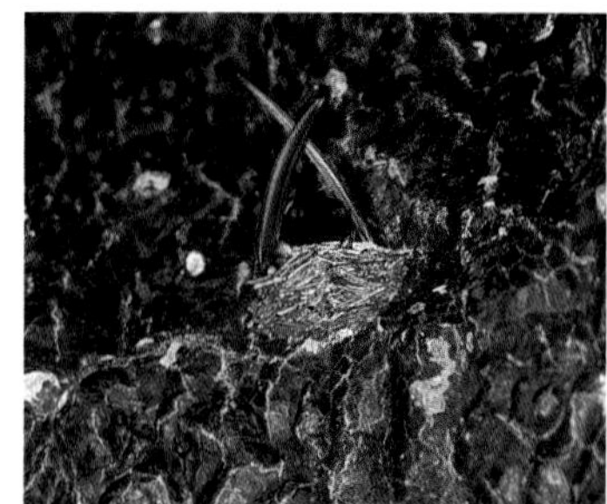

courtship behaviour is not well known. One regular feature is mutual preening in the nest, and courtship feeding has also been observed. Swifts are monogamous and the pair bond seems to be maintained by fidelity to the nest site in successive breeding seasons. A White-rumped Swift's nest was used for nine consecutive years and a ringed pair bred in it for at least three years. A ringed Horus Swift was recaptured in the same breeding area after an interval of 13 years, and Alpine Swifts were found breeding in the same cave 38 years after the small colony had first been discovered.

Copulation usually takes place in the nest or at the entrance. In the case of the Palm Swift the male descends from a position above the nest and copulates with the female in a vertical position, once every seven minutes in one observation. Remarkably, swifts have also been seen to copulate in flight, and African species that have been known to do so are the Alpine and White-rumped Swifts and the Mottled Spinetail. The birds glide together for a brief period with wings held in a 'V' position.

Nesting is either colonial or solitary depending on the species and often there is fierce competition for a suitable site. The most important requirement is that it should have a drop below for ease of access and departure, as the birds cannot take off horizontally. Species such as the Little Swift have taken to nesting on man-made structures and are now rarely found breeding in ancestral situations such as cliffs. As a result, they have greatly expanded their range into areas that previously lacked nest sites. Similarly the Palm Swift has been able to spread into new areas thanks to the planting of introduced palms.

Because they cannot alight to pick up nest material, swifts use whatever they can gather opportunistically in flight, often after an updraft of air or a dust-devil. Feathers, grass awns, pieces of dry leaves and any other suitable airborne fragments are taken, the smaller items being chewed in the birds' mouths and the larger pieces being carried in their bills. Saliva is used to cement the pieces together to form the nest, and in the breeding season swifts' salivary glands become much enlarged. During the construction process the swift clings onto the surface of the nest site by means of its incredibly strong feet and very sharp claws. Once the nest is complete and in use, an unpleasant aspect of it is the large number of parasites it contains. The parasites of swifts are a study in themselves, with several species being endemic to swifts and, in some cases, even to a particular part of their bodies.

FEATHERED NESTS AND GLUED EGGS

Palm Swifts attach their nests of feathers to palm fronds, under road and rail bridges, and rarely to a leaf clump in a eucalypt. Nesting on man-made structures was recorded as early as 1947 in South Africa, and in Zimbabwe the use of bridges has been known since 1960. Sometimes these sites are used even when there are palms available nearby, and they may be chosen because they are more secure. Usually several pairs breed in the same palm, or under the same bridge, in a small colony, but from time to time solitary nests may be found.

Like other swifts, Palm Swifts collect feathers as they fly, but are apparently quite selective about the size and texture of those they gather. They then glue the feathers into position with saliva to form a small vertical pad which has a lip at the bottom. The pair roost and copulate on this pad, and when the female lays she clings firmly to it, then slides down until her breast rests against the egg. What follows is remarkable: she vomits copious amounts of saliva round the egg and moves her body from side to side while pressing against it. This process may be repeated several times, even later in the incubation period, and has the effect of gluing the egg to the nest. Two eggs are laid, their small ends pointing downwards against the lip, and the parent broods in a vertical position. The eggs hatch successfully despite not being turned during incubation as are those of most birds.

When the eggs hatch the parent pushes against the shells to break them and the fragments fall to the ground below. The chicks have very long sharp claws with which they grip onto the nest. They are restless even when small, and from the age of about four days they clamber up the nest and even off it onto the palm frond itself. From the age of six days they develop a coat of white down that resembles the feathers of the nest and camouflages them very effectively .

The nestlings are brooded for the first week and then both parents fetch food for them, making 16 to 27 visits in a day, but feeding only one chick per visit. The young leave the nest after between 31 and 35 days and may return to roost on the nest for the first night or two, but not thereafter ●

A Palm Swift has attached its nest to the underside of a palm frond and glued the eggs into it with saliva.

Alpine Swifts have nested in cracks in the roof of this cave for some 40 years.

The four large species of swifts – the Black, Mottled, Alpine and Bradfield's – breed mainly in cracks in high cliffs or in cave roofs. The first three prefer vertical crevices whereas Bradfield's Swift uses horizontal cracks. They all breed colonially and sometimes Alpine and Black Swifts nest together. The former have recently been observed nesting on man-made structures and Bradfield's Swifts have been seen breeding in introduced palm trees in towns and also occasionally on buildings. The nests of these four species are shaped like a half cup with a shallow depression on top. Attached firmly with saliva, they are extremely difficult to dislodge.

The White-rumped Swift utilises a variety of sites such as holes in cliffs or spaces under the eaves of buildings, but it prefers swallows' nests, specifically those that are enclosed and bowl-shaped. As the swallows have increased their range thanks to the availability of man-made structures for their nests, so too have the swifts. If the swallow's nest is in use the swift will evict the eggs or chicks, then line it with feathers which protrude from the nest entrance, betraying the usurper's occupation.

The Horus Swift is the only African swift that nests in tunnels in banks that have been excavated by hole-nesting species such as bee-eaters, Ground Woodpeckers, martins and Pied Starlings. There is no evidence that they evict these species, and statements that they excavate their own burrows have yet to be supported by direct observations. Sometimes just a few birds may nest in association, but where numerous holes are available, for example in a bee-eater colony, several hundred may nest colonially. The nest, a shallow pad of aerial debris and feathers cemented round the rim with saliva, is placed at the end of the tunnel.

Little Swifts usually nest in colonies, some very large, on tall man-made structures such as grain silos and water towers, but also under bridges and occasionally on cliffs. Sometimes they make use of the bowl-shaped nests of swallows as do White-rumped Swifts, or adapt cup-shaped swallows' nests to their own design. Their nests are untidy, bowl-shaped structures with a side entrance and are usually so close that they touch each other. In the initial stage of construction the swift clings to the surface of the site and outlines the position of the nest with saliva before adding more solid material. By the time the structure is complete it is, as a rule, inaccessible and secure from predation. However, the Gymnogene is one species that regularly exploits Little Swift colonies, hanging upside down from nests and tearing them open.

The nests of the Mottled and the Bat-like Spinetails differ from those of other swifts in that they are small and bracket-shaped, and comprise twigs and leaf ribs cemented with saliva. The spinetails collect the twigs by snapping them off with their feet in flight. Very little is known about the breeding of these species in southern Africa, but Mottled Spinetails nest inside hollow baobabs and Bat-like Spinetails nest underground in old mine shafts and wells. When incubating the spinetail faces inwards with its wings projecting over the edge of the nest.

Swifts lay elongate white eggs, usually at intervals of about 48 hours. The four large species lay two eggs, or sometimes one, and the three white-rumped species – White-rumped, Horus and Little – usually lay two or three. Five eggs occasionally found in White-rumped Swift clutches may be the product of two females. The Palm Swift lays two eggs and spinetails usually two or three, although the Mottled Spinetail has been known to lay up to four eggs.

When the eggs hatch the shells may be removed in the mouth of an adult, but the Palm Swift breaks the shells into fragments with its body. Both parents share incubation, which normally begins with the completion of the clutch, but the extent to which eggs are covered is very variable. If poor weather conditions force the parents to range far afield to find food, the eggs may be left unattended for several hours. Remarkably, swift eggs can cool completely without the embryos being harmed.

The incubation period for most species is not known, but as a rule is long relative to the swift's size. The White-rumped Swift incubates for 21 to 26 days, the Horus Swift for 28 days, the Little Swift for 21 to 24 days, and the Palm Swift for 18 to 22 days. The Bat-like Spinetail's incubation period is given as 17 to 18 days, but may possibly be as short as 14 days.

Swift chicks are not creatures of beauty, hatching naked and blind with extremely distended bellies. However, their legs are strong and soon after hatching they are able to crawl about the nest. Both parents take turns to brood them while they are still naked. They feed them with foodballs carried in the throat, regurgitating part of the foodball initially, but later the chick is able to swallow the entire mass whole. It feeds by completely enfolding the adult's bill in its own wide gape. Some foodballs may contain 900 items or more. The rate of growth is rapid and a newly-hatched Little Swift may treble its size in two days. A three-quarter grown Horus Swift nestling weighed 41 grams compared with its parent which weighed only 30 grams. Once the chicks are large enough, they maintain sanitation by defecating over the nest edge or out of the entrance hole.

When the adult swifts have to fly considerable distances in search of food the chicks, like the eggs, may be left unattended for long periods. The young birds' entire metabolism slows down and they fall into a state of torpor; in one case a newly-hatched European Swift chick survived for two days before its parents returned with food. In another instance a nestling removed from its nest survived for 21 days without food before it died, during which time its weight dropped from 57 grams to 21 grams.

In some species, such as the White-rumped and Horus Swifts, the adults roost in the nest with their offspring even when they are fully grown. The nestling period, like the incubation period, is long, and it can be extremely variable. Little Swifts leave the nest after 36 to 40 days, and as a rule White-rumped Swifts leave after between 41 and 53 days, although in one observation the first chick left at 47 days and the second at 69 days. The nestling period of the Palm Swift is 31 to 35 days, and that of the Bat-like Spinetail about 40 days. However, the young of both these species as well as the Mottled Spinetail may clamber off the nest before they are ready to fly, and they cling nearby. Young swifts have been observed to leave their nests mainly in the early part of the morning. After a long period in a cramped nest, where wing exercises were restricted, they launch themselves into space and fly off, usually never to return.

Mousebirds

The mousebirds belong to the order Coliiformes, the only order of birds that is endemic to Africa. Of its six species, three – the Speckled, White-backed and Red-faced Mousebirds – occur in southern Africa where their ranges overlap considerably. They are gregarious species and their highly integrated social life extends into their breeding, which includes communal and co-operative behaviour. One of the more unusual demonstrations of their social interaction is when up to 14 birds roost together in a hanging cluster.

Mousebirds' courtship behaviour is unusual and not always consistent. Apart from regular mutual preening, which may not necessarily be confined to a pair, the most commonly described display is 'bouncing'. Perched crosswise on a branch and horizontal to it, a bird bounces up and down, either holding onto the branch or releasing it and rising about 3 centimetres above it in a series of jumps. In various observations the bouncing bird, on the evidence of copulation, has been either male or female. Mating ensues irrespective of which sex 'bounces'. In one observation a male White-backed Mousebird grasped the female's bill in his during copulation and may have fed her. In another record for this species copulation took place on the ground, the male approaching the female with raised and spread wings. After mating the pair often preen each other. Mousebirds are considered to be monogamous, but some polygamous situations may also occur.

They may breed throughout the year, but there are relatively few records for the winter months from April to July. Breeding peaks vary depending on the locality, and in the south-western Cape, for example, Speckled Mousebirds breed mainly from August to October.

1

Although they prefer to nest in indigenous thorny trees and shrubs, mousebirds make use of a variety of trees, including introduced species such as pines. Creeper-covered trees or bushes are also chosen. They nest either solitarily or in loose clusters. Both birds build the nest, sometimes assisted by helpers, and they construct an open bowl on a base of twigs or soft material such as grass. As a rule the Red-faced Mousebird uses twigs, often thorny ones, and the other two species soft material. Sometimes the nest is built on top of an old nest of another species such as a Cape Sparrow or on another mousebird's nest, and it is also not unusual for mousebirds to re-use a nest. They line the bowl well with miscellaneous soft materials such as feathers, cobweb, wool, string and pieces of rag and paper, and sometimes add green leaves and sprays of flowers, making additions during incubation and occasionally even in the nestling period.

The clutch usually comprises 2-4 eggs, the average being three. However, up to eight eggs have been recorded, and it has been estimated that a third of all mousebird clutches are the product of two or more females nesting jointly. The eggs are dull creamy white with a chalky texture, and only those of the Red-faced Mousebird are marked, with sparsely distributed spots and scrolls of dark red. A mousebird egg is very small, weighing no more than 5 per cent of the adult's body weight. It is also unpalatable: in tests involving the eggs of 212 bird species, those of mousebirds were amongst the seven most distasteful.

Eggs may be laid at intervals ranging from 15 to 48 hours, with those deposited more quickly undoubtedly being the products of two females laying in the same nest. Incubation begins when the first egg has been laid, which results in young of different sizes in the nest and sometimes the death of smaller chicks. Both sexes incubate, as do helpers too, and they sit extremely tight, sometimes so tight that a sitting bird can be caught. The incubation period may be as short as 11 days, but averages 13.

The flesh of newly-hatched Red-faced and Speckled Mousebirds is dark pink, and that of White-backed Mousebird nestlings is greenish. The chicks hatch blind with a sparse covering of down, and the parents and their helpers, usually offspring from a previous brood, are extremely attentive towards them, feeding them by regurgitation. The chick's soft faeces are eaten by an adult which nibbles at its anus to stimulate it to defecate. Subseqently the chick may receive a regurgitated meal which is mixed with its own faeces.

The parents seldom leave the chicks unattended, brooding them throughout the nestling period and even refusing to move when an observer tries to inspect the contents of the nest. The chicks grow quickly and by ten days old may clamber off the nest into the tree. Because they often leave the nest prematurely it is difficult to establish when they first fly, but it appears that they do so at 17 to 20 days old. The young continue to be fed after leaving the nest and join the social group, interacting with them through various play activities.

2

1 *A White-backed Mousebird broods its chicks in a nest that is warmly lined with sheep's wool.* ***2*** *A Speckled Mousebird may add lining to its nest throughout the incubation period.*

The Narina Trogon

The trogons are a colourful group of forest birds, of which the Narina Trogon is the only representative in southern Africa. The male's green upperparts and crimson belly make him one of the most attractive of all birds and the female appears drab only when compared with her mate. Despite their bright underparts, trogons are extremely difficult to locate, especially when they sit immobile for long periods watching for insects. According to numerous reports, they deliberately sit with their backs to the observer, thus rendering themselves even more inconspicuous.

To make locating a trogon in the forest canopy even more difficult, the male's deep hooting call has a ventriloquial quality and is not easy to pinpoint. He uses it for both territorial advertisement and courtship, puffing out the blue flesh under his throat and depressing his tail to a vertical position as he makes it. He also emits a rapid series of low *kuk* notes during courtship and when he relieves his mate at the nest.

During the breeding season, which extends from October to January, several males may assemble and chase each other while calling. Two observed competing for the attentions of a female were seen flying towards her with a whirr of wings and displaying their crimson bellies. Other courtship behaviour involves the male chasing the female in a slow 'butterfly flight', with wings raised high to reveal the white underwing. When perched he fluffs out his crimson belly feathers, flicks his tail up and down and fans it to reveal the white outer feathers. In one observation a male, his crimson feathers puffed out, approached in slow 'butterfly flight' a female perched on a branch, landed on her back and copulated. Copulation occurs frequently while a pair is prospecting for a nest site. It appears that the male selects the site and the female inspects it thoroughly before accepting it.

The Narina Trogon nests in a natural hole in a vertical limb, either a living branch or dead stump, and often has to compete with other hole-nesting species for a suitable site. The entrance, which may be at any height between 1 and 12 metres, is just large enough to admit the trogons. No lining is used and the eggs are laid on whatever may be lying at the bottom of the hole, usually decomposed wood chips. Nest sites that prove to be secure may be re-used for several seasons.

The clutch comprises 1-4 rounded and slightly glossy eggs which are laid at 24-hour intervals. Both sexes incubate over a period of 16 to 17 days, the male from mid-morning to late afternoon and the female for the rest of the time. They sit very tight and are reluctant to leave the nest. The tail feathers of both birds may become very worn by the end of the incubation period.

During the nestling period the male shares duties with the female and broods the small chicks. He appears to bring more food than the female, and at a nest watched by the author the feeding rate was very erratic: the single chick was not fed for 3½ hours and then the male brought four insects in ten minutes. On another day the male brought food eight times in two hours and the female did not visit at all. Stick insects, some of them large, were the main food items, comprising 11 of the 18 insects recorded. Faeces are not removed and accumulate in the nest.

The nestlings make a repeated ventriloquial *weee-er* begging call that sounds like a distant Helmeted Guineafowl and is difficult to pinpoint, and when an adult arrives with food they wheeze excitedly. When alarmed they hiss like snakes, open their bills and move their tongues laterally – all behaviour that would presumably deter a predator – and also snap their bills. However, at one nest this behaviour did not deter a boomslang which ate the chick. The nestling period in southern Africa has not been established, but in Kenya it has been recorded twice, at 25 and 28 days. The young are fed until about two months old, and they may remain with their parents for several months.

1

2

***1** Narina Trogons nest in natural holes in trees.* ***2** During courtship the male flaunts his bright underparts.*

KINGFISHERS

AS A GROUP KINGFISHERS ARE SOME of the most attractive of all birds. Most are colourful or strikingly marked, many are extremely vociferous, and some have spectacular courtship displays. There are ten species in southern Africa, all of which breed in the region. They range in size from the Pygmy Kingfisher, which is 12 to 13 centimetres long and weighs 19 grams, to the Giant Kingfisher, 43 to 46 centimetres long and weighing about 380 grams. The latter is the largest aquatic kingfisher in the world and is exceeded in size only by the dry-land Laughing Kookaburra of Australia.

The southern African kingfishers may be divided into two groups: typical and woodland kingfishers. The typical kingfishers include the Pied, Giant, Half-collared, Malachite and Pygmy Kingfishers, and the Woodland, Mangrove, Brown-hooded, Grey-hooded and Striped Kingfishers belong to the woodland group. Four species – the Pied, Giant, Half-collared and Malachite Kingfishers – are aquatic, and the other six are dry-land kingfishers that feed mainly on arthropods. The Pygmy, Woodland and Striped Kingfishers are all intra-African migrants. The little-known Mangrove Kingfisher moves between 5 and 50 kilometres inland to breed, but the remaining species are resident.

Pied and Giant Kingfishers are sexually dimorphic, as is the Striped Kingfisher, a fact which has only recently been discovered; the male shows a darker band on the wing. In general kingfishers are monogamous and territorial, but two species are co-operative breeders: the Striped Kingfisher may breed in a group comprising two males and a female in which both males mate with her, and the Pied Kingfisher has a number of male helpers.

As a generalisation, the typical kingfishers have less demonstrative courtship behaviour than the woodland group, but one feature that is common to all of them is courtship feeding. The Half-collared, Malachite and Pygmy Kingfishers are quiet species and no special courtship displays have been described. Giant Kingfishers, on the other hand, are extremely noisy and indulge in chasing flights in the breeding season. Pied Kingfishers, too, are noisy and up to a dozen birds may interact in chasing flights.

Duetting forms part of the courtship of the woodland kingfishers. The display of the Woodland Kingfisher is fairly typical of the group: the male and female face each other, raise and spread their wings, and then pivot on their perch so as to reveal both upper and underside wing patterns. The performance is accompanied by loud calling and is spectacular to watch. In another impressive display they dive down and weave through the treetops.

In general kingfishers breed during spring and summer, from August to December, but some may breed as late as March or April. An exception is the Pied Kingfisher which breeds in most months. Several species have been known to raise two broods in a season. Most kingfishers nest solitarily except, again, the Pied Kingfisher as it may also breed in loose colonies.

The five typical kingfishers and the Brown-hooded and Grey-hooded Kingfishers all nest in a hole in a bank, or sometimes in the side of an antbear's burrow. The hole is usually placed at a height that makes it difficult for predators to reach it. Both sexes excavate the tunnel, initially with the bill and later with the feet too. Sometimes considerable quantities of earth may be ejected from the entrance; one Giant Kingfisher kicked out soil to a distance of a metre from the hole and reached a depth of 2 metres in a week. Tunnels range in length from 30 to 60 centimetres for smaller species to a prodigious 8.5 metres in the case of one Giant Kingfisher, although the average length for this species is 1.8 metres. At the end of the tunnel the kingfishers excavate a chamber, sometimes with a slightly raised lip where tunnel and chamber join. The tunnels of some species, such as the Malachite Kingfisher, may slope slightly downward to the entrance to assist the drainage of excreta. The egg chamber is unlined, but as the nesting cycle progresses it becomes filled with fish bones or the exoskeletons of insects from regurgitated pellets.

The Woodland, Mangrove and Striped Kingfishers nest in holes in trees, usually in the old nest of a barbet or woodpecker, but also in natural cavities. Sometimes they make use of nest-boxes, and nests have been found too in unusual situations: Striped

A Pygmy Kingfisher approaches its nest in a low roadside bank. Both sexes feed the young.

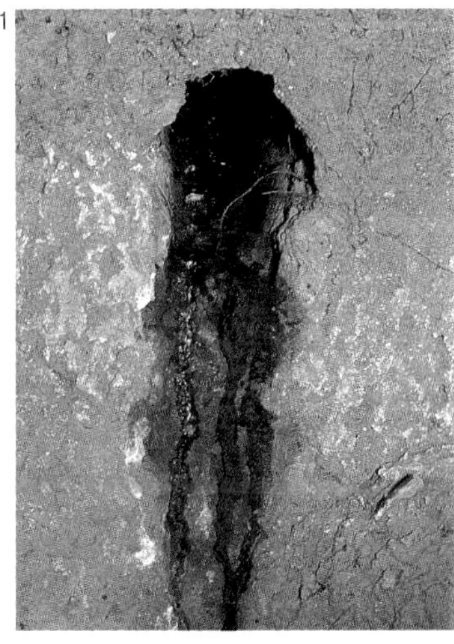

__1__ Liquid excreta ooze from the entrance to a Malachite Kingfisher's tunnel. Sanitation poses a problem for the tunnel-nesting species. __2__ A young Striped Kingfisher peers out of its nest-hole, which was originally excavated by a Crested Barbet. __3__ A Woodland Kingfisher feeds its chicks in a disused barbet's nest. __4__ 'Tram lines' at the entrance to a Grey-hooded Kingfisher tunnel in a roadside embankment indicate that it is occupied.

CO-OPERATIVE BREEDING OF PIED KINGFISHERS

Conspicuous, noisy and quite often gregarious, Pied Kingfishers sometimes nest in loose colonies, and recent research in Kenya has shown that they may also breed co-operatively. It appears that there are more males than females (in a ratio of 1.8 males:1 female), and that one in three pairs has at least one helper. These helpers fall into two categories, primary and secondary. Primary helpers are year-old males raised by one or both members of the pair being assisted. Secondary helpers are unrelated non-breeding males, or males whose breeding cycle has failed.

A breeding pair may have either one primary helper or up to three or four secondary helpers, and a pair with a poor food supply seems to accept more secondary helpers than one with plentiful food. A primary helper feeds both members of his adopted pair during the courtship and incubation period, and he also feeds the young. Secondary helpers only participate after the chicks have hatched; initially they are driven away, but later they are accepted and feed the nestlings, although less effectively than a primary helper does.

It emerges that a primary helper contributes more to the success of the breeding cycle than a secondary helper, but at cost to himself. He expends more energy, lowers his chance of survival and also reduces his own chance of breeding. A secondary helper on the other hand contributes less but has a better chance of survival. His reward is that he may get to breed in the same place where he helped the previous season, sometimes with the female he assisted.

What are the benefits for the primary helper? His behaviour results in no direct benefit to himself, but the effort he invests results in a better chance of survival for both the parents and offspring who are his kin ●

Some Pied Kingfisher pairs are assisted in their breeding attempts by one or more helpers.

Kingfishers have been recorded using nests of Lesser Striped Swallows and a hollow beam in the roof of a garage, and Woodland Kingfishers have nested in Little Swift nests and in holes under the eaves of houses. Secure nest sites may be used for several breeding seasons.

Clutches range from 2-6 eggs which are rounded and white with a slight gloss. Occasional atypical Brown-hooded Kingfisher eggs have been described as being marked with black dots. Because kingfisher nests are difficult to observe, details on egg-laying intervals are scant, but differences in the size of chicks in some broods indicate that incubation may begin before the clutch is complete.

Incubation is shared by both sexes, although the female sits overnight. The number of change-overs during the day varies from one species to another: Giant Kingfishers change over three or four times a day, smaller species more often, and Half-collared Kingfishers alternate every 1 to 2 hours, incubating the eggs 92 per cent of the time. The incubation period of the Giant Kingfisher is the longest at 25 to 27 days, while other species incubate for between 14 and 18 days.

When the chicks hatch the eggshells are dropped a short distance from the tunnel entrance. At one Giant Kingfisher's nest the shells floated downstream after they had been dropped and were dive-bombed by the parent until they sank! The chicks hatch naked and blind but are soon covered with a mass of spiky quills which encase the feathers to prevent them from becoming fouled. Nest sanitation is a problem and the burrows of species such as the Malachite and Half-collared Kingfishers can become extremely foul, with liquid excreta oozing down the tunnel. The parent birds often find it necessary to clean themselves off after a feeding visit to the burrow and they dive into water, sometimes repeatedly, to do so. At the same time they rid themselves of parasites that collect in the tunnel. The older chicks of species which nest in tree holes void their liquid excreta in a strong jet through the entrance, thus keeping the interior relatively clean. They also regurgitate pellets of undigested food out of the nest-hole, whereas the pellets of bank-nesting species usually accumulate in the brood chamber.

Both parents feed the young and brood them while they are still small. In one observation a Half-collared Kingfisher pair visited the nest 37 times in eight hours, feeding the chicks an estimated 50 grams of food each. The nestling period ranges from 18 days for the Pygmy Kingfisher to 37 days for the Giant Kingfisher; most of the other species leave the nest at between 21 and 28 days old. The young signal their readiness to leave by moving to the nest entrance to be fed, and in observations of several different species it was noted that the feeding rate decreased, sometimes ceasing altogether, when they were ready to fly. Sometimes the parents even perched in view nearby with the deliberate intention of inducing their offspring to leave the nest. The young may catch food for themselves quite soon after leaving the nest – a Malachite Kingfisher has been recorded doing so within a week – but generally they remain dependent on the adults for several weeks.

Bee-eaters

ATTRACTIVE BIRDS NOT ONLY FOR their marvellous colours but also for their streamlined shape and graceful flight, bee-eaters look like colourful darts as they fly up to catch insects. A colony of breeding Carmine Bee-eaters numbering some 2000 brightly hued birds must rank as one of the most spectacular sights in the avian world.

Seven species of bee-eaters breed in southern Africa. The Carmine Bee-eater is an intra-African migrant, as is the European Bee-eater although there is also a separate population that visits from the Palaearctic region. The situation regarding the Olive Bee-eater is less clear; it is described as an intra-African migrant but there is a resident population on the Bazaruto Island group in Mozambique. Bee-eaters normally have a long-term fidelity to their breeding areas, and the Olive Bee-eater is also unusual in that in Zimbabwe it breeds sporadically in its traditional area. The remaining four species – Böhm's, White-fronted, Little and Swallow-tailed Bee-eaters – are resident, although the last-mentioned may wander locally at times.

Bee-eaters are monogamous and pair for life. They are also territorial and confront intruders with threatening postures. Courtship feeding, to the accompaniment of chattering or twittering calls, is common to all species and often serves as a prelude to copulation. In some species it is accompanied by 'head bobbing', shivering movements of the wings and spreading of the tail. The White-fronted Bee-eater female performs looping flights, bumping the male as she alights.

Breeding takes place from August to December and several species have a September/October laying peak. This is timed to coincide with an abundance of insects which emerge at the onset of the rainy season. All bee-eaters excavate a burrow which comprises a tunnel with a chamber at the end, often with a slight lip to prevent the eggs from rolling out. The length of the tunnel varies from 50 to 75 centimetres in the case of the Little Bee-eater, to 1 to 3 metres in larger species such as the Carmine Bee-eater. Even when holes exist from the previous breeding season, the

1 *The European Bee-eaters that breed in southern Africa are intra-African migrants, and nest colonially in riverbanks.* ***2*** *Carmine Bee-eater colonies, usually located in riverbanks, may number up to 1000 pairs, and are a spectacular sight.* ***3*** *Occasionally this species excavates holes in flat ground, and when they do so the nests are well spaced.*

Little Bee-eaters nest solitarily, but excavate a succession of holes over several breeding seasons.

birds prefer to excavate new ones. Digging seems to be part of the breeding stimulus and both sexes take part, initially with the bill but later the earth is shovelled out with a bicycling action of the feet.

Böhm's, Little and Swallow-tailed Bee-eaters nest solitarily and the other species breed in colonies that are either tightly packed or loose. The colonies of Carmine Bee-eaters are large, sometimes comprising as many as 1000 holes or more, with a density of 60 nests per square metre. Most are situated in riverbanks but occasionally nests are excavated in flat ground, in which case the density drops dramatically to only 3 nests per square metre. Carmine Bee-eaters may use the same locality for a long period; a site 100 kilometres north-west of Bulawayo in Zimbabwe was in use for at least 80 years. A feature of their colonies prior to egg-laying is a sudden panic when the birds fly off simultaneously and then return. We do not yet know why these panics occur, but they do so sometimes as many as eight times in an hour.

European and White-fronted Bee-eaters nest in banks, the latter species habitually using steep sites along rivers or erosion gullies and on mine dumps. The Olive Bee-eater breeds on level ground, in low banks and in the sides of dunes. Swallow-tailed Bee-eaters nest in low banks and on level ground, and Böhm's Bee-eaters excavate in flat ground. Little Bee-eaters often use low banks but they prefer the wall or roof of an antbear burrow where this is available. Such a site has the distinction of being the safest of all bee-eater nests, although it does not prevent parasitism by honeyguides.

All bee-eaters lay rounded white eggs that are smooth and glossy. Clutches comprise 2-6 eggs which are laid at intervals of one or two days, and within a colony there is a high degree of laying synchrony. Incubation begins when the first or second egg is laid, with the result that the chicks are of varying ages and the younger ones will die if food is short. Both sexes brood, the females overnight, and the incubation period is close on 20 days for those species where it is known.

The chicks hatch blind and naked, with a pink skin. After about a week their eyes open and their skin turns grey. When the quills break out the feathers initially remain encased for protection against fouling, in a 'hedgehog' stage that is typical of hole-nesting birds. Both parents participate in feeding the chicks on insects, and in the case of White-fronted Bee-eaters they are regularly assisted by helpers. There may also be helpers at the nests of Carmine and European Bee-eaters; four birds have been recorded at a nest of the latter species. As the breeding cycle progresses the nest chamber and tunnel become filled with chitin from decomposed pellets, and perhaps even the body of a dead chick. This debris is an ideal breeding ground for parasites and this may be one of the reasons why bee-eaters excavate new tunnels each season.

The nestling period varies from 23 or 24 days in the case of the Little Bee-eater to 30 days for the European Bee-eater. After leaving the nest the young remain dependent on their parents for at least 21 to 42 days and perhaps longer, since Carmine Bee-eaters have been seen feeding begging chicks far from known breeding colonies. A fairly long period of parental care is required because young bee-eaters need to learn the art of removing the venomous stings of bees and wasps that make up much of their diet.

SOCIAL ORGANISATION IN A WHITE-FRONTED BEE-EATER COLONY

On the surface, life at a colony of White-fronted Bee-eaters appears to be straightforward. The birds go about their business in typical bee-eater fashion: they feed their mates as part of courtship, they drive off intruders with loud chattering, they excavate nest burrows and they sunbathe on the riverbank with wings and tails spread and heads hanging limply to one side. Appearances are deceptive however, for White-fronted Bee-eaters have one of the most complex social organisations recorded for any bird.

Detailed long-term studies of marked birds in Kenya have revealed some amazing facts about the lives of White-fronted Bee-eaters. They live in highly gregarious colonies comprising extended social units known as clans, each of which is made up of three, sometimes four, generations of individuals. Although the nests of the various clan members may be spread throughout the colony, the clan itself forms a cohesive unit that is held together at a defended feeding territory, the focal point of clan activity. This may be some distance from the colony. The clan remains intact year-round and continues to use the colony as a roosting site.

Within each clan the monogamous pairs may have up to five helpers which are drawn from their own kin. About half the nests observed had helpers, which were non-breeding birds or those whose own breeding attempt had failed. Not all the pairs in a colony breed and sometimes birds that bred in a previous season became helpers in the subsequent one. The helpers assisted in all aspects of the breeding cycle: defending nests, excavating, incubating and feeding young. Their presence resulted in a marked difference in productivity: pairs without helpers raised an average of 0.53 offspring whereas those with two or more helpers averaged 1.34 young.

Among other aspects of White-fronted Bee-eater behaviour revealed in the Kenya study it was found that rape was a regular occurrence, whereby several males attacked an unaccompanied female. To avoid this situation males were careful to accompany their mates to protect them. Parasitism of nests also occurred, when unpaired females or those whose own breeding had been disrupted at the time of laying deposited their eggs in another bird's nest. These parasitic eggs had to be laid at just the right time if they were to be accepted: if deposited too early they would be evicted by the nest's legitimate owner, and if laid too late the young would not be able to compete with older chicks. As in the case of cuckoos, the 'window of vulnerability' was while the host's clutch was being laid.

The question again arises, why do some birds help others in their breeding attempts? The answer lies in the principle of kin selection which, in essence, means that if a bird is unable to breed it can at least compensate for its own lost reproductive output by aiding its relatives. Another advantage may be that the helper enhances its own future breeding ability by serving an apprenticeship as a parent ●

White-fronted Bee-eaters are gregarious, living in colonies that are made up of a number of cohesive units, or clans.

Rollers

Five rollers occur in southern Africa, of which one, the European Roller, is a non-breeding migrant from the Palaearctic. The Lilac-breasted, Racket-tailed and Purple Rollers are resident, sometimes making nomadic movements, and the Broad-billed Roller is a breeding intra-African migrant. Although these birds are conspicuous and colourful, their breeding habits have been little studied, and details on the incubation and nestling periods of the Lilac-breasted Roller in this account are drawn from the author's own unpublished observations.

Rollers are monogamous and aggressively territorial, with impressive aerial displays which have given rise to their name. Except for the see-saw chasing flights of the Broad-billed Roller, their display flights are very similar, and that of the Lilac-breasted Roller is typical. The displaying roller tumbles about the sky to the accompaniment of harsh grating calls, then shoots upwards with closed wings and, at the point of stall, loops over and dives. Other aerobatics include a fast flight during which the bird rocks from side to side, a manoeuvre akin to 'rolling' in pilot's parlance. These displays are used for territorial advertisement and courtship, and to intimidate intruders near the nest.

Courtship feeding is part of the pre-breeding ritual and is often a prelude to copulation. Although some pairs visit their nest-holes throughout the year, nest inspection and display activity intensify as the breeding season approaches. All four species breed from September to December, with a peak in October before the rainy season. In Namibia the Purple Roller may breed until June, depending on the rainfall.

A typical roller nest site is a hole in a tree that may be either natural or that of a barbet or woodpecker. The nest is unlined and the birds may even clean out debris from the bottom of the hole. Both Lilac-breasted and Purple Rollers have been found breeding in large-diameter vertical pipes, and in Namibia the latter species also breeds in holes in cliffs. A nest-box may also be used if its entrance is large enough. The birds may return to secure nest sites in successive breeding seasons.

The clutch comprises 2-4 eggs which are rounded and white. At Lilac-breasted Roller nests in the author's garden in Zimbabwe, eggs were laid at intervals of two or three days. Incubation began in one case after the second egg of a three-egg clutch had been laid, and in another observation only after a four-egg clutch was complete. The birds were very secretive when laying and during the rest of the incubation period. Both sexes

Like other rollers, the Purple Roller nests in trees, but it has also been found nesting in vertical pipes and holes in cliffs. Long-horned grasshoppers form part of the chicks' diet.

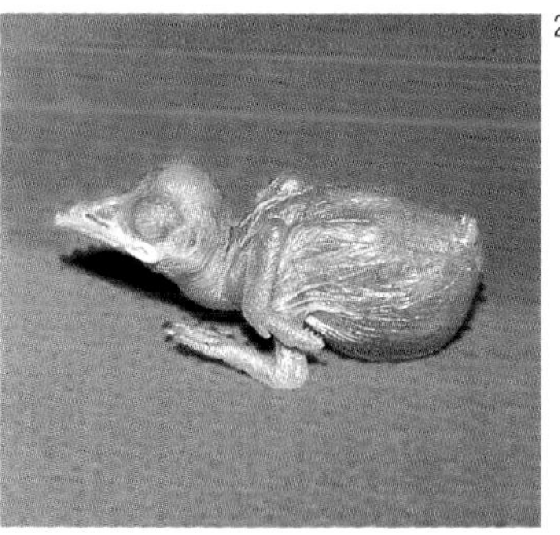

1 *A Racket-tailed Roller at the entrance to its nest, a natural hole in a tree.* **2** *At five days old a Broad-billed Roller chick is still naked and blind and has a distended abdomen.* **3** *A Broad-billed Roller takes food to its chicks.*

incubated and sat very tight, even remaining on the nest when the inspection hatch on a nest-box was opened. The incubation period, observed on six occasions, varied from 18 to 23 days. The incubation periods of the other roller species have yet to be recorded.

In a Lilac-breasted Roller's nest the eggshells were not removed and were eventually crushed at the bottom of the hole. The newly-hatched chicks were blind, pink and naked, with distended abdomens. By the time they were nine days old their eyes had opened and quills were sprouting rapidly, transforming them into 'hedgehogs' before the feathers emerged. Both parents fed and brooded during the day for about the first week, and overnight until the chicks were at least two weeks old. During a 13-hour dawn-to-dusk watch at a nest with two week-old chicks the parents brought food 40 times. Visits started to increase only from 09h00, as it became warmer. However, feeding rates were variable: at one nest ten and 13 feeds were brought in two four-hour watches, whereas at another there were 14 visits in only nine minutes.

The author's observations on the nestling periods of Lilac-breasted Rollers showed that they were very variable. At one nest the two young left at between 25 and 27 days; at another with four offspring the first two left after 28 days and the remaining two after 30 to 31 days; the first two in a nest with three young left after 28 to 30 days and the last chick after 34 days. Other nestling periods were 34 and 37 days. At these nests the young dispersed with their parents from the immediate nest area and it is not known for how long they remained dependent. The nestling periods of the Racket-tailed Roller and Purple Roller are unknown, but a Broad-billed Roller reared in captivity by the author was able to fly weakly at 30 days old.

The Hoopoe and Woodhoopoes

The Hoopoe and the three woodhoopoes are closely related, and as there are many similarities in their breeding biologies for convenience they will at times be referred to collectively as hoopoes. The Red-billed and Violet Woodhoopoes were once considered to be the same species but are now treated as separate ones. Unpublished research on the latter, which occurs in northern Namibia, has established that its breeding biology, including co-operative behaviour, is similar to that of the Red-billed Woodhoopoe. The details of the Scimitar-billed Woodhoopoe's breeding biology have been drawn largely from the author's own unpublished information.

The Hoopoe and the Scimitar-billed Woodhoopoe live in pairs – although the Hoopoe has occasionally been observed with a single helper at the nest – whereas the Red-billed and Violet Woodhoopoes associate in groups. All four species are territorial and monogamous, even when they live in groups. They have distinctive calls and the voices of the male and female are different.

The male Hoopoe emits his monotonous *oop, oop, oop...* for long periods to proclaim his territory. Courtship feeding is regular and during the course of it the Hoopoe pair may pass a grub back and forth to each other 25 to 30 times before eating it. The male also chases the female both in flight, raising and lowering his crest in excitement, and on the ground with tail spread and raised wings flapping slowly. Copulation takes place on the ground.

In the woodhoopoes weaving movements of the body and the flicking of the tail from side to side may be part of courtship behaviour. When the breeding season approaches the male Red-billed Woodhoopoe keeps the female apart from other members of the flock, guarding her from their attentions. He preens her and initiates copulation by scratching on her back with his foot, then mounts her with pedalling movements of his feet. Copulation lasts for a minute or more.

The breeding season falls mainly in spring and early summer, from August to December with a distinct peak in September and October, although nests have also been found at other times of the year. All species nest in holes, and will use a nest-box if one is available. The Hoopoe uses any suitable cavity such as a hole in a tree, wall or termite mound. It appears that the male prospects for a nest site and leads the female to it. The woodhoopoes prefer holes in trees, either natural or those of barbets or woodpeckers. Prospecting for a nest site may begin long before eggs are laid, a wise precaution as suitable holes are often scarce and competition for them is fierce. The most important feature of a nest is that the opening should be narrow, and sometimes the female can enter only with difficulty.

The eggs are laid on wood chips or whatever may be lying in the hole, or a shallow scrape is prepared for them. The birds sometimes remove debris from the hole and rarely, in the case of the Scimitar-billed Woodhoopoe, they may bring a few pieces of lining to the nest. The holes are usually poorly ventilated and foul-smelling, and develop a characteristic 'hoopoe odour' as the breeding cycle progresses. In this respect Hoopoe nests are the least savoury.

The Hoopoe lays 2-6 eggs, or rarely seven, the Red-billed Woodhoopoe lays 2-5 and the Scimitar-billed Woodhoopoe lays 2-4. The eggs, produced at daily intervals, are an attractive blue or turquoise when fresh with minuscule white pores, but they

Once a Hoopoe chick is old enough to come to the nest entrance, the parent may feed it without alighting.

THE SOCIAL SYSTEM OF THE RED-BILLED WOODHOOPOE

The communal social system of Red-billed Woodhoopoes has been intensively researched in Kenya and in the Eastern Cape in South Africa. There are many similarities in the two studies but also marked differences, a fact which underlines the value of comparative research in different localities.

In southern Africa the size of a group of Red-billed Woodhoopoes ranges from a single breeding pair to as many as 12 birds. Where more than two birds occur the flock comprises a core breeding pair with helpers which are usually, although not necessarily, related. Within this group there is a strict hierarchy: the breeding male and female are the oldest and dominant members of the group, whereafter males dominate females and older birds have a higher status than younger ones.

When a breeding vacancy occurs the highest-ranked bird of the required sex fills the gap so that there is no disruption of the breeding unit. Some dispersal takes place, and in Kenya up to five birds of the same sex may leave the group; in South Africa it is usually a single individual of either sex. Incestuous relationships are regular and Red-billed Woodhoopoes have one of the highest incidences of inbreeding among group-territorial birds with no apparent deleterious effects.

Within the group cohesion is maintained in various ways. Feeding and grooming of other birds occurs regularly and is linked to an individual's status, and group displays are performed. One of these, 'rallying', is highly vocal, the whole flock giving vent to loud cackling calls as they sway their bodies about and flick their tails up and down. In the display known as 'flag-waving' a member of the flock, usually one of the breeding pair, picks up a piece of bark or lichen or a twig or berry and, supported by the entire group, waves it at another flock of Red-billed Woodhoopoes or towards a predator. In addition to maintaining cohesion, both these displays may have the effect of reducing tension within the group.

Group cohesion is especially evident when the birds are breeding, and the helpers come into their own when the young hatch. They supply food to the female and then, as the chicks grow, they try to feed them directly. When the young leave the nest they are easy to distinguish by their black bills which gradually change to red by the time they are eight months old. These juveniles are tended in various ways and are then integrated into the flock, becoming helpers themselves.

Why do Red-billed Woodhoopoes live in groups? The answer may be found in the availability of roost cavities. Every night a secure roost-hole – sometimes two if the group is large – is required. To deceive predators some groups pretend to roost at one spot, then slip in quietly to the actual roost-hole. In Kenya it was found that males and females sometimes roosted separately, mainly because the larger males could not fit into holes occupied by females. As there was a high level of predation several birds of the same sex would sometimes be eliminated.

A Red-billed Woodhoopoe at its nest entrance. The social organisation of this species is linked to the availability of roost-holes.

In the South African study the sexes did not roost separately and predation was not high. It was also revealed, by direct observation as well as experimentally by increasing and decreasing the availability of roost cavities, that lack of roosts led to the formation of larger flocks. When this happened non-breeding birds were forced to remain in their natal flocks because both space and resources were limited; in essence there was an environmental constraint on dispersal. In a Red-billed Woodhoopoe's world secure roost cavities are a critical resource; without them it is dangerous to forage too far afield for fear of having nowhere safe to sleep.

Contrary to expectation, it was found that the presence of helpers made no difference either to the woodhoopoes' overall breeding success or to the survival of the breeding pair, even though their energy expenditure was reduced. The usual argument pertaining to co-operative breeding, whereby the helped or the helpers derive an advantage, does not apply to Red-billed Woodhoopoes. Instead, it appears that helping constitutes an example of misdirected parental care stimulated by the begging of the young or of the breeding female. This conclusion has been confirmed by experimental manipulations in the South African study ●

discolour rapidly as incubation progresses. An unusual feature of hoopoe clutches is the frequency with which infertile eggs are recorded. The female Scimitar-billed Woodhoopoe sits overnight even when she has not started brooding during the day, which may explain why the last chick of a brood often lags behind the others. Only the female hoopoe incubates and she rarely leaves the nest; when she does so it is usually in the morning and late afternoon. She is fed by the male, and in the case of the Red-billed Woodhoopoe sometimes also by helpers. The incubation period of the Hoopoe is 17 days, that of the Red-billed Woodhoopoe 18 days, and that of the Scimitar-billed Woodhoopoe 14 days.

When the eggs hatch the shells are not removed and they disintegrate in the nest. The female continues to sit in hornbill-like incarceration, brooding the chicks. She is fed assiduously by the male, and passes the food on to the young. In the case of the Red-billed Woodhoopoe helpers assist in feeding the chicks and attempt to by-pass the female if they can. Female Hoopoes and Scimitar-billed Woodhoopoes stay in the nest for about a week after the chicks have hatched, but Red-billed Woodhoopoes stay for about two weeks.

Feeding the young may involve considerable expenditure of energy and in one record it was calculated that a pair of Hoopoes covered 11 kilometres during 21 feeding visits to the nest. A day-long watch from first to last light at a Scimitar-billed Woodhoopoe's nest with three chicks recorded 122 feeding visits, 42 by the male and 80 by the female, and on 22 occasions the birds came to the nest together. It is worth noting that in woodhoopoes males are heavier than females and have sturdier and longer bills. This enables the pair (and helpers) to forage on different food items so that a wide range of available food resources is utilised.

Newly-hatched chicks are blind, pink and naked, except for a wispy covering of white down. They have prominent white flanges on either side of the gape which

A female Scimitar-billed Woodhoopoe brings food to the nest. Having stayed with the chicks for about a week, she then helps the male to feed them.

appear as white circles when they are begging with mouths open, enabling the female to locate them in the dark hole. The spiky 'hedgehog' stage begins when the chicks are a week old, with quills sprouting in all directions, and the feathers remain encased in their long sheaths for another two weeks. Then the sheaths are suddenly shed, the last sheaths to be lost being those on the head.

Once young hoopoes become aware of their surroundings, they defend themselves against predators in a number of different ways. They sway from side to side like snakes and hiss (a ploy also used by the female in the nest) and poke at the source of alarm with their bills. They also spray liquid excreta, and exude a vile-smelling fluid from the preen gland.

All hoopoe chicks beg with a characteristic squeaky call similar to that of the female in the nest when she is soliciting for food. Excreta are not removed and the chicks merely defecate against the side of the hole. Conditions are far from sanitary and it is of interest that one of the first things that a young Hoopoe was observed to do on emerging from the nest was to take a dust-bath. As the time to leave approaches the chicks move up to the entrance to be fed, and it is not uncommon to see a Hoopoe hovering to feed its offspring.

The nestling period of the Hoopoe is 26 to 32 days, that of the Red-billed Woodhoopoe 30 days, and that of the Scimitar-billed Woodhoopoe 21 to 24 days. However, the smallest chick of a brood sometimes leaves the nest several days after its siblings. Once Hoopoe offspring have left the nest they can feed themselves after only a week, and their parents may lay a new clutch within days of their departure. Scimitar-billed Woodhoopoes remain with their parents for at least a month, and in one observation a pair laid a second clutch 70 days after the young had left the nest. Red-billed Woodhoopoes usually raise only one brood per season in southern Africa. The young are integrated into the flock where helpers compete with each other to lavish attention on them, feeding, preening and guarding them, and even guiding them to their roost-hole at night.

Hornbills

Hornbills are well known for their unique nesting behaviour, notably their habit of sealing the female into the nest-hole. The nine species that occur in southern Africa can be divided into three groups: six members of the genus *Tockus*; the Trumpeter and the Silvery-cheeked Hornbills of the genus *Bycanistes*; and the large, terrestrial Ground Hornbill.

The *Tockus* hornbills inhabit a wide range of savanna, bushveld and woodland, the most widespread species being the Grey, Red-billed, Yellow-billed and Crowned; Bradfield's Hornbill has a range extending from north-western Zimbabwe westwards into northern Namibia; and Monteiro's Hornbill, inhabiting the most arid habitat of any hornbill, is found in northern Namibia. The Silvery-cheeked Hornbill, a true forest bird found mainly in eastern Zimbabwe and Mozambique, and the Trumpeter Hornbill have the same nesting habits as *Tockus* hornbills, except for some minor variations. The Ground Hornbill differs from other hornbills in that it does not seal its nest and it lives in social groups and breeds co-operatively.

Hornbills are often gregarious and may form large flocks. *Tockus* hornbills are monogamous and territorial and appear to maintain a long-term pair bond. Depending on the species, there are two distinct types of territorial display. The Grey, Crowned and Bradfield's Hornbills forage mainly in trees and have a whistling or 'piping' display in which the bill is pointed vertically upwards and the body is jerked slightly backwards and forwards at each whistle. The Red-billed, Yellow-billed and Monteiro's Hornbills all forage on the ground and are 'cluckers'. They emit their calls while perched on a branch with their heads dropped down to the level of their feet and wings raised high above their backs. In all *Tockus* hornbills both sexes perform these displays, but the males are more active. The displays also serve to strengthen the pair bond. Trumpeter and Silvery-cheeked Hornbills apparently do not execute territorial displays and defend only the immediate vicinity of the nest. Courtship feeding is common to all species, and the male hornbill feeds his mate diligently before breeding so that she has reserves both for laying and for replacing her feathers after she has moulted in the nest.

In general hornbills breed in the early summer months, but the start of nesting is determined by the onset of the rains and may vary from year to year. The rains result in mud being available for nest-sealing and, more importantly, in there being a good food supply which enables the female to build up her reserves prior to laying. In common with many other birds, hornbills in Namibia breed later than elsewhere in southern Africa because the rains are later.

The nest of a hornbill sets it apart from all other birds. Most nests are in holes in trees, sometimes in holes in rocks, but all, except for those of the Ground Hornbill, are sealed. The nest and sealing behaviour of *Tockus* hornbills is fairly typical. The ideal hole in a tree is 3 to 4 metres above the ground, and quite often it is a site that has been previously used. The average internal diameter of the cavity is 20 centimetres and the floor is usually about 10 centimetres below the lower lip of the entrance. Often it is a tight squeeze for the bird to fit into the hole, which may be no more than 2.5 centimetres across and 3.5 centimetres high, and there have been cases of females trapped inside a nest when the tree's growth has closed the entrance slightly. An important requirement above the nest chamber is a chimney, or 'funk hole', in which the female – and later the chicks – can hide if danger threatens.

The female does all the sealing and to begin with she uses mainly mud. Not only does she reduce the size of the entrance, but she also carefully seals off any holes or cracks leading to the nest cavity. Finally, when she can just enter the nest, she begins the last stages of sealing from within, using a mixture of her own droppings and nest lining which comprises bark chips, dry grass and leaves. Quite frequently the male brings millipedes to the nest and it is thought that their sticky exudations are used for sealing purposes rather than food. The material is held in the bill and

A male Bradfield's Hornbill passes food to his incarcerated mate.

***1-4** A Crowned Hornbill chick peers from its nest which has been sealed to a slit. After pecking at the seal it is able to get its head through the opening and then manoeuvre its way out of the narrow hole. **5** Yellow-billed Hornbill eggs, like those of other hornbills, are plain white with a pitted surface.*

patted from side to side so that a thin layer is applied. The final closure of the nest is usually accomplished quickly, sometimes in four hours.

Several *Tockus* hornbills occasionally use rock holes or crevices for breeding, but the only species to do so regularly is Monteiro's Hornbill because of the absence of suitable trees in its arid environment. The disadvantage of a rock hole is that its configuration often means that there is no chimney. These holes are sealed in the same way as tree holes, but sometimes a great deal of material is required; at one nest an area of 600 square centimetres was closed off. Because the task is sometimes prodigious, the male also helps by bringing mud. However, previously used nests generally require little additional sealing.

Male Trumpeter and Silvery-cheeked Hornbills regularly assist with nest sealing and regurgitate pellets of mud for their mates to use. Once the female has finished sealing only a narrow vertical slit remains, through which she can be fed by the male. The slit is smoothed off, the material blending so well with the tree or rock that it is not easily detected unless the male is seen going to the nest with food. It should be noted that several hornbill species have been induced to use nest-boxes, but they still seal the entrance. A box with an inspection panel is one of the best ways to study their breeding biology.

In *Tockus* hornbills the average time from the female's entry into the nest until the laying of the first egg is six days. The male feeds her often at this time and also provides nest lining, continuing to do so during the incubation and nestling periods. At a number of hornbill nests males have brought snail shells to the female and, as this is not suitable lining material, it is possible that they provide a source of calcium for the female while she is laying.

Tockus hornbills normally produce 2-5 eggs, but in good rainy seasons larger clutches may be laid; eight eggs have been recorded in a nest of Monteiro's Hornbill. In this species there are records of five young being raised in years of good food supply, and in one case six chicks were successfully reared. The Trumpeter Hornbill lays 2-4 eggs and the Silvery-cheeked Hornbill, which appears to be an irregular breeder, lays one or two eggs.

The eggs of all species are oval and white with a pitted surface. Incubation starts with the first egg, and it appears from the differing ages of the chicks that the intervals between eggs become longer as the clutch progresses. A clutch of five would probably take at least nine days to complete.

Once the female has sealed herself in, one of the most remarkable features of a hornbill's life takes place. She starts to shed her wing and tail feathers, and by the time the clutch is complete they have all been dropped. The body feathers are moulted gradually and the female is never naked as is sometimes erroneously stated. When a female *Tockus* hornbill is ready to break out of the nest, on average 46 days after laying the first egg, she has regrown her flight feathers. We have no details about the Trumpeter and Silvery-cheeked Hornbills in this respect, but it appears that the former does moult its flight feathers while the latter does not.

The average incubation period for *Tockus* hornbills is 24 days and the eggs hatch in the order laid. The incubation period of the Trumpeter Hornbill has yet to be recorded and that of the Silvery-cheeked Hornbill is approximately 40 days. Hornbill chicks are pink, naked and blind when they hatch and the upper mandible is 2 millimetres shorter than the lower. Within a short time, and before they are seven days old, extensive air sacs develop on their backs and shoulders and these are retained until they are about 14 days old. The function of these sacs is not known, but they may protect a chick from being trampled by the female or its siblings.

By the time *Tockus* hornbill chicks are ten days old their quills are breaking out and their legs are so well developed that they appear out of proportion to the rest of their bodies. Feathers start emerging rapidly after about 21 days and the young are able to fly capably, without prior wing exercises, by the time they leave the nest at about 45 days old. However, the nestling period ranges from 39 to 50 days, depending on the food supply.

A young hornbill characteristically holds its tail up at right angles to its body to protect it from damage and to conserve space in the nest. As soon as it is able it clambers to the entrance and defecates forcibly out of the slit, as the female has done during her incarceration. This maintains sanitation and the nest, in contrast to those of many other hole-nesting species, remains remarkably clean throughout the breeding cycle. The excreta are not ejected only when they are required for sealing the nest-hole.

The *Tockus* female breaks out of the nest once the chicks are half-grown, at about 21 days old. The oldest chicks then reseal the entrance as the female had done and although the result may not be as expert, it is nevertheless quite effective. When the oldest chicks are ready to leave a conflict of interests sometimes occurs; while they attempt to break out the smaller chicks are trying to seal the slit again! The females of Trumpeter and Silvery-cheeked Hornbills remain sealed in the nest until the young are ready to leave.

Throughout this time the male hornbill continues his task of feeding which started even before the female entered the nest. In most *Tockus* species it is a laborious process, as only one food item at a time is brought in the bill tip. Monteiro's Hornbills, however, carry up to three items at a time, and this may reflect the difficulty of finding food in an arid environment. In this species the female may provide more food than other *Tockus* females for the same reason. As a rule the male continues to be the main provider and once the female has left the nest her contribution varies. The chicks beg with a characteristic call and watch through the slit for food deliveries. They greet these with a piercing shriek, as did the female while she was in the nest. Trumpeter and Silvery-cheeked Hornbills, both of which are mainly frugivorous, carry many fruits in their gullets at a time and regurgitate them to the female or young. It was calculated that at a Silvery-cheeked Hornbill's nest 1600 visits were made, and 24 000 fruits delivered, during the entire breeding cycle of about 120 days.

Young hornbills fly strongly from the nest and then remain hidden in the foliage of trees where they are fed by the adults. They do not return to the nest once it has been vacated, remaining initially in family groups. In *Tockus* species they can usually feed themselves about 18 days after they have left the nest.

The hornbills' sealed nests provide security from predation and breeding success is usually high, with most of the eggs hatched producing young. In times of food shortage smaller chicks die of starvation and this serves as a means of brood reduction. The major disadvantage of the sealed nest is that during the long breeding cycle the death of the male could prove to be disastrous, and it is notable that hornbills are extremely wary when they are approaching and leaving the nest. It has recently been observed that some Trumpeter Hornbills breed co-operatively and up to three males have been seen bringing food to the nest. The question remains, however, why other hornbills do not appear to have evolved a similar co-operative breeding system, whereby the female and chicks would still be fed in the event of the male's death.

Ground Hornbills are stately birds and a group of them plodding their way steadily and methodically across the veld in search of food is a fascinating sight. Sadly the pleasure of watching them is often restricted to large conserved areas such as the Kruger National Park; they have declined considerably outside these refuges and are classified as 'Vulnerable' in the South African *Red Data Book* for birds.

Although occasionally found in pairs, Ground Hornbills usually occur in groups of four or five birds. This social unit consists of a dominant pair and helpers, which are usually immatures of various ages. The dominant female is recognisable by the patch of blue on her throat and young birds by the drab colour of their facial skin. By the time the latter are three years old this skin begins to turn red, but they do not acquire full adult coloration until they reach sexual maturity at the age of six. At this stage females would leave the group and wander alone or join up with other females.

A group of Ground Hornbills defends

Recent observations record that some Trumpeter Hornbills breed co-operatively, with up to three males feeding the female and young.

a territory of approximately 100 square kilometres against incursion by other groups. A resonant, booming *boo boo boo-boo*, often delivered in duet and frequently in the early morning, is used as a means of territorial advertisement. On a still day it can be heard 4.5 kilometres away, like the roar of distant lions.

Breeding, which is very irregular, is linked to the rainy season, as is the case with most hornbills. Apart from courtship feeding by the male, as well as other members of the group, there is little by way of display. Only the dominant male mates with the female and copulation takes place on a branch, the male grasping the back of her neck with his bill. Mating often induces such excitement in other males in the group that they may try to interfere and have to be driven off.

The nest is usually in a large cavity in a tree, occasionally in a hole amongst rocks or in an earth bank. In trees it is usually about 4 metres above the ground in a cavity averaging 40 centimetres in diameter. The depth of the cavity varies but usually the female cannot be seen from below. A secure nest site may be used for several breeding seasons.

When the female is ready to breed she may sit in the nest for several hours at a time while her mate and the helpers bring bundles of leaves and grass, mixed with food items, and pass them to her. Once the first egg has been laid she rarely leaves the nest except for half an hour in the early morning, mid-morning and late afternoon. She does not defecate in the nest, waiting instead until one of these breaks from incubation. The male and the helpers continue to feed her and bring lining throughout the incubation period.

Ground Hornbills lay one or two eggs which, like those of other hornbills, are white with a pitted shell. When two eggs are laid, which is the case in three-quarters of the clutches recorded, the second egg is invariably smaller than the first and is laid three to five days later. The incubation period is close on 40 days. Because incubation starts with the first egg, the second chick is much smaller and weighs about 60 grams when it hatches, by which time its sibling has reached 250 grams. It is unable to compete and invariably dies of starvation.

The newly-hatched chick resembles other hornbill chicks, but after a few days its pink skin darkens to blackish and only the gape remains pink. After 14 days the chick's eyes are fully open and its body is covered with spiky quills. Not noted for its beauty, the young hornbill at this stage bears a strong resemblance to a leathery black gargoyle. By the time it is a month old the feathers have emerged from the quills and cover the body, and until the chick is ready to leave the nest at the age of 86 days, its wing and tail feathers develop.

The female remains in the nest with the chick initially and feeds it with food supplied by the male and the helpers. It is not known when she first leaves the nestling alone in the nest, but by the time it is a month old it is left totally alone. She joins the others in searching for food, although the extent of her contribution is not known. The nestling defecates on the lining in the nest, which is so copious that it does not become fouled until the end of the nestling period. The young hornbill has a voracious appetite and devours anything brought for it, greeting the arrival of food with a braying nasal shriek.

When it leaves the nest the young hornbill is integrated into the group and may still receive food from the others until it is two years old, although it can usually feed itself from the age of about six months. Gradually its behaviour evolves from a state of dependence until, when it is about two years old, it is an active helper.

A study of Ground Hornbills in the Kruger National Park revealed that breeding productivity there was low and that on average a group raised one chick every six years. It was estimated that the hornbills needed to have an average lifespan of 28 years to maintain a stable population. Because they are birds that invest in a long-term breeding strategy, any persecution such as shooting or poisoning can result – and has resulted – in local extinctions.

1

2

3

***1**, **2** Ground Hornbills usually nest in a large hole in a tree or dead stump such as this, laying the eggs on a thick bed of leaves. **3** Monteiro's Hornbill regularly nests in rock crevices in its arid habitat, still sealing the entrance in much the same way as other hornbills seal tree-hole entrances.*

Barbets

Barbets are well represented in Africa and occur in a range of habitats from acacia thornveld to evergreen forests. Several of the ten species that occur in southern Africa – such as the Black-collared, Pied and Crested Barbets, and the Yellow-fronted Tinker Barbet – are widely distributed and common in the region. Others, such as the Red-fronted and Golden-rumped Tinker Barbets, are more localised in the forested eastern areas, and the Green Barbet is found only in the Ngoye Forest in Zululand. The Green Tinker Barbet has only been recorded in Mozambique and its nest has yet to be discovered anywhere in its African range.

All barbets are vocal and some of their calls, like the trilling notes of the Crested Barbet and the tin trumpet-like call of the Pied Barbet, are familiar sounds of the bush. Others, such as the Black-collared Barbet, perform regular antiphonal duets, and the tinker barbets deliver monotonous clinking notes, like a distant hammer on a small anvil, for long periods at a rate of 100 to 130 clinks per minute. The slow *chop-chop-chop...* of the Green Barbet gives an indication of where to locate it, but trying to pinpoint the calling bird in the forest canopy can be frustrating.

Barbets are strongly territorial and aggressive, advertising their territories by calling and by swaying, bobbing, flicking their wings and wiping their bills on a branch. Some of these movements may form part of courtship, but the birds do not perform elaborate courtship displays. Calling to each other, and in some cases duetting, strengthens the pair bond, and courtship feeding has been observed in several species.

Although barbets are monogamous, three species – the Black-collared, White-eared and Whyte's Barbets – are usually gregarious and may have helpers when breeding. They roost communally and as many as 11 birds have been seen emerging from a hole. The Green Barbet also roosts communally but it is not known whether there are helpers at the nest. The other species are encountered solitarily or in pairs.

Barbets breed mainly in spring and summer, although some species continue into autumn and a few breed in most months. Sometimes as many as four or five consecutive broods may be raised, in which case the season is inevitably prolonged.

All barbets excavate a hole in a tree for their nests. Usually it is in a dead vertical or slightly sloping trunk or branch, but tinker barbets often prefer the underside of an almost horizontal branch. The diameter of the entrance varies from one species to another; in the case of tinker barbets the hole is a mere 2 centimetres across. The depth of the nest also varies, with larger species making a cavity at least 20 centimetres deep, and often much more. In those species which have been observed, both male and female excavate, taking turns. After the initial horizontal entry has been made the birds excavate downwards with their strong bills until they reach the required depth, removing wood chips in their bills and dropping them away from the nest. Barbets can be attracted to nest in gardens if they are provided with a log of soft wood or a sisal stem which has been attached to a tree in a suitable position. A small hole cut into the wood often stimulates them to start pecking.

Holes are used for both roosting and breeding, and there is often fierce competition for nesting sites which must be defended against other hole-nesting species. White-eared Barbets are tolerant of neighbours and several pairs of this species may nest in the same trunk. The Pied Barbet has diversified and has been recorded roosting in the enclosed nests of swallows, in a Brown-throated Martin's burrow, and in a Sociable Weaver's nest. However, of these, the only evidence of breeding was found in the nests of Greater Striped Swallows. This species has also extended its range into the fynbos habitat of the south-western Cape this century, apparently following the spread of alien tree species, especially Australian acacias, which now provide nest sites. In tandem with the Pied Barbet, the Lesser Honeyguide, which parasitises this species, has also expanded into the south-western Cape.

Barbet clutches are usually three or four eggs, although they may be 2-6. The white eggs are laid on wood chips at the bottom of the nest and no lining is added. Both sexes incubate, sometimes assisted by helpers, and quite often the bird on duty sits for long periods with its head protruding slightly from the entrance. Presumably the temperature in the hole is maintained by this means and the nest is guarded at the same time. Both members of the pair, and helpers too, may roost in the nest, even when there are eggs or young. The incubation period varies from about 12 days for the Golden-rumped Tinker Barbet to close on 18 days for the Black-collared and

1

2

3

1 *Pied Barbets have extended their range into areas pioneered by the alien vegetation in which they nest.* ***2*** *A Crested Barbet keeps its nest clean by removing a mixture of wood chips and droppings.* ***3*** *A Red-fronted Tinker Barbet carries food to its nest.*

White-eared Barbets. The Crested Barbet incubates for 14 days and the Pied Barbet for 14 to 16 days. The periods for the remaining species are not known.

The young hatch blind, naked and pink, with a pronounced thickened 'heel' pad at the bend in the leg, as in woodpeckers. Both parents (and helpers in the gregarious species) brood and feed the chicks initially, but once they are about seven days old they are left alone in the nest, except at night when the entire family occupies the hole. As the end of the nestling period approaches the young come up to the entrance and take turns at watching out of the hole where they are fed by the adults. Faeces are removed and dropped away from the nest, either in a sac or in a mixture of wood chips from the bottom of the hole. As the nestling period progresses the hole may become several centimetres deeper as the bottom is pecked to clean it of droppings.

The nestling period lasts 20 to 25 days for the small tinker barbets and 35 to 39 days for the larger species. The young join their parents to form a family group, and some may become helpers in subsequent breeding attempts. In most species they initially return to the nest to roost, but the duration of their attachment varies.

Honeyguides

Honeyguides have some extraordinary habits, of which the most widely known – and the most misunderstood – is the guiding behaviour from which their name derives. The Greater Honeyguide is the only species which guides on a regular basis, leading humans to bees' nests with a chattering call like a half-empty matchbox being shaken lengthwise. The Scaly-throated Honeyguide is also said to guide, but evidence for this is scanty and the habit is probably poorly developed. Another remarkable aspect of the biology of honeyguides is their ability to digest wax derived from honeycomb or, in the case of the Sharp-billed and Slender-billed Honeyguides, from the waxy exudations of scale insects. Wax is not the only constituent of a honeyguide's diet, however, and nestlings in particular eat whatever is provided.

The breeding behaviour of honeyguides is just as fascinating as other aspects of their biology. Like cuckoos, honeyguides are brood parasites, laying their eggs in the nests of other birds and taking no further part in the breeding cycle. Six honeyguide species, all resident, breed in southern Africa. Four – the Greater, Scaly-throated, Lesser and Eastern Honeyguides – belong to the *Indicator* group, and two small flycatcher-like species, the Sharp-billed and Slender-billed Honeyguides, belong to the *Prodotiscus* group. The Eastern Honeyguide occurs in eastern Zimbabwe and adjacent Mozambique, but virtually nothing is known about its breeding habits.

In *Indicator* honeyguides the call-site is the focal point of courtship activity in the breeding season and, apart from the Eastern Honeyguide, their behaviour is well documented. The male calls for long periods – sometimes up to eight hours – from regular perches, some of which have been in use for more than 20 years by a succession of birds. The best-known call is the Greater Honeyguide's *whit-purr* or *vic-torr*, with the emphasis on the first syllable. The Scaly-throated Honeyguide emits a frog-like trill which rises in pitch, and the Lesser Honeyguide makes a series of 15 to 40 piping *klew* notes, rests for a while, and then starts another series. In some observations three different males of the same species have been known to occupy the same call-site during a breeding season.

The Greater and Lesser Honeyguides also perform display flights, usually stimulated by the presence of a female, in which the winnowing of their wings makes a resonant sound. Chasing behaviour has been observed and presumably forms part of courtship. It appears that in *Indicator* honeyguides females are attracted to the call-site of the male and copulation takes place there. The male may mate with several females on a promiscuous basis.

Courtship of *Prodotiscus* honeyguides takes the form of aerial displays during which the white outer tail feathers and the white patch on the lower back are flaunted. The Sharp-billed Honeyguide makes a metallic *zink* call, and the Slender-billed Honeyguide emits a series of harsh *skeee-aa* calls, sometimes while gliding with raised wings and fanned tail. Both species are usually unobtrusive but their presence is betrayed by their display flights.

The breeding season of honeyguides falls mainly between September and January, matching that of their host species. *Indicator* species parasitise a variety of mainly hole-nesting birds, with the Greater Honeyguide's range of hosts being the widest. The Sharp-billed Honeyguide parasitises small warblers and cisticolas, and the Slender-billed Honeyguide lays in the nests of Yellow White-eyes. The known hosts of honeyguides recorded in southern Africa are summarised opposite, but there can be little doubt that further hosts remain to be discovered.

Honeyguides normally lay plain white eggs but the Slender-billed Honeyguide may lay either blue or white eggs, as does its host, the Yellow White-eye. In a study in Zimbabwe the eggs of the honeyguide were a perfect match to those of the host in 11 out of 12 nests: seven had blue eggs and four had white ones. Only in one nest was there a mismatch; a white egg lying with two blue ones was deduced to be that of the honeyguide. The eggs of the Sharp-billed Honeyguide have yet to be described.

Usually only a single egg is deposited in the host's nest, but sometimes two or more may be found. It is estimated that a Greater Honeyguide would lay 4-8 eggs in a breeding season and a Sharp-billed Honeyguide five eggs. When laying takes place an egg of the host is often punctured or broken, and in some cases removed. An observer saw a Sharp-billed Honeyguide remove an egg of its host on two occasions. If the honeyguide lays its egg before the host has started its clutch, the egg may be removed or broken by the host. As in the case of cuckoos, the best time to lay is when the host species has begun laying.

It appears that when the female lays she works alone, unassisted by the male. Having watched for evidence of her host's breeding activity, she seeks an opportunity to slip in and lay, an operation that is not without danger, especially considering the formidable bills of barbets. One observation relates how a Lesser Honeyguide was caught in a Black-collared Barbet's nest-hole by one barbet and attacked by its mate as it tried to escape. Fortunately for the honeyguide some of its tail feathers came out in the barbet's bill and it was able to break free. Sometimes, however, the honeyguide is killed.

The incubation periods of honeyguides are difficult to obtain with accuracy, but generally, like those of cuckoos, they are shorter than the host species'. Of the incubation periods known, that of the Scaly-throated Honeyguide is about 18 days, that of the Lesser Honeyguide has twice been recorded as 12 days and once as 16½ days, and that of the Slender-billed Honeyguide is 13 days or less.

Newly-hatched honeyguides are pink, naked and blind, and like barbets and woodpeckers, they have 'heel' pads. Their bills are stubby and they have raised nostrils.

1 *Although blind, this Greater Honeyguide chick grasps anything it feels with its sharp bill hooks.* ***2*** *A well-grown Lesser Honeyguide nestling calls for food from its Pied Barbet host.*

Their most characteristic feature, however, is the sharp bill hooks at the tips of the upper and lower mandibles, which are used to kill the young of their host. Remarkably, where two honeyguide chicks of the same species hatch in the nest together, both survive. This occurrence has been observed on two occasions, once when two Slender Honeyguide chicks hatched together, and again when two Greater Honeyguide chicks were found in a bee-eater's nest. In *Indicator* species the hooks are retained until the nestling is about 14 days old, but in *Prodotiscus* honeyguides they disappear by the time it is seven days old.

In a bee-eater's nest a single Greater Honeyguide was heard to produce a begging call that mimicked *two* bee-eater chicks to provide an extra stimulus for its foster parents to feed it. Greater Honeyguide nestlings have also been known to mimic the calls of Red-billed Woodhoopoe offspring. Towards the end of the nestling period the young of *Indicator* honeyguides may come up to the entrance hole to be fed, and in doing so may lean right out. At a Pied Barbet's nest there was evidence that the barbets reduced their visits when this happened to induce a Lesser Honeyguide to leave the nest.

The nestling period is relatively long for all honeyguide species. The Lesser Honeyguide leaves the nest after 38 days, staying longer than would the chicks of one of its main hosts, the Black-collared Barbet. The Greater Honeyguide remains in the nest for about 30 days, the Scaly-throated Honeyguide for between 27 and 35 days, the Slender-billed Honeyguide for about 22 days, and the Sharp-billed Honeyguide for 17 to 21 days.

Information about the post-nestling period is scanty, but it appears that the young honeyguide does not return to the nest of its host to roost and may become independent within a short while. Greater Honeyguides may be independent ten days after emerging from the nest, and at three Olive Woodpecker nests young Scaly-throated Honeyguides were fed by their foster parents on the day they left and not thereafter.

It requires dedication and patient observation to study the breeding habits of honeyguides, but in time further aspects no doubt will be elucidated. Why, for example, did a Lesser Honeyguide investigate a nest several times when it contained a Lesser Honeyguide chick about to leave? Was it checking on progress with a view to parasitising the Pied Barbet's subsequent breeding attempt, or was it showing an interest in its own chick?

HOST SPECIES OF HONEYGUIDES	
GREATER HONEYGUIDE	Woodland, Brown-hooded, Grey-hooded, Striped and Pygmy Kingfishers; Little, White-fronted, Carmine, Olive, Böhm's and Swallow-tailed Bee-eaters; Hoopoe; Red-billed and Scimitar-billed Woodhoopoes; Black-collared, Pied and Crested Barbets; Golden-tailed and Knysna Woodpeckers; Red-throated Wryneck; White-throated, Red-breasted and Greater Striped Swallows; Banded Martin and Brown-throated Martin (unconfirmed); Southern Black Tit; Ant-eating Chat; Glossy and Pied Starlings; Scarlet-chested Sunbird; Grey-headed and Yellow-throated Sparrows
SCALY-THROATED HONEYGUIDE	Black-collared Barbet; Cardinal, Golden-tailed and Olive Woodpeckers
LESSER HONEYGUIDE	Striped Kingfisher; Little Bee-eater; Red-billed Woodhoopoe; Black-collared, Pied, White-eared and Crested Barbets (other barbets likely); Golden-tailed Woodpecker; Red-throated Wryneck; White-throated Swallow; Glossy and Plum-coloured Starlings; Yellow-throated Sparrow
EASTERN HONEYGUIDE	No hosts confirmed; circumstantial evidence of White-eared Barbet.
SHARP-BILLED HONEYGUIDE	Bleating Warbler; Wailing Cisticola and Neddicky (other warbler/cisticola species likely)
SLENDER-BILLED HONEYGUIDE	Yellow White-eye

Woodpeckers and the Red-throated Wryneck

Specialist feeders on insects and their larvae which, in many cases, they excavate from dead or rotten wood, woodpeckers have a number of adaptations which enable them to obtain their food. Their skulls are specially designed to withstand the shock of pecking against wood, and they have remarkably long tongues which in some species are barbed to impale insects and their grubs, and in others are sticky to mop up ants and their eggs and larvae. They also have strong claws and feet, with two toes pointing forwards and two backwards, which enable them to clamber on tree trunks with impressive agility.

There are nine species of woodpeckers in southern Africa, of which the Speckle-throated Woodpecker is often considered a subspecies of Bennett's Woodpecker, and the Little Spotted Woodpecker is found only in eastern Zimbabwe and adjacent Mozambique. The Cardinal Woodpecker, however, is widely distributed throughout the region. Most species occur in woodland, but only the Olive Woodpecker is a regular forest inhabitant. The Ground Woodpecker, one of only a few wholly terrestrial woodpeckers in the world, is endemic to southern Africa and, not unexpectedly, its breeding behaviour differs in several respects from that of other woodpeckers. Wrynecks are closely allied to woodpeckers, but only one species, the Red-throated Wryneck, is found in our region.

With the exception of the Ground Woodpecker, woodpeckers are sexually dimorphic with, among other differences, the males having more red on the head. They are monogamous, aggressively territorial and highly vocal. Some species also drum on hollow branches as a means of proclaiming their territory, or to keep in contact with their mates. As a rule they are found in pairs, or sometimes in small family groups after breeding. The Ground Woodpecker again differs in that it often occurs in small groups of three to six and, because its terrestrial habits expose it to predation, a sentry keeps watch while the rest of the group forages. When breeding, however, it is

Like most woodpeckers, Cardinal Woodpeckers excavate their own nest-holes, usually in dead wood.

***1** A Ground Woodpecker at the entrance to the tunnel it has dug into a roadside embankment.* ***2** A Red-throated Wryneck feeds its chick. Unlike woodpeckers, this species does not excavate a nest-hole, using instead a natural hole or one created by a barbet or woodpecker.*

strongly territorial, and there is no evidence to support statements that there are sometimes helpers at the nest.

Little is known about the courtship behaviour of woodpeckers, but the male and female call to each other to maintain the pair bond. Bennett's Woodpeckers display by raising and flapping the wings, probably to threaten other birds in territorial encounters, but possibly also as part of courtship. Observations of the male running around the tree trunk in front of the female and calling as he does so may relate to courtship. The male Ground Woodpecker approaches the female with wings spread and quivering, then he assumes an upright posture and spreads them wide; she responds with quivering wings and copulation ensues.

Most woodpeckers breed in spring and summer from August to December, but some, like the Cardinal Woodpecker, may breed in most months. In South Africa the Bearded Woodpecker nests in winter, from May to July, but in Zimbabwe it continues until as late as November.

Both sexes of arboreal woodpeckers take turns to excavate a nest-hole in a tree trunk or branch, usually in dead wood. Bennett's Woodpecker is unusual in that it frequently makes use of an existing hole, either a natural cavity or the hole of a barbet or another woodpecker. The entrance is usually round, but that of the Bearded Woodpecker is oval and larger than other species' entrances. For this reason its cavities are often used by other hole-nesters such as rollers, starlings and the African Scops Owl. The eggs are laid on wood chips at the bottom of the hole.

The Ground Woodpecker excavates a tunnel 0.5 to 1 metre long into a bank and at the end of it a chamber about 15 centimetres in diameter. Several holes may be found in the same bank, but only one is used at a time. Rarely a tunnel may be excavated in a termite mound.

A woodpecker clutch comprises 2-4 eggs which are plain white. At a Cardinal Woodpecker's nest under observation they were laid in the early morning on consecutive days. Both sexes incubate but the male always sits overnight. The Cardinal Woodpecker incubates for 13 days and the Olive Woodpecker for 15 to 17 days. Little is known about the incubation periods of other species, although it seems that most are about 14 days long.

When the young hatch they are naked, pink and blind and have prominent 'heel' pads. The eggshells are not removed and disintegrate in the nest. Both sexes take turns in brooding the chicks until they are about seven days old, the male again taking the overnight shift. At a Cardinal Woodpecker's nest the male continued to brood overnight until the two chicks were 21 days old. Both parents feed, stimulating the small chicks to beg by tapping on the side of their gapes. Ant-eating species pass food to their young by regurgitation, whereas others bring individual food items, such as grubs, one at a time. The adult pecks the anus of the chick to stimulate it to defecate and then removes the faecal sac, dropping it away from the nest.

Once their feathers start to develop the chicks clamber up to the nest entrance to be fed, and may even sleep clinging to the inside wall of the cavity. For those species where information is available, the nestling period is close on 27 days, shorter than that of most barbets, which also excavate their own nest-holes. Having left the nest, the young seldom return to roost in it as barbets do. They usually remain with their parents in a family group and, in a brood of two, each parent may take over the care of one chick. Ringed Cardinal Woodpeckers were observed accompanying their parents two months after leaving the nest. In some species family groups may remain together until the following breeding season. It is a feature of the Cardinal and Bearded Woodpeckers that the young, irrespective of sex, have red on their heads and resemble the male, whereas the young of other species look like the female.

Red-throated Wrynecks resemble woodpeckers superficially but are cryptically coloured, apart from the red throat. They have very long, sticky tongues which they use to feed on ants and their eggs, larvae and pupae, the main component of their diet. Their breeding biology is similar to that of woodpeckers and they, too, are monogamous, territorial and vocal. The name 'wryneck' derives from one of their main displays, in which the bird leans forward, with breast exposed and tail cocked, and sways its head from side to side. This display is used both in aggressive territorial encounters and as part of courtship.

Red-throated Wrynecks breed between August and February, with a peak in October, and use a natural cavity or the old nest of a barbet or woodpecker instead of excavating their own holes. One of their main competitors for nest-holes is the Crested Barbet, which may evict breeding wrynecks from a hole that it originally excavated. Sometimes wrynecks nest in a hollow metal fence post, and they readily accept a suitably placed nest-box.

Usually three, but up to five, cream-coloured eggs are laid. Both sexes incubate, although at one nest observed one of the pair did the greater share. They begin incubating before the clutch is complete – with the result that chicks may be different sizes – and continue for 13 to 15 days. The chicks hatch naked, pink and blind, and are brooded continuously until they are ten days old. The adult birds take turns at brooding and bringing food, removing faeces after visits to the nest. After ten days the young are brooded only at night by one parent of unknown sex.

From the age of 21 days they peer out of the nest-hole and may emerge for short excursions, even onto the ground. Large nestlings threaten predators with an intimidating display that resembles a striking snake. They extend the head upwards with partially open bill, then suddenly retract it, hissing loudly through the bill which is by now fully open. After another four or five days the young leave the nest and are able to fly strongly within a day or two. They are not dependent on their parents for long and may be feeding themselves within two weeks. The parents may start a second brood a little over a week after the first has departed.

A male Olive Woodpecker perches at its nest-hole, its strong claws evident. Both male and female woodpeckers excavate the hole, incubate the eggs and tend the chicks.

The African Broadbill

The only broadbill species in southern Africa, the African Broadbill inhabits evergreen and riparian forests and deciduous thickets. In areas of Kwazulu-Natal where coastal forest has been decimated, for example around Durban, its numbers have been severely reduced. It is a flycatcher-like bird that hawks insects from a perch, but also forages on branches and sometimes on the ground. Small, drab and unobtrusive, it would be easily overlooked were it not for its territorial display. Performed solitarily by both sexes but mainly the male, the display is remarkable; the broadbill flies out from its perch on an elliptical course for about a metre and returns to its starting point. As it flies it produces a vibrating *purrrup* sound with its wings and raises its back feathers to reveal a white patch which is concealed when it is perched. It may make flights about once a minute and continue doing so for some time, especially in response to displaying neighbours. The sound, which carries for a considerable distance, is not easy to pinpoint, but a displaying bird may be attracted to a tape-recording of it.

Another display described, which was presumed to be courtship, involved two birds perched about 45 centimetres apart and flicking their wings like Familiar Chats. Then both hung upside down from their perches for about 30 seconds, continuing to flick their wings, and emitted 'loud klaxon notes'.

The breeding season extends from September to February, but with a marked November peak in Zimbabwe. The nest, built by both birds, is an upright oval in shape with a triangular side entrance over which there is a slight 'porch'. It is characteristically suspended from a thin horizontal branch and is not usually more than 2 metres above the ground in thick vegetation. The material is slung over the branch, and at the bottom of the nest long strands hang down so that the entire structure resembles debris caught in a tree. In this respect it is very like the nest of another forest species, the Olive Sunbird. Constructed with 'old man's beard' lichen or with other dry matter such as grass, fibres, leaves or bark strips, the nest is lined with soft dry material.

Up to three white, elongated and slightly glossy eggs are laid. The incubation and nestling periods are unrecorded and, apart from the fact that the female incubates, virtually nothing is known about the breeding cycle of this elusive species.

__1__ An African Broadbill's nest made entirely of 'old man's beard' lichen has strands hanging below it that make it look like debris caught in the tree. __2__ The African Broadbill usually builds its nest low down in the understorey of its forest habitat.

The Angola Pitta

Only one pitta, the Angola Pitta, occurs in southern Africa and it is an intra-African migrant found in Zimbabwe and adjacent Mozambique, with occasional vagrants in South Africa. This colourful but elusive bird inhabits lowland evergreen forest and deciduous thicket, including such habitats in the middle Zambezi Valley. Extremely secretive, it is seldom seen, although its habit of migrating at night has resulted in dead or injured birds being found after they have flown into lighted buildings. In Zimbabwe the earliest arrivals are in late October and the birds depart again in April. The only nests recorded there have been found in November and December.

The Angola Pitta resembles a stocky, short-tailed thrush and feeds on the ground as a thrush does, but its displays, nest and nestlings are quite different. Like that of the African Broadbill, its display is impressive, and could be for either territorial or courtship purposes, or both. A pitta uses several regular display perches on stout branches about 5 metres above the ground where there is no overhanging foliage. It calls and displays simultaneously, mainly in the early morning and evening, and is most active in the fortnight before laying. From its perch it flies 30 to 40 centimetres up with a trilling *ffrrueeep* call, then parachutes back to its perch with shallow wing-beats, fluffing out its brilliant scarlet belly feathers as it does so. It also displays these feathers by swaying from a horizontal to a vertical position while perched.

The pitta's nest, like that of a coucal, is an untidy, ball-shaped structure with a side entrance. It is made of twigs and dead leaves, but the exterior is encased with thin sticks. The roof of the nest is flat and fringed with similar sticks in such a way that they resemble the upturned brim of a hat. The entrance is neatly formed and the interior is lined with soft dry material such as rootlets, tendrils and leaves. The nest is placed 2 to 8 metres above the ground, often in a thorny tree, and from a distance it resembles a cluster of debris caught on a branch.

The Angola Pitta's nest is an untidy, ball-shaped structure of twigs and leaves, with a side entrance.

Usually three, but sometimes one or two, eggs are laid, and they are round and creamy white with scattered dark blotches and underlying scrolls of grey. A week-old nestling had pitch-black skin with wispy black down and was difficult to detect in the dark interior of the nest. The most remarkable feature of the chick was its bill, which had a 1-millimetre-wide band of orange running from its base to the tip and orange swellings at the base of the mandibles. The bill was encircled in the middle by a band of black and the gape was orange. There was also an orange ring around the cloaca when it was distended. The colour of the bill and gape and the orange ring around the cloaca presumably enable the parent to feed the chick in the dark nest and also to locate the faecal sacs for removal. Nothing is known of parental care during the breeding cycle and the incubation and nestling periods are unrecorded.

Larks and Finch-larks

Southern Africa is rich in lark species, several of which are endemic and some – Rudd's, Short-clawed, Dune, Red, Botha's and Gray's Larks – have highly restricted distributions. Relatively little is known about the Short-clawed, Botha's and Red Larks, whose nests have been discovered as recently as the 1980s, whereas the breeding behaviour of the Dune Lark, which is confined to the dunes of the Namib Desert, has been thoroughly studied despite the bird's inhospitable environment. Of the 25 species only one, the Dusky Lark, is a non-breeding intra-African migrant. The breeding species occur in habitats such as grassland and bushveld, as well as the semi-desert or desert environments of the Karoo, Kalahari and Namib. In fact, the greatest diversity of species is to be found in the arid western regions.

Generally cryptically coloured and difficult to identify, larks are also very variable in coloration, and their plumage usually matches the soil type on which they live. The sexes of most species are similar, with the exception of the finch-larks, the males of which appear striking in contrast to the drab females. Many species are resident but others, especially the granivorous species such as finch-larks, are highly nomadic and appear suddenly in an area after rain.

Because they are terrestrial and normally unobtrusive, larks need to advertise their presence with territorial displays. In some species displays may occur throughout the year, but all larks intensify their display behaviour in the breeding season. In this respect, and for the same reasons, they are similar to the cisticolas. Most larks display aerially as well as from prominent elevated perches, and vocalisations are a major part of their behaviour.

Each species has its own distinctive means of territorial advertisement. The Melodious Lark flies overhead for 25 minutes or more, having ascended to a height of about 20 metres on fluttering wings like a bumble-bee. In its song it incorporates imitations of other birds, of which 57 species have been identified. Imitations of birds, and even once of a rock hyrax, are a feature of the calls of several lark species, notably the Sabota and, to a lesser extent, the Rufous-naped Larks. The Monotonous Lark, normally difficult to see, repeats the same phrase over and over, even continuing into the night. The Clapper Lark, which is also active day and night, has one of the most characteristic of all displays; it flies high into the air, producing a clattering sound with its wings, then drops down with a drawn-out whistle that rises in pitch. The display of Gray's Larks was undescribed until an observer in the Namib Desert used a night scope to observe the birds performing in the sky in darkness, mainly for about an hour after sunset and for two hours before dawn. The finch-larks and species such as Pink-billed, Sclater's and Stark's Larks, being gregarious and nomadic, do not have displays that are as striking as many of the resident species. Male Grey-backed Finch-larks flutter overhead with dangling legs and emit soft chirping notes, and male Black-eared Finch-larks hover with a butterfly-like flight without dangling their legs.

Little is known about the courtship behaviour of larks. In some species the male postures in front of the female; a pair of Spike-heeled Larks has been observed in a crouched position with quivering wings and tails spread. Stark's Larks raise their crests, presumably as a part of courtship. It is possible that the territorial displays of larks serve a dual function as part of courtship, but this requires further investigation.

In some areas, for example the winter-rainfall region of the south-western Cape, larks breed regularly in spring and early summer. However, in the arid regions which many of them inhabit, breeding is triggered by rainfall and sometimes occurs within a week of a good downpour. As rainfall is highly variable from year to year and from region to region, it is virtually impossible to define breeding seasons. There are also some anomalies: the Chestnut-backed Finch-lark, for example, breeds mainly in the dry season, and the Dune Lark's breeding season peaks in January and February, independent of rainfall.

Larks' nests are cup-shaped and set into the ground so that the rim is level with, or very slightly higher than, the surrounding substrate. They are built in a natural hollow such as an animal's hoofprint, or the lark excavates the hollow itself. The cup is made of grasses and lined with finer material and flowering grass heads. A large group of species – Melodious, Monotonous, Rufous-naped, Clapper, Flappet, Fawn-coloured, Sabota, Rudd's, Long-billed, Karoo, Dune and Red Larks – build a dome-shaped bower over the back of the nest, fashioning it from living grass surrounding the nest or from loose pieces of grass. Some domes are substantial while others may be rudimentary or even absent altogether, and the Melodious Lark adds a raised 'doormat' of grass at the entrance to its nest. Originally all these species were classified in the genus *Mirafra*, but some have recently been reclassified, and the presence of a dome is no longer thought to be diagnostic of this genus.

The Long-billed Lark frequently has no dome, but it places stones – often quite large ones – around the front part of its nest. Many other species build a rim of small sticks or stones at the front of the nest, including finch-larks, whose nests are a small

1 

 2

***1** The Karoo Lark is one of several lark species that builds a grass hood over the back of its nest. In this case the nest is well lined with downy plant material. **2** The Red Lark also builds a hood over its nest, as was discovered when this nest, the first ever found, was located in 1986.*

THE LARK THAT LAYS ONE EGG

Sclater's Lark is a nomadic species which has a restricted distribution in the Karoo, Bushmanland and southern Namibia. Until recently its biology was little known, but observations in the Brandvlei district in Bushmanland have revealed some fascinating aspects of its life.

This species breeds on areas of exposed shale, placing its nests completely in the open. The birds excavate a hollow in the hard ground and build a small neat cup measuring 5 centimetres across and 3 centimetres deep. The rim is level with the ground and the perimeter is surrounded by stones, sometimes four concentric rings; at two nests observed the larks had collected 130 and 175 stones.

Breeding is highly synchronised and all recorded clutches contained a single elongated egg. The reason for a single egg being laid has not yet been established. The chick is no more advanced than other larks and leaves the nest after ten to 11 days, like many other species. Perhaps the single egg enables the birds to breed opportunistically, although other species such as finch-larks also do so and lay two or three eggs. The emphasis seems to be on the production of quality rather than quantity, but why Sclater's Lark should be alone in this strategy remains puzzling.

The exposed situation of the nest means that the larks have to cope with heat that is often intense. Unlike most other larks, the parents share incubation equally, changing over about every 15 minutes. When the chick hatches they take an equal share of brooding and feeding it, never leaving it unattended in the heat. The bird at the nest stands over the chick with its bill half open, wings slightly away from its body and feathers raised, all in an effort to keep cool. The main food is harvester termites and the parents alternate at the nest each time food is brought. If there is a breeze the chick stretches up its neck and opens its gape to keep cool, behaviour that is commonly adopted by most lark nestlings.

A particularly remarkable aspect of the breeding behaviour of Sclater's Larks is their probably unique habit of placing stones in the nest when the chick hatches. The effect of this is to disrupt the rounded outline of the nest. The chick's tufts of down also contribute to its camouflage, making it appear more like a clump of grass seeds caught in a hollow than a bird.

Sclater's Lark occurs in very similar habitat to that of the Double-banded Courser, and it is interesting that both species lay a single egg and employ similar strategies to counteract the harsh conditions in which they breed ●

1

2

__1__ A bare gravel plain in Bushmanland is the typical nesting habitat of Sclater's Lark. __2__ This species characteristically builds a rim of stones around its nest. It is the only southern African lark which invariably lays a single egg. __3__ When the chick hatches its camouflage is sometimes enhanced by stones being placed in the nest to break up its round outline. __4__ The chick needs to be constantly shaded in its exposed nest site, and the parents take turns to feed it and shield it from the sun.

3

4

neat cup. The Chestnut-backed and Grey-backed Finch-larks often build a rim of earth clods or stones, and the nest of the Black-eared Finch-lark is characterised by having earth mixed with spiderweb around its rim.

Most larks place their nests against a tuft of grass, rock or small bush and in arid areas orientate them so that they face east, south-east or south and are in shade most of the day. In contrast some species, notably Sclater's Lark, place their nests completely in the open. It appears that the females do all or most of the nest building, although male finch-larks may assist by bringing material. The Black-eared Finch-lark brings offerings of earth and spiderweb for his mate to place in position, even during the incubation period.

The normal clutch is two or three eggs, sometimes four, which are laid at daily intervals, and there is evidence that the two-egg clutch of species such as finch-larks may increase to three in years of good rainfall. The Dune Lark never lays more than two eggs, and Sclater's Lark is unique amongst southern African larks in that it invariably lays a single egg. Most eggs are white with brown spots and blotches of varying intensity, sometimes concentrated in a zone around the large end. Occasionally aberrations occur, as in the case of a Thick-billed Lark's nest which contained two pure white eggs.

For those species where information is available, it appears that only the females incubate. However, finch-larks are again an exception, and the males, especially the Black-eared Finch-lark, do a substantial share. This is something of an anomaly as males are more conspicuous than females. For most species where the incubation period has been recorded it averages 12 days. The Dune Lark incubates for 13 to 14 days, and Sclater's Lark for no less than 13 days.

When the eggs hatch the eggshells are removed. Newly-hatched chicks have pink or blackish skin depending on the species, the gape is orange or yellow and there is a pattern of spots on the tongue. Tufts of down cover the back, serving very effectively as disruptive camouflage. Later, when the feathers develop, they have pale buff edges, so that the chicks are still well camouflaged in the nest.

Both parents feed the chicks, either by regurgitating seeds or by carrying insects in the bill, depending on the diet. Quite often species that are usually granivorous change to a diet of insects when they are feeding chicks. The adult birds are very cautious about betraying the whereabouts of their nest and usually approach it by landing some distance away and then walking to it. A pair of Red Larks were so wary that they approached from a different spot each time they brought food, and it took two experienced observers three days to find the nest. Some species have been seen performing injury-feigning displays to lure predators from the nest, and in one instance a Spike-heeled Lark dragged its wings and spread its tail to reveal the white pattern. Both parents remove faecal sacs and drop them away from the nest. Larks were not known to be co-operative breeders, but there are two cases on record of Spike-heeled Larks feeding chicks in a nest with the aid of a single helper. This species is often encountered in small groups and it is possible that the presence of a helper, or helpers, will prove to be the norm.

The nestling period of larks is usually brief because the chicks leave the nest well before they can fly. Some species leave when only seven to ten days old, often having been called out of the nest by their parents. An early departure from the nest may help to reduce predation since the chicks are no longer at a single spot where the whole brood could be eaten. Very little is known about how long chicks remain dependent on their parents, but Dune Larks, which leave the nest after 12 to 14 days, remain with them for a month. It is estimated that young larks would probably fly at the age of 14 to 21 days.

1 *The Long-billed Lark builds up the front of its nest with stones that are often quite large.* ***2*** *Typically of larks, the Spike-heeled Lark fashions a cup-shaped nest set into the ground, but does not add a hood.* ***3*** *The first Botha's Lark nest was discovered as recently as 1982.* ***4*** *A Stark's Lark nest has been placed against a grass tuft so that the eggs are shaded for much of the day.* ***5*** *Chestnut-backed Finch-larks are unusual in that they breed mainly in the dry season.* ***6*** *Grey-backed Finch-larks often build up a rampart of stones around their nests.* ***7*** *The male Black-eared Finch-lark takes on a substantial share of incubation, despite being more conspicuous than the female.*

Swallows and Martins

Fifteen species of swallows and martins breed regularly in southern Africa, and the House Martin, a migrant from the Palaearctic, has been recorded breeding sporadically in five scattered localities since 1892, the last record dating from 1969. Seven of the 15 breeding species are intra-African migrants, and the Pearl-breasted Swallow is resident in parts of its range but is a migrant to the south-western Cape.

Swallows are territorial and the majority are solitary breeders, the exceptions being the South African Cliff Swallow and Brown-throated Martin, as well as the House Martin when it occasionally breeds. They are also monogamous, although there is a record of a male and two females at a Red-breasted Swallow's nest, each female having contributed four eggs to the clutch. The birds may return to the same nest sites over a period of many years; Blue Swallows have been known to breed on the same building for 20 years. In all species except the Blue Swallow the sexes are similar.

There are usually no obvious displays of courtship, although pairs have been observed perching together and calling, the male puffing out his throat feathers as he does so. Some aerial manoeuvres may form part of courtship and male saw-wing swallows pursue females with stiff wing-beats. The male Blue Swallow, identifiable by his longer outer tail feathers, may puff out his white flank feathers while perched. In flight he flutters after the female, then glides above her and bends his head down until he almost touches her bill with his. In most species occupation of the nest site establishes the pair bond.

The breeding season is not clearly defined but in general lasts from about August to March or April. It may be protracted by the raising of multiple consecutive broods, up to three or even four in some species. In the south-western Cape the breeding season is more defined and is mainly in spring and summer from about September to January. In the tropics species such as the Wire-tailed and Lesser Striped Swallows may breed in all months. The Grey-rumped Swallow is an exception in that it nests

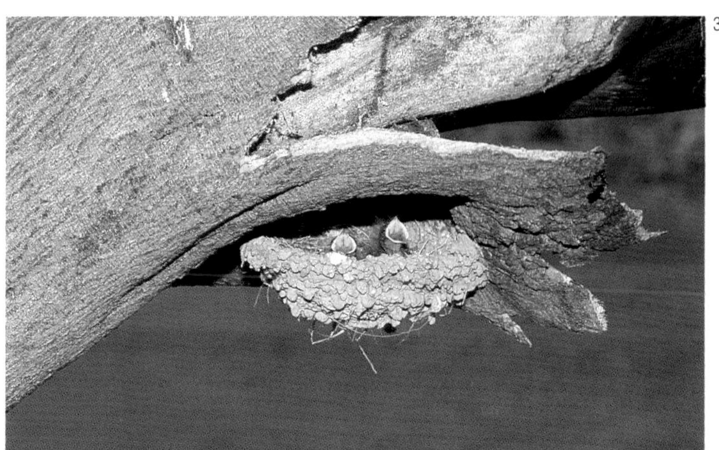

1 *A Pearl-breasted Swallow feeds its chicks, a duty carried out by both sexes.* ***2*** *Before the advent of man-made structures natural sites such as the underside of branches were undoubtedly used by many swallow species, including the Lesser Striped Swallow.* ***3*** *A Wire-tailed Swallow's nest has also been placed under a branch, an unusual site as this species has adapted well to using man-made structures.*

1 The White-throated Swallow is one of the species that builds a cup-shaped mud nest on a vertical surface, in this case the side of a road culvert. The nest is well lined with soft materials. ***2*** *The Blue Swallow still builds mainly in natural sites and often underground, such as on the ceiling of an antbear's burrow.* ***3*** *Secure nest sites may be used again and again, as this one on a farmhouse verandah has been by Rock Martins for many years.*

mainly in the dry winter months so as to minimise the risk of its nest burrow being flooded. Nest-building may often be triggered by rain which makes mud available.

There are three main nest types: a half cup of mud pellets placed against a vertical or sloping surface; a hemispherical, retort-shaped nest of mud pellets under a horizontal surface; and a hole in a bank or in the ground, usually excavated by the birds themselves. Cup-shaped nests are built by White-throated, Blue, Wire-tailed and Pearl-breasted Swallows, and by the Rock Martin. Hemispherical nests, with spouts of varying lengths, are constructed by Red-breasted, Mosque, Greater Striped, Lesser Striped and South African Cliff Swallows, and by the House Martin. Grey-rumped Swallows utilise rodent burrows in flat ground and Brown-throated and Banded Martins, and Black and Eastern Saw-wing Swallows nest in holes in banks.

Swallows that build mud nests traditionally nested on cliffs, on eroded riverbanks, in antbear burrows and under horizontal branches of large trees. These days the proliferation of man-made structures, especially road bridges and culverts, has resulted not only in a switch from natural sites but also to an expansion into areas where nest sites were previously unavailable. The South African Cliff Swallow, for example, is now found as far south as the Calvinia district in the Karoo.

Both sexes build the mud nests,

applying the moist pellets of mud, sometimes mixed with grass, one at a time. Construction usually starts with a bracket of pellets and then expands gradually outwards in layers. Those species that build retort-shaped nests start by constructing the bowl and then progress to the tunnel. Finally the nest is lined with soft materials such as feathers, grass, wool and hair.

The rate of construction varies considerably. A Rock Martin's nest that probably contained close on 1000 pellets took 40 days to build, Pearl-breasted Swallows make a new nest in 21 to 28 days, and two Red-breasted Swallow nests took 13 and 16 days to complete. South African Cliff Swallows may use between 1300 and 1800 pellets in their nests and complete them in as little as five to seven days, a prodigious feat in which pellets are sometimes added at a rate of one every minute. However, because swallow nests often last well, they may be re-used for successive seasons and merely repaired as necessary.

Nests are built in a wide variety of situations such as under bridges and in culverts, in outbuildings on farms, under the eaves of houses and on water towers. The Blue Swallow is one of the species which has more specific requirements and it breeds mainly below ground in sink-holes, antbear burrows and old mine workings, but also in eroded banks of dongas and streams and occasionally on buildings. It constructs its nest with a mixture of mud and grass which is applied evenly and not in pellets as in the case of other swallows. Sadly this species has lost much of its habitat to afforestation in the last 50 years and the only place where its future now seems secure is on the Nyika Plateau in Malawi. Another species that uses mainly natural sites is the Mosque Swallow which breeds in holes in trees, mainly baobabs.

The Red-breasted Swallow breeds almost exclusively in culverts less than a metre high and its enclosed nest is thus rarely usurped by the fast-flying White-rumped Swift which finds access to such situations difficult. This swift is an aggressive species which sometimes takes over the enclosed nests of swallows and even evicts half-grown chicks of the victimised species. Little Swifts, Ant-eating and Mocking Chats, and House and Cape Sparrows may also usurp enclosed swallow nests.

Grey-rumped Swallows nest in level ground in rodent burrows and line the chamber at the end with soft material. Brown-throated and Banded Martins build their

A CONDOMINIUM OF SWALLOWS

South African Cliff Swallows are highly gregarious and nest in tightly packed colonies of a few dozen to 900 or more nests. Breeding activity commences soon after the swallows arrive at their nest sites in August, depending on the availability of mud after rain. An unmated male builds a small platform on which to cling and then attracts a mate by quivering his wings. Once paired, the birds recognise each other by their voices, a feature which remains one of the marvels of colonial breeders.

The swallows build their retort-shaped nests mainly under road bridges but also on buildings, water towers and cliffs, often siting them over water. Those birds that have undamaged nests from the previous season breed earliest, while others are still repairing their nests or building new ones. Although the nests are so close to each other that they touch, each spout points in a different direction. Because of the tightly packed nature of colonies, pairs protect only the immediate area of the nest entrance, so that spacing is based on pecking range.

The eggs are laid at night or in the early morning at 24-hour intervals. The clutch is usually three eggs, but nests sometimes contain four as a result of parasitism by another female in the colony. In one sample of 177 nests 16 per cent were found to be parasitised. However, parasitism is counter-productive, because the swallows normally are unable to raise more than three young. The male shares incubation despite the fact that he lacks a brood patch, and each bird sits for an average of nine minutes. The incubating swallow leaves the nest only when its mate clings to the entrance, thus indicating its willingness to take over. The young are fed by both parents and when they first fly they do not return to the nest for four days.

One disadvantage of colonial breeding is the presence of ectoparasites, especially in nests which are intact from the previous season. A tick, a flea and a louse-fly are specific to South African Cliff Swallows, and the mite *Macronyssus bursa* may cause the deaths of many nestlings.

After breeding the swallows migrate to Zaïre, returning to their colonies the following season to repeat the cycle of squabbling over nest sites, building nests and rearing the young that will perpetuate succeeding generations in their condominiums ●

The tightly packed nests of South African Cliff Swallows under a road bridge are pecking distance apart.

***1** A Lesser Striped Swallow constructs its hemispherical nest under the roof of a tent. **2** Sometimes nest architecture goes awry, and this double nest of a Greater Striped Swallow has three entrances.*

1 *Brown-throated Martins nest colonially in sandbanks, excavating tunnels up to 60 centimetres long.* ***2*** *Grey-rumped Swallows, on the other hand, breed solitarily, utilising rodent burrows in flat ground.* ***3*** *Black Saw-wing Swallows are also solitary breeders, but they nest in holes in banks.*

nests in banks, excavating tunnels about 45 to 60 centimetres and up to a metre long respectively. Occasionally they use the abandoned burrow of another bird such as a kingfisher, bee-eater or Pied Starling. The martins line the chamber at the end of the tunnel with soft materials such as grass, rootlets and feathers. Whereas the Brown-throated Martin nests in colonies of between ten and 100 or more pairs, the Banded Martin's nest is usually solitary, although a few pairs may occasionally breed in the same bank. The two saw-wing swallows nest solitarily in banks, often roadside cuttings, and line their nests mainly with lichens picked off branches.

Some retort-shaped nests have interesting architectural deviations with two or even three tunnels, or a tunnel that is curved to give the birds access in an enclosed situation. Some species show no lack of ingenuity and may be remarkably persistent in their attempts to build in a difficult site. Occasionally aberrant nests are found, such as that of a Rock Martin that bred successfully on top of a flattened Cape Sparrow's nest on girders supporting a roof. No mud pellets were used, and the top of the sparrow's nest was merely lined with soft materials.

The clutch usually comprises 2-4 eggs, mostly three, but species such as the Rock Martin and Red-breasted Swallow may lay up to six eggs. Burrow-nesting species all lay plain white eggs, whereas those that lay in cup-shaped nests lay eggs that are speckled with rust-red, often in a zone around the large end. The exception in this group is the Pearl-breasted Swallow which lays white eggs. Of the species that build retort-shaped nests, some lay white eggs and others speckled ones. Eggs are laid on successive days, and incubation usually begins on completion of the clutch, although partial incubation may occur sooner as the nest is occupied overnight from the laying of the first egg.

Both sexes of the Rock Martin and South African Cliff Swallow incubate, and in the other species only the female does so, although the male may sleep in enclosed nests overnight with the female. The female Red-breasted Swallow may spend remarkably little time (only 14 per cent) on the nest during the day. The thick-walled nest, long tunnel and soft lining all help to maintain a constant temperature, and often lining is added throughout the incubation period. Most species incubate for between 14 and 17 days, but occasionally for longer.

Newly-hatched chicks are blind and have pink skin with a sparse covering of fine down. Prominent yellow flanges at the side of their gapes enable their parents to see them when feeding in a dark nest. In all species both parents feed, often assiduously, and there is a record of three feathered Wire-tailed Swallow chicks receiving 40 feeds in 90 minutes. Food is brought in the bill or mouth, not in a large food ball in the manner of swifts. Initially the parents remove faecal sacs, but once the chicks are older they defecate over the edge of the nest or out of the entrance. When old enough to do so, the chicks of hole-nesting species come to the entrance of the burrow to be fed.

As the young develop they begin to exercise their wings. In an enclosed nest they do so as best they can, but in a cup-shaped one they cling to the edge, facing inwards, and flap vigorously. By the time they make their first flight it is generally a good one, and they soon become proficient on the wing. The nestling period for most species is 20 to 25 days, and sometimes as much as 30 days; the South African Cliff Swallow averages 24 days.

Young swallows remain dependent on their parents for 14 to 21 days, returning to roost in the nest during this period. Initially they are still fed, even sometimes in flight, but soon they catch insects for themselves and join the ranks of some of the most graceful masters of the air.

Cuckoo-shrikes

Cuckoo-shrikes are unobtrusive birds and are usually found solitarily or in pairs, or in small family parties after breeding. Of the three species found in southern Africa, the Black Cuckoo-shrike occurs in a variety of woodland habitats including exotic plantations; the White-breasted Cuckoo-shrike is most common in miombo woodland; and the Grey Cuckoo-shrike is a forest species whose habits are the least well known of the three.

The sexes of the Black Cuckoo-shrike are strikingly different, but the differences in the other two species are slight. Cuckoo-shrikes are monogamous and territorial, the White-breasted Cuckoo-shrike displaying its presence by flying above the canopy of its territory. Courtship behaviour is poorly known, but the male Black Cuckoo-shrike may follow the female with a fluttering, moth-like flight, his rump feathers raised and mouth open to reveal his red gape.

These species breed in spring and summer, but mostly from September to November. The nest, placed high in a tree on top of a horizontal or slightly sloping branch where there is a fork, is a small, flattened pad of leaf ribs and fine twiglets and is often covered with lichens. Some nests appear to be made almost entirely of lichens and are bound with cobweb for strength. Being small and well camouflaged, and situated on top of a branch, the nests are remarkably difficult to locate. Only the female Black Cuckoo-shrike builds, but the male accompanies her to and from the nest as she collects material. Observations of the White-breasted Cuckoo-shrike are conflicting: in one account the female was said to do most of the building, whereas in another only the male was reported to build, with the female merely testing the nest for size from time to time. The latter nest was completed in six days.

The clutch is normally two eggs, sometimes one, and they are greenish in ground colour with brown spots or, in the case of the Grey Cuckoo-shrike, with dark olive spots and streaks. The Black Cuckoo-shrike is said to lay three eggs, but it is doubtful whether three grown nestlings would fit into its nest. The eggs are laid at daily intervals. At a White-breasted Cuckoo-shrike's nest the male incubated for the first day and during the morning of the second until the female laid the second egg. Thereafter they shared incubation equally and the eggs were left unattended only for short periods. Grey Cuckoo-shrikes also share incubation, but in the Black Cuckoo-shrike only the female incubates while the male is usually on guard nearby. The sitting bird looks somewhat incongruous on its small nest, and from a distance appears to be merely squatting on a branch. The incubation period of the Black Cuckoo-shrike is 20 days, that of the White-breasted Cuckoo-shrike is 23 days, and the Grey Cuckoo-shrike's is unrecorded.

Newly-hatched cuckoo-shrikes have blackish skin and are covered with white down which renders them inconspicuous in the nest. As the feathers emerge they are tipped with black so the chicks are still remarkably well camouflaged, appearing more like a growth on a branch than birds. If there are two chicks they soon fill the nest and sometimes one falls out to the ground below. Small chicks are brooded, and a female Black Cuckoo-shrike has been known to brood overnight until the young were 14 days old. Both parents feed the chicks and remove faecal sacs.

The nestling period of the Black Cuckoo-shrike is 21 to 23 days and that of the White-breasted Cuckoo-shrike somewhat longer at 30 days. The young remain dependent on their parents for some weeks after leaving the nest, and for as long as 70 days in the case of the White-breasted Cuckoo-shrike. The young of this species, as well as those of the Grey Cuckoo-shrike, remain with their parents until the following breeding season.

In an unusual incident at a White-breasted Cuckoo-shrike's nest the male demolished the structure as soon as the single chick had left it, tearing it to pieces until it became dislodged and dropped to the ground.

1 

 2

1 *Placed high in a tree and made almost entirely of 'old man's beard' lichen, a White-breasted Cuckoo-shrike's nest is difficult to see from the ground.* ***2*** *The Black Cuckoo-shrike chick's black-tipped feathers camouflage it well on its lichen-covered nest.*

Drongos

The two drongos which occur in southern Africa are the Fork-tailed and the Square-tailed, the latter a species confined to forest and thus found only in the eastern part of the region. The conspicuous Fork-tailed Drongo, on the other hand, is widespread in a variety of habitats, including plantations of exotic trees (in which it breeds), and it has recently expanded its range into the south-western Cape.

Both species are monogamous and aggressively territorial, but the behaviour of the Square-tailed Drongo is seldom observed. Fork-tailed Drongos chase raptors, crows and various other birds relentlessly, even clinging to their backs and pecking them in flight. They imitate bird calls to perfection, and their repertoire usually includes the calls of diurnal raptors and owls, especially the Pearl-spotted Owl.

The breeding season for both species is from September to January, or occasionally later. In Zimbabwe the peak period for the Fork-tailed Drongo is September to November and it may raise two or sometimes three broods in a season, whereas the peak for the Square-tailed species is in October and November.

The Fork-tailed Drongo constructs a saucer-shaped nest of fine twigs, leaf ribs, rootlets, tendrils and similar material, binding it with cobweb and slinging it hammock-like in a fork at the end of a horizontal branch. Some nests are so flimsy that the eggs can be seen from the ground. They are rarely less than 3 metres above the ground and some are as high as 12 metres or more. The nest of the Square-tailed Drongo is similar, except that it is deeper and has thicker walls, and the exterior is covered with lichens. It usually hangs at the end of a slender branch of a tree or sapling. There are several records of Square-tailed Drongos destroying their nests after use, and in one observation the material was used in the construction of a new nest.

The clutch of the Fork-tailed Drongo is 2-4 eggs and that of the Square-tailed Drongo two or three. The former's eggs are extremely variable, with five main types having been identified. Some eggs are plain white, others are white with speckles, and yet others are various shades of pink with speckles and blotches of red-brown. The African Cuckoo parasitises this species exclusively (see page 105), and faces the dual problem of matching such variable eggs and laying in the nests of alert and aggressive birds. The eggs of the Square-tailed Drongo may also be plain or speckled.

The incubation period of the Fork-tailed Drongo is 16 to 17 days and the nestling period slightly longer at 17 to 18 days. At one nest both parents were seen feeding the chicks and they removed and dropped faecal sacs. Considering that the Fork-tailed Drongo is so widespread and conspicuous, there is remarkably little information about its breeding cycle, and none at all about the incubation and nestling periods of the Square-tailed Drongo.

1 

 2

__1__ The Fork-tailed Drongo slings its nest in a horizontal fork at the end of a branch. __2__ The Square-tailed Drongo builds a similar nest but decorates it with lichens.

Orioles

The orioles form another group whose breeding biology is little known. Four species occur in southern Africa, of which the European Oriole is a non-breeding migrant from the Palaearctic. The breeding species are the African Golden Oriole, an intra-African migrant; the resident Black-headed Oriole; and the Green-headed Oriole which in southern Africa is represented by an isolated population on Mount Gorongosa in Mozambique. The nest of this last species has not been discovered in the region, but elsewhere it is similar to those of other orioles.

Orioles are notable for their clear liquid calls which probably serve as territorial advertisement. Few observations have been made of their courtship behaviour, but in one instance a male Black-headed Oriole hung almost upside down near a female, his tail stiffly fanned. He emitted a miscellany of warbling notes and she responded by quivering her wings. The interlude lasted about two minutes before another male appeared and was chased off.

The breeding peak for the Black-headed and African Golden Orioles is from September to November but they may breed throughout spring and summer. Their nests are deep basins slung like a hammock in the fork of a slender horizontal branch, some distance from the main trunk and rarely less than 5 metres above the ground. The Black-headed Oriole usually constructs its nest from 'old man's beard' lichens, often concealing it amongst them, whereas the African Golden Oriole uses a mixture of grass, rootlets and lichens. The nest is bound to the branch with cobweb.

Black-headed Orioles lay two or three attractive eggs in their nests, which are usually made of 'old man's beard' lichens.

The clutch comprises two or three eggs which are very pale pink when freshly laid, with brown spots concentrated at the large end. Both sexes of the Black-headed Oriole share incubation, and it and the African Golden Oriole incubate for close on 15 days. The nestlings are fed by both parents and they leave the nest after about 15 days.

Crows and Ravens

Collectively known as corvids, crows and ravens are among the most intelligent of all birds. Most species are gregarious when not breeding and roost communally. They utilise a wide range of food sources and are adept at scavenging, with the result that species such as the Pied Crow have become a familiar sight in many urban environments throughout southern Africa. Four corvid species occur in the subcontinent: the Black, Pied and House Crows and the White-necked Raven. The House Crow, a newcomer, was first sighted in Durban in 1972 and appears to have become well established, despite attempts to eradicate it. It has recently been seen in East London and Cape Town where its arrival may have been ship-assisted.

All our corvids nest solitarily and are monogamous and territorial. They appear to pair for life and White-necked Ravens especially are always seen flying in pairs, although they may collect in numbers at carrion. Renowned for their vocal abilities, most species have a repertoire of caws, croaks and gurgles which are intermingled with imitations of a variety of bird calls and other sounds. Vocalisations are undoubtedly used as part of courtship and for maintaining the pair bond, as are preening and the feeding of a mate, but courtship displays are seldom described. Sometimes the pair will stand together with head and neck feathers raised, or assume a bowing posture and flick the wings and raise the tail. Copulation is often initiated by the female who crouches in a horizontal position, her wings quivering.

The breeding season for all species extends from early spring into summer, with a peak in September and October. The White-necked Raven builds on a ledge or in a pothole on cliffs, or very rarely in a tree. Crows, on the other hand, often breed in trees, especially in very tall eucalypts, but the Black Crow also nests on large aloes such as *Aloe ferox*. Other favoured nest sites of Black and Pied Crows are telegraph poles, high-tension power pylons and windmills, and they also occasionally nest on cliffs. In Harare Pied Crows have been recorded nesting on a tall office block and on a radio mast. The use of man-made structures as nest sites has enabled crows to expand their range, and they have been followed into new areas by the Greater Kestrel which in some localities uses only old crows' nests for its own breeding. In the Karoo and other treeless regions crows' nests on telegraph poles are regularly spaced, giving a good indication of the size of a pair's territory. Even where Black and Pied Crows breed in the same area, the spacing is maintained.

Crow and raven nests are sturdy, basket-shaped structures made of stout sticks. One belonging to a Black Crow was found to contain 392 sticks which weighed 2.7 kilograms of the 3.2 kilograms total weight. Quite often nests incorporate wire, such as pieces of discarded fencing, and some are constructed entirely of this material. One wire nest weighed 20 kilograms. The cup is deep and thickly lined with a wide range of soft materials such as wool, hair, fur, feathers, grass and dung, as well as with rags, string, paper and pieces of cloth. Irrespective of the materials used, the cup is always very neatly finished by the female as she turns round and round, shaping it with her breast. The nests of the Pied and Black Crows and the White-necked Raven are 40 to 60 centimetres across with a cup approximately 20 centimetres in diameter; those of the House Crow are smaller, about 30 centimetres across.

Usually four eggs are laid, but the clutch may vary from three to six eggs, or rarely seven. The eggs of the Pied and House Crows and the White-necked Raven are greenish and spotted and streaked with shades of brown and sepia, whereas those of the Black Crow are pink in ground colour with rust-red speckling. All four species lay their eggs at intervals of 24 to 48 hours; in one observation a Black Crow completed its clutch of five within seven days. Pied and Black Crows are regular hosts of the Great Spotted Cuckoo (see page 105).

***1** Three well-grown White-necked Raven chicks fill their nest. Even after they have left the nest they remain dependent on their parents for some time.* ***2** The White-necked Raven is a cliff-nesting species, siting its nest on a ledge or in a pothole.*

__1__ The Black Crow is one of a number of species that have extended their ranges by nesting on man-made structures. __2__ Black Crow eggs differ from those of other corvids in that they are pink in ground colour. The cup in which they are laid is invariably deep and neatly lined. __3__ The House Crow's nest is smaller than that of other crows.

Because the sexes are alike, observations on the share of parental duties at the nest are difficult. In most corvids the female incubates and is fed by her mate. However, the male Pied Crow has been seen taking a 20 per cent share of the duty, and both sexes of the Black Crow have also been seen incubating. Little is known about the breeding biology of the White-necked Raven because it is so wary and nests in remote places. Corvids invariably begin incubating before the clutch is complete, usually after the second or third egg has been laid, although there is no set pattern. Crows sit so low in the deep cup that often it is difficult to see from below whether there is a bird on the nest. The incubation period of the Black and Pied Crows is close on 18 days and that of the White-necked Raven is 19 to 21 days. In India the House Crow incubates for 16 days and both sexes are said to share this parental duty.

When the eggs hatch the shells are removed and eaten. The newly-hatched chicks are naked and have pink gapes that become scarlet as they grow older. Because the eggs hatch at different times, some members of the brood are always smaller and may die if the food supply is inadequate. The young grow rapidly; by ten days old their quills are sprouting and at the age of 14 days their feathers emerge. Initially the parents maintain sanitation by eating the faeces, but later the chicks defecate over the edge of the nest.

Both parents feed the chicks, mainly by regurgitation, but White-necked Ravens have been observed carrying pieces of meat in their bills for the young; at one nest the female fed them approximately 30 times in three hours. Pied Crows have been known to mob humans climbing to the nest, and Black Crows fly around with rapid fluttering wing-beats, presumably as a form of distraction. Young Pied Crows remain in the nest for an average of 38 days, but will leave the nest after 35 days if they have been disturbed. The Black Crow's nestling period of 36 to 39 days is similar, but that of the House Crow (in India) is much shorter at 21 to 28 days. The nestling period of the White-necked Raven is not known. After they have left the nest young corvids remain dependent for some considerable time; in the case of the Black Crow they may be fed for up to three months and then remain with their parents for a further three months.

Tits

Six tit species – the Southern Grey, Northern Grey, Ashy, Black, Carp's Black and Rufous-bellied Tits – occur in southern Africa, in a wide range of habitats from the arid Karoo to the miombo woodlands of Zimbabwe. They are noisy, gregarious birds, and some species regularly join mixed bird parties.

Only the Black Tit has been studied in any detail, and it was found to be territorial throughout the year and to breed co-operatively. For about three weeks before laying the female is fed by the male and she accepts his offerings by adopting a submissive posture with quivering wings. At this stage, if there are helpers in the group (and these are always males) the dominant male tries to prevent them from feeding his mate by chasing them. In a case where there was only one helper the male chased him away 28 times over a period of three hours.

In all species noisy calling is used for territorial advertisement and also probably serves as part of courtship. The breeding season is mainly during spring and early summer, although species in arid regions also breed opportunistically after rain. The nest is always hidden from view in a cavity, usually a natural hole in a tree but sometimes in woodpecker or barbet holes, and nest-boxes are also readily accepted. Other sites include hollow metal fence posts, holes in derelict buildings or amongst rocks, or a hole in a bank excavated by another bird species. The nest is a thick pad of soft material such as grass, hair or wool. In the Black Tit, the only species for which observations are available, only the female builds the nest.

The clutch size is usually 3-5 eggs, although 2-6 may be laid, and they are

The Southern Grey Tit, which inhabits dry areas where few trees occur, is more likely to nest in a hole among rocks than in a tree.

white or pale pink in ground colour with red speckles. As is often the case with birds that nest in cavities, accurate details of incubation and nestling periods are difficult to establish. In the Black Tit only the female incubates and she is fed by the male as well as by helpers if they are present. She incubates for periods of 47 minutes on average and then leaves the nest for a break. If disturbed while she is in the nest she sits tight, emits a snake-like hiss and lunges at the predator; similar behaviour is adopted by large young. The Black Tit's incubation period is 15 days and that of other species would presumably be similar.

The newly-hatched chicks are blind, naked except for a few faint wisps of down, and pink, with prominent yellow flanges on the gape which enable their parents to locate their mouths in the dark nest cavity. In Black Tits they are brooded by the female while still small and food is brought by the male and up to three helpers. Later the female leaves the nest and helps to feed the young. It was found that at nests with helpers more offspring were reared than at those without them. The nestling period of the Black Tit is 22 to 24 days.

Once they have left the nest young Black Tits are fed by their parents (and helpers where present), but after 14 days they start to feed themselves and by the time they have been out of the nest for seven weeks they are no longer fed. Sometimes dependent young leave their own family and join another group, but most young remain in their natal group for many months.

The basic breeding biology of other southern African tits is likely to be very similar to that of the Black Tit, and the Ashy and Carp's Black Tits have also been recorded breeding co-operatively. It remains to be seen whether this applies to all species.

A Black Tit takes food to its nest, a natural hole in a tree.

Penduline Tits

The penduline tits, of which there are two species in southern Africa, are notable for two reasons: they are the smallest birds in the region, and they construct a remarkable nest. Although their ranges overlap in some areas, the Cape Penduline Tit inhabits the more arid western regions, and the Grey Penduline Tit is found mainly in woodland in the north-east and north. Only the Cape Penduline Tit has been well studied, but from the information available there appear to be no major differences in the nests and breeding behaviour of the two species.

These very active little birds live in groups and feed on insects and their larvae. They roost communally in their own small nests, but also utilise weavers' nests for this purpose. In most respects they are like true tits, but have been placed in their own family mainly because of their unique nests. Their breeding season varies from one region to another, and in arid areas the birds may breed opportunistically after rain. In general, however, most breeding occurs between the months of August and February, and in Zimbabwe the Grey Penduline Tit shows a marked peak from September to November.

It is not easy to observe the tits' behaviour because they are so active, and the only description of apparent courtship behaviour was when a male pursued a female during the nest-building period, accelerating as he caught up with her and simultaneously emitting an abrupt call note.

The oval nest is a remarkable structure with great strength. Two features are distinctive: the hollow below the entrance, often known as the false entrance, and the tubular entrance 'spout' which the bird closes using a special technique. The nest is usually suspended from a side branch or sprig of a tree or bush, often a thorny species, at a height varying between 0.5 and 5 metres above the ground. In arid regions it tends to be lower because of the lack of large trees, and in the Karoo one was found hanging from a strand of fence wire. A wide variety of soft materials is used in its construction – including cotton, cobweb, especially that of the social spider *Stegodyphus*, the downy seeds of *Eriocephalus* species known as 'kapokbossie', and hair of wild and domestic animals – but sheep's wool is preferred where it is available. Other birds also find the soft material attractive and sometimes steal it for their own nests, on occasion causing the penduline tits to desert and move to a new spot. The nest is usually white when newly constructed but it darkens with weathering, and in areas where karakul sheep are farmed it may be blackish.

Detailed observations of a Cape Penduline Tit's nest being constructed record that the birds started by forming a ring as weavers do (see page 210), and from this base extended the nest to the back and front and substantially thickened the floor. By the tenth day the shell was complete and two days later the entrance spout was being formed. At this stage the birds sheltered in the nest from a heavy thunderstorm, closing the entrance vertically, not horizontally as is usual. On the seventeenth day, once they had finished the spout, the birds began work on the false entrance, the construction of which had the effect of pulling the real one into a horizontal slit instead of a round opening. By the twenty-second day the nest was fully formed and the first egg was laid, but the birds continued adding material to the interior throughout incubation and into the nestling period.

Both birds took part in building, bringing large wads of material in their bills and sometimes storing them temporarily on twigs near the nest. During construction the nest was loosely put together and had a lacy appearance, and it was only by constantly pulling, teasing and jabbing the soft material with their bills that the tits created the strong felting; no weaving was involved. The completed nest was waterproof and subsequent weathering appeared to strengthen the felting.

The clutch of the Grey Penduline Tit comprises 3-6 eggs, but as many as ten have been recorded in nests of the Cape Penduline Tit. The suggestion that these larger clutches are laid by two females has not been proved. The minuscule, pure white eggs are laid on successive days. In one observation the eggs were covered with soft material during the laying stage, but in view of the fact that lining is constantly being added to the nest interior, it is difficult to know whether this covering was deliberate or incidental. The eggs were no longer covered once the clutch was complete. In the Cape Penduline Tit both sexes share incubation, which lasts for close on 15 days.

The tiny chicks are naked when they hatch, with orange skin and yellow gapes, and they are fed assiduously by both parents. As yet there is no direct evidence that helpers also feed the young. Near the end of the nestling period they move to the nest entrance to be fed, and at the age of 23 or 24 days they depart. However, they remain in a compact family unit, returning with their parents to roost in the nest.

The Cape Penduline Tit regularly produces two broods in a season, sometimes laying a second clutch a month after the previous brood has left. At night the young of the first brood roost with their parents while the second is being incubated, and eventually all the young of both broods roost in the nest. As many as 18 birds have been recorded sleeping in a nest, a remarkable feat in an area only 7 centimetres high and 6 centimetres across. As the young of both broods and their parents could not account for the total number, a few 'strangers' must have joined in. An interesting question is how the birds know when to get up, for despite the fact that it is dark inside the nest, they coming streaming out at sunrise, even on overcast days. At the end of the breeding season the nest is still used as a roost, for three to four months in one observation. The incubation and nestling periods of the Grey Penduline Tit have yet to be established, but they are unlikely to differ much from those of the Cape Penduline Tit.

EXITS AND ENTRANCES – HOW DO PENDULINE TITS OPEN AND CLOSE THEIR NESTS?

No aspect of a penduline tit's nest is more misunderstood than the way in which the bird opens and closes the entrance spout. On arrival at the nest it lands on the false entrance's substantial rim and with its bill, or with one foot if it is carrying nest material or food, it pulls down the lower lip of the spout. Grasping the lip with both feet, it levers itself inside and turns around to close first the corners and then the centre of the entrance. Thereafter shaking movements of the nest indicate that it is closing the tube along its length by pulling the floor up against the ceiling. The floor and ceiling are lined with coarse cobweb so they stick together like strips of Velcro.

On leaving the nest the bird merely pushes its way out along the tube and the spout opens. Once outside it swings down to perch on the rim of the false entrance and prods the spout with its bill and forehead to close it, using extremely rapid movements that have led some observers to believe, quite incorrectly, that it pinches the entrance closed with its bill.

There can be no doubt that when the true entrance spout is closed the pocket beneath it could easily deceive a predator such as a snake into trying to gain entry to the nest at that point, and being defeated by the thickness of the wall there. However, this bonus may be incidental to the structural importance of the false entrance, and to its use as a perch when the bird enters the nest and closes it on leaving ●

1

2

3

4

1*, *2 *A Cape Penduline Tit opens the entrance spout to its nest with one foot, then grasps the lower lip before entering.* ***3*** *The false entrance has a substantial rim on which the bird perches when it opens and closes the real entrance.* ***4*** *The tit closes the entire length of the spout by pushing up against it with rapid movements of its forehead and bill.*

The Spotted Creeper

The Spotted Creeper, the only representative of its family in southern Africa, occurs mainly on the central plateau in Zimbabwe, but also in adjacent Mozambique. In its preferred habitat of miombo woodland it occurs singly or in pairs and, being cryptically coloured, is not easily seen. Its habits are woodpecker-like, although it does not use its tail as a brace against the tree trunk.

Little is known about its behaviour, and nothing at all about courtship. However, the breeding season is well defined, extending from August to October with a marked September peak that coincides with the flush of insects emerging in the hot weather before the rains begin. The nest is placed at a height of between 4.5 and 7.5 metres in the vertical fork of a branch or on a horizontal limb where there is a kink to conceal it. Like a batis nest it is a cup made of leaf ribs, lichens, chips of bark and similar material bound together with cobweb and plastered externally with pieces of lichen. It blends perfectly with the lichen-covered branch, and even when there is a bird on the nest its cryptic plumage merely serves as an extension of the camouflage.

The normal clutch is three eggs which are greenish in ground colour with dark brown spots and grey clouding concentrated mainly at the large end. The incubating bird relies on its camouflage for protection and sits tight, often allowing an observer to approach closely. In such circumstances it points its bill skywards at an angle of 75° so that it resembles a twig and enhances the camouflage. Once the danger has passed the bird relaxes again and holds its bill at the normal angle. The sitting bird often gives its *sweepy-swee-seepy* call while on the nest and experienced observers have learnt to locate its position by this means. The sexes are similar, and reports that only the female incubates have still to be confirmed by sustained observation.

The newly-hatched chicks have dark skin which camouflages them in the nest cup. They are fed by both parents, which approach the nest by clambering up the branch from below in the same way that they normally forage on a tree trunk. The parents also remove faecal sacs and drop them away from the nest. At a nest containing large young an adult has been observed performing a distraction display in which it fluttered weakly to a tree and hung out each wing in turn as if injured. Then it hung upside down with its wings flopping loosely over its back. There are no records of either incubation or nestling periods.

1 *A Spotted Creeper's nest, bound together with cobweb and covered with lichen, is difficult to discern on lichen-covered bark.* ***2*** *The sitting bird is already well camouflaged but, aware of an intruder, it points its bill skyward to resemble a twig.*

BABBLERS

OF THE FIVE SPECIES OF BABBLERS that occur in southern Africa, the Arrow-marked Babbler is the most widely distributed. The Black-faced, Bare-cheeked and Pied Babblers are found in more arid habitats, the first two being restricted mainly to northern Namibia where little is known about their breeding biology. The White-rumped (also known as Hartlaub's) Babbler occurs in the moist habitats of the Okavango Delta and adjacent Caprivi, and in northern Namibia.

Known for their sociability (as demonstrated by their habit of preening each other) and co-operative breeding behaviour, babblers are gregarious, territorial and very noisy. Their harsh chattering calls rise to a crescendo when opposing groups meet on the boundary of a territory and face each other with spread tails and quivering wings. In Zimbabwe the size of 28 flocks of Arrow-marked Babblers ranged from three birds to 13, with an average of six. Sometimes flocks are larger, and a Pied Babbler group may comprise as many as 15 birds. Group cohesion is closely maintained, and in one incident five Pied Babblers vigorously attacked a Gabar Goshawk which had captured one of their young. The hawk was later found dead.

Babblers are presumed to be monogamous and, besides other adults, a flock usually includes immatures from a previous brood. However, as there are no direct observations of either the pair bond or courtship behaviour, it can only be assumed that the group comprises a core pair assisted by helpers. Most species breed during the summer months. The Arrow-marked Babbler, however, may breed in all months, with a September to November peak in Zimbabwe.

The nests of the different species are similar, comprising an untidy bulky exterior of small twigs, leaves and dry grass, and a neat bowl lined with rootlets and fine fibres. They are built in trees – thorny ones are favoured in arid regions – at varying heights up to about 5 metres and may be hidden in creepers or amongst debris caught in the tree. In the case of the Arrow-marked Babbler they are occasionally placed in holes in trees. In one observation all seven birds in a group of Arrow-marked Babblers helped to build the nest.

Usually three eggs, sometimes 2-4, are laid, and they are plain turquoise or greenish blue, or pale blue in the case of the Pied Babbler. The eggs of the Black-faced and Pied Babblers have small nodules on the surface, and it is of interest that these are the only two babblers which have not been parasitised by the Striped Cuckoo (see page 105). The Arrowed-marked Babbler is the main host of this cuckoo, which lays perfectly matched eggs in its nest.

The Arrow-marked Babbler is the only species which has been studied in any detail at the nest, but it is unlikely that the breeding biology of the other babblers would differ to any great extent. The eggs are incubated for close on 14 days, but it has still to be established to what extent helpers may assist. At a nest where the members of the group were colour-ringed, four adults and two immatures fed the chicks, although the contributions varied; one immature contributed very little. Of the meals delivered, three adults brought three-quarters of the total. An adult sat on the nest all the time until the chicks were nine days old, after which they were left alone. Faecal sacs were eaten at the nest and members of the group sometimes competed for this task. In one instance the chicks left at the age of 12 days, before they could fly properly. In another observation the nestling period was 14 days. Once they leave the nest the young join the flock and become helpers when subsequent breeding takes place.

1 *An Arrow-marked Babbler has built its nest at the base of a thorn tree, an unusually low position for a babbler.*
2 *Pied Babbler eggs, like those of Black-faced Babblers, have small nodules on the shell. Neither species is parasitised by the Striped Cuckoo.*
3 *Immature Arrow-marked Babblers (distinguished by their duller eye colour) apparently help to tend the chicks of subsequent broods.*

The Boulder Chat and Rockjumpers

The Boulder Chat and the two rockjumper species were traditionally classified in the family Turdidae (which includes thrushes, robins and chats) but the former, albeit correctly placed, belongs to the distinct genus *Pinarornis* and is very rockjumper-like in its build and habits. The rockjumpers, also in their own genus, *Chaetops*, are now considered to belong to the family Timaliidae, with the babblers. Because of the similarity between the two genera, however, it is convenient to discuss them together.

In southern Africa the Boulder Chat occurs from eastern Botswana across the granitic shield of Zimbabwe and then extralimitally into Zambia and Malawi. It is common among the wooded granite koppies of the Matobo Hills south of Bulawayo, and may be seen running over rocks and gliding from one outcrop to another just as rockjumpers do, or foraging in the leaf litter for insects.

Nothing is recorded about its courtship behaviour but it is monogamous and territorial. The usual song is a pleasant series of high-pitched warbles, but the alarm call is a distinctive triple-noted *tsi-tsi-tsi* which is repeated frequently and sounds like the squeaky wheel of a pram. There is no direct evidence of co-operative behaviour, but on one occasion three adults scolded an observer who was near young that had recently left the nest. In Zimbabwe Boulder Chats breed from September to December, with a clearly defined peak in October and November. They place the nest against or partially under a rock, in a cleft in a boulder or sometimes against a log or amongst fallen branches, always concealing it well. The neat cup, 9 centimetres across and 4 centimetres deep, is lined with leaf ribs and rootlets and set into an accumulation of material such as bark pieces, twigs, dry leaves and small earth clods, below the level of the surrounding leaf litter.

The clutch comprises three eggs which are greenish white in ground colour with numerous red-brown speckles often concentrated in a cap at the large end. The Boulder Chat is regularly parasitised by the Red-chested Cuckoo, whose eggs match those of the chat very closely and can only be distinguished because they are slightly smaller and rounder (see page 105). In a survey of 12 nests found in two consecutive breeding seasons in the Matobo Hills, half had been parasitised. The incubating birds are extremely wary and after giving vent to their squeaky alarm call at the slightest sign of danger, they are reluctant to return to the nest. It is not known whether both parents incubate, and the period of incubation has still to be recorded.

When the eggs hatch the shells are removed and dropped away from the nest. With their yellow gapes, dull pink skin and tufts of dark grey down, the newly-hatched chicks are readily distinguishable from the Red-chested Cuckoo's offspring which has black skin and no down. The chats' grey down serves as camouflage in the nest, and the brooding bird too, with its dark plumage, is well concealed. Both parents share feeding duties, approaching the nest by a regular route, and they eat the faecal sacs at the nest. When there are young in the nest the intensity of alarm calls increases, and on one occasion a bird tried to distract an observer away with a 'rodent run' display which developed into a full 'broken wing' act with wings and tail spread out on a rock.

The young birds leave the nest after 16 to 20 days. Not yet able to fly, they hide in crevices among rocks when their parents give the alarm, their black plumage making it very difficult to detect them. It is not known for how long they remain dependent on their parents.

Once considered a single species, the Cape and Orange-breasted Rockjumpers have similar habits. The former is found in the mountains of the south-western and southern Cape as far east as Port Elizabeth and occurs from the lowest rocky slopes at sea level to the tops of the highest peaks. The range of the Orange-breasted Rockjumper extends from the vicinity of Graaff-Reinet to Lesotho and the Natal Drakensberg.

Rockjumpers, which live in pairs or small groups, may be located by their far-carrying piping calls as they run rapidly among rocks, appearing and disappearing like a magician's trick. Males and females are easily distinguished by the former's much brighter colours, but courtship behaviour has not been recorded. The birds are presumably territorial because they are found in the same area from one season to the next; indeed, one nest was in the same hollow as that used the previous year.

Both rockjumpers breed from September to December, placing the nest on the ground against or partially under a rock, in a cleft between rocks, in a grass clump or sometimes in a patch of berg palmiet. The fairly bulky structure is built by both sexes and comprises coarse grass and other plant material, with finer fibres lining the spacious bowl. Two, occasionally three, plain white eggs are laid. In the case of the Cape Rockjumper both male and female incubate, but the duration of the incubation period is unrecorded.

The chicks are pink with tufts of grey down when they hatch, and they are brooded and tended by both parents. The Cape Rockjumper often becomes confiding when there are young in the nest and will allow observers to sit nearby and photograph them as they go back and forth by a regular route. At one nest the birds even accepted insects that were offered to them. It is not known for how long the young remain in the nest, but they probably leave it before they can fly. The frequency with which small groups are observed suggests that the family stays together for some while. It is possible that the extra male found assisting with the feeding and brooding of chicks at some nests may be drawn from the parents' own kin.

1

2

1 *A Boulder Chat's nest is typically set in leaf litter at the base of a rock.*
2 *A rock slab provides protection for a Cape Rockjumper chick.*

Bulbuls, the Bush Blackcap and the Yellow-spotted Nicator

Bulbuls are well represented in Africa, but of the 52 species that occur on the continent, only nine are in its southern part. These may be divided into four groups: the Cape, Red-eyed and Black-eyed Bulbuls of the genus *Pycnonotus*; the Terrestrial, Yellow-streaked and Slender Bulbuls of the genus *Phyllastrephus*; the Sombre and Stripe-cheeked Bulbuls of the genus *Andropadus*; and the Yellow-bellied Bulbul of the genus *Chlorocichla*. The blackcaps and nicators are each represented by a single species.

The ranges of the Cape, Red-eyed and Black-eyed Bulbuls are, for the most part, mutually exclusive, although a degree of overlap does occur. The drably coloured Terrestrial, Yellow-streaked, Slender, Sombre and Stripe-cheeked Bulbuls occur in evergreen forest or thicket, and the Yellow bellied Bulbul occupies a wide range of habitats ranging from thick bush to evergreen forest. Only the Terrestrial Bulbul regularly lives in groups – usually of three to six birds – but Yellow-bellied Bulbuls may sometimes be seen in small groups too.

Although the nests of all nine species have been found and their eggs described, there is generally a lack of information about the breeding biology of bulbuls, especially the more elusive forest denizens. Only the Cape Bulbul has been comprehensively studied in southern Africa. Bulbuls are monogamous and territorial, and all have distinctive calls varying from the cheerful liquid phrases of the Cape, Red-eyed and Black-eyed species to the chattering notes of the Terrestrial Bulbul and the monotonous ringing call of the Sombre Bulbul. Calling, especially in forest species, serves as territorial advertisement and to maintain the pair bond.

Little is known about courtship behaviour, except that of the Cape Bulbul. In this species the male flutters around the female with spread tail, or he perches above her to display his yellow vent and bends forward with flattened crest, spreading his wings and tail and fluffing up the feathers of his lower back and rump. In an observation of Sombre Bulbul pre-copulatory behaviour, the male and female faced each other on a branch, the male with his bill slightly open and the feathers of his crown and nape erect. He raised and lowered his head in an elliptical arc and, without stopping the head movements, rotated slowly on a horizontal plane until he had completed a full circle. The female responded by rotating her head, and both emitted a soft, keening call. After about five minutes mating took place. Similar head-moving behaviour has been described for the Yellow-bellied Bulbul.

The breeding season for most species extends from spring into late summer (March). In Zimbabwe, however, the Black-eyed Bulbul breeds in most months, with a marked peak from September to December, and especially in October. In regions of marginal rainfall breeding may be opportunistic after rain.

In those species where observations are available, both sexes build the cup-shaped nest. The Cape, Red-eyed and Black-eyed Bulbuls construct a relatively sturdy structure of small twigs, rootlets, leaf ribs and tendrils which is neatly lined with finer material. The nests of the forest bulbuls are often flimsy, sometimes to such an extent that the eggs may be seen from below. Some species, for example the Yellow-streaked Bulbul, incorporate leaves into the structure, making it look like a clump of debris and thus difficult to detect. The nests are situated in bushes, trees and sometimes in creepers, and are usually well hidden, except those of the Sombre Bulbul which are often more easily seen. They may be found at varying heights, although those in forest are often low down in saplings in the understorey.

Each species is usually consistent in the number of eggs it lays – for example, Cape, Red-eyed and Black-eyed Bulbuls generally lay three eggs, and Sombre and Yellow-bellied Bulbuls lay two – but a clutch may vary from one to four eggs. Those of the Cape, Red-eyed and Black-eyed species are white or pale pink in ground colour with numerous rust-red speckles, usually in a zone at the large end, whereas other species lay white eggs with dark brown or sepia markings. Most beautiful of all are the eggs of the Yellow-streaked Bulbul: rich salmon-pink in colour, with dark red scrolls concentrated at the large end. A plain white egg found in a Cape, Red-eyed, Black-eyed or Sombre Bulbul nest is almost certain to be that of the Jacobin Cuckoo, a regular parasite of these species (see page 105).

Eggs are laid at daily intervals and incubation usually, but not invariably, begins when the clutch is complete. In those species for which information is available it appears that only the female incubates and that she leaves the nest to feed, although sometimes the male brings food to her. Some forest species – notably the Sombre and Yellow-streaked Bulbuls – sit extraordinarily tight, relying on their drab coloration for camouflage. Cape, Red-eyed and Black-eyed Bulbuls incubate for 12 to 14 days, the Yellow-bellied Bulbul for 14 days, and the Sombre Bulbul for 15 to 17 days.

The naked chicks are brooded for the first few days, mainly by the female

1

2

3

4

1 *The Cape Bulbul builds a sturdy, neatly lined nest and usually lays three eggs.* ***2*** *The Yellow-streaked Bulbul incorporates dry leaves in its nest so that it looks like debris in the tree. Its beautiful salmon-pink eggs are distinctive.* ***3*** *The Terrestrial Bulbul's nest is characteristically built with black rootlets.* ***4*** *A single egg in this Sombre Bulbul's nest had been freshly laid and the clutch was probably incomplete.*

***1** The forest-dwelling Stripe-cheeked Bulbul often builds its nest in a sapling in the understorey. Both birds tend the chicks. **2** The Yellow-spotted Nicator's flimsy nest of twigs is very similar to the nests of bush shrikes.*

although the male also does a share. Both birds feed the chicks and remove faeces, which are usually eaten at the nest but may be carried away and dropped. There are a number of accounts of Cape, Red-eyed and Black-eyed Bulbuls performing injury-feigning displays in which they drop to the ground and drag their outspread wings and tail, over a distance of 10 metres in one observation. Similar behaviour has been recorded for the Yellow-bellied Bulbul. The chicks of this species remain in the nest for 16 to 18 days, Sombre Bulbul nestlings stay for 14 to 16 days, and the offspring of Cape, Red-eyed and Black-eyed Bulbuls leave any time between ten and 15 days after hatching, but usually after 12 or 13 days. Young Cape Bulbuls are fed for about 14 days after they have left the nest and only become fully independent after another 28 days, even though their parents may have laid a second clutch before this time. The same apparently applies to young Red-eyed and Black-eyed Bulbuls.

Although bulbul-like in appearance and voice, the attractive Bush Blackcap is normally classified with the babblers. It bears no resemblance to the southern African babblers however, and is thus included in this account. A species of the evergreen mist-belt and montane forests of eastern South Africa between Uitenhage and the Transvaal escarpment, it moves down to the coast in winter.

Little is known about this species' breeding biology. It is monogamous and territorial, often singing melodiously for sustained periods during its breeding season, which lasts from November to January. Its nest, a neat cup of twigs, roots and moss lined with bark strips and rootlets, is usually less than 2 metres above the ground in the fork of a tree or bush, often at the edge of the forest or a clearing within it. The two eggs are white in ground colour with brown and slate markings concentrated at the large end.

Another taxonomic enigma is the Yellow-spotted Nicator: some authorities place it with the bulbuls, whereas others believe that it belongs with the bush shrikes because its nest is so similar to theirs. Again, details of its breeding biology are meagre: it breeds in summer, from November to January, and builds a shallow, saucer-shaped nest of twigs, using finer ones for the lining. The nest is placed low down in a bush in a thicket and contains usually two eggs which are creamy in ground colour with a heavy overlay of dark brown and sepia blotches. The nestling is remarkable in that when its feathers are well grown, it still has a bare neck and only a couple of patches of down on the head.

Thrushes

THE NINE THRUSHES THAT OCCUR in southern Africa fall into three groups: the 'true' thrushes – Kurrichane, Olive and Groundscraper – of the genus *Turdus*; the 'ground' thrushes – Spotted and Orange – of the genus *Zoothera*; and the rock thrushes – Cape, Sentinel, Short-toed and Miombo – of the genus *Monticola*. Most species are sedentary, although some populations of the Olive and Orange Thrushes and the Sentinel Rock Thrush may move to lower altitudes in winter, and the Eastern Cape population of the Spotted Thrush is thought to move to Kwazulu-Natal at the same time of year.

With the exception of the rock thrushes, male and female thrushes are alike. The breeding biologies of all species are very similar and all are monogamous. However, co-operative breeding has been observed only in the case of the Groundscraper Thrush where, at one nest, four birds were seen to feed the chicks. All species are also territorial and, although they lack the vocal skills of robins, thrushes sing pleasantly and advertise their territories by this means. Little is known about their courtship behaviour, but displays involving quivering wings and the tail raised almost vertically presumably relate to courtship.

Most breeding activity is concentrated in spring and summer, and there are marked peaks in different regions. In Zimbabwe, for example, the Kurrichane and Groundscraper Thrushes breed mainly from September to November, with most records in October. In the south-western Cape Olive Thrushes may breed in all months, but around Cape Town the main peak is between August and November, with a mini-season from February to April. Several species have been recorded raising up to three broods in a season.

Thrush nests have a characteristically bulky and rather untidy exterior which is constructed with a wide range of plant material such as dry leaves, twigs, rootlets and moss. Near human habitation miscellaneous items such as pieces of string, plastic, cotton waste and paper are incorporated, and Kurrichane and Olive Thrushes use mud, often mixed with plant material, when building their nests. In contrast to the untidy exterior, the spacious cup is always neatly finished and is lined with fine fibres such as rootlets. The nest of the Orange Thrush is constructed mainly of moss, which continues to grow even after the birds have finished breeding, and it is always lined with black rootlets.

It is thought that only the females are responsible for construction, but a male Kurrichane Thrush was once seen with nest material. A female of the same species, with a history of failed breeding attempts, was seen to build seven nests between August and November, completing each one in a day or, at most, two. However, it is unlikely that nests are usually built so quickly. Sometimes the same nest is used for a subsequent brood, or the same site is occupied the following season. There are also records of nests having been built on the old nest of another species such as a dove or a Fiscal Shrike.

Except for the rock thrushes, most species build their nests in trees, either in a main vertical fork or, less frequently, on a side branch. Both the Orange and the Spotted Thrushes usually choose a low site, about 2 or 3 metres above the ground, and the latter characteristically uses saplings in the understorey of the forest. The fact that the nest is relatively conspicuous in this position may explain the very high level of predation on Spotted Thrushes, which continue to nest until they raise a brood. Orange Thrush nests have also been found on top of a tree fern, in a hollow on an earthen bank and amongst ferns on a boulder on the slope of a gully, and Olive and Kurrichane Thrushes have been recorded nesting in gutters.

The rock thrushes nest in crevices in rock walls, under rocks or under a tuft of grass on a hillside. The Cape Rock Thrush, which becomes tame around human habitation, occasionally nests on buildings, and the Short-toed Rock Thrush of the more arid regions quite often nests among the roots of a fig tree growing on a rock face. The Miombo Rock Thrush, as its name suggests, inhabits mainly miombo woodland and nests in an open hole in a tree, varying the structure of the nest according to the site and in some cases furnishing it with only a sparse lining of rootlets.

Most thrushes lay two or three eggs, sometimes as many as four, but the clutches of some species, such as the Orange Thrush, are invariably restricted to two. The eggs, laid in the early morning of consecutive days, have an attractive ground colour of

2

3

1 *Both male and female Kurrichane Thrushes feed the nestlings, and are very protective of their brood.* ***2*** *Predation on Spotted Thrushes is high, possibly due to their nests being relatively conspicuous in the forest understorey.* ***3*** *Typical of thrush eggs, the Spotted Thrush's eggs are greenish blue with red-brown markings.*

1

1*, *2 *The nest of the Orange Thrush is difficult to detect. Made of moss which continues to grow, it is lined with black rootlets.* ***3*** *True to its name, a Cape Rock Thrush has jammed its untidy nest into a cleft in a rock wall.* ***4*** *The male Miombo Rock Thrush shares incubation and feeding duties with his mate. This species nests in an open cavity in a tree in its miombo woodland habitat.*

greenish blue or turquoise with spots and blotches of red-brown. Those of the rock thrushes are plain as a rule, although both the Cape and Miombo Rock Thrushes sometimes lay eggs with some spotting. It is thought that only the females of most species incubate, an exception being the Miombo Rock Thrush. Where known, the incubation period is close on 14 days.

Naked except sometimes for a few wisps of down, the newly-hatched chicks are brooded by the female. In the case of the Orange Thrush, which lives in moist montane forest, they are brooded even when well feathered to protect them from the cold. Both parents feed the chicks and remove faeces, either eating the sacs at the nest or carrying them off and dropping them. The Kurrichane and Groundscraper Thrushes are protective of their offspring and may even attack humans near the nest. Nestling periods of the different species vary from 13 to 20 days.

There is little information about the duration of parental care once the young have left the nest. A young Groundscraper Thrush was still with its parents six weeks after leaving, although it was feeding itself. Sometimes a new clutch may be laid as soon as ten days after the young have left, and there are records of young from a previous brood coming to the nest and begging for food when the chicks of the next brood hatch. When a new brood is started soon after the previous one the male looks after young that are still dependent.

Chats

The chats fill a wide spectrum of habitats in southern Africa and several are typical inhabitants of arid regions. They are sedentary or make minor movements, with the exception of the Capped Wheatear which is an intra-African migrant to some areas and resident or nomadic in others. Twelve species are included in this account, of which one, the Herero Chat, is a taxonomic enigma. Another species, the Boulder Chat, also has uncertain taxonomic affinities and is treated with the rockjumpers (see page 157).

The sexes of the various species are either similar or, in the cases of the Stonechat and Buff-streaked, Mountain and Mocking Chats, they have strikingly different coloration. All species are monogamous and territorial, and the Ant-eating Chat is the only one for which co-operative breeding behaviour has been observed.

Most chats breed in spring and summer, but those in arid regions nest opportunistically after rain. Often several broods are raised in a season. Courtship behaviour, where it is known, always involves the display of white patches in the plumage which are characteristic of even the drab species. The male Capped Wheatear flexes his wings, fans his tail to reveal the white rump, and runs round the female. He also makes short fluttering flights to a height of about 3 metres, with his tail fanned and singing as he does so; these song-flights are interspersed with snatches of mimicry. The Stonechat postures in front of the female to reveal his white neck, rump and wing patches, and hovers above her with dangling legs. The Ant-eating Chat also hovers with dangling legs, displaying his white shoulder patches and singing vigorously. The male Mocking Chat chases the female at a crouching run, wings half opened and tail cocked. In addition, he makes a zigzagging song-flight through trees and amongst rocks, and in both these displays reveals his white shoulder patches to striking effect.

The Stonechat and the Mountain, Buff-streaked, Familiar, Tractrac, Karoo and Sickle-winged Chats place their nests on the ground under a rock, bush or tuft of grass, in a hollow in an embankment or in a crevice amongst rocks. In arid regions they often orientate them so that they are in shade most of the day. Very occasionally species such as the Familiar Chat may nest in a hole excavated by another species such as a bee-eater, or even in a disused chamber in a Sociable Weaver nest. The Mountain and Familiar Chats become tame around human habitation and may locate their nests in or on buildings. As well as readily occupying nest-boxes, the latter in particular shows great ingenuity in its choice of nest situations. Perhaps the most unusual site on record was the end of the hollow boom at a border post between Zimbabwe and Botswana. Every time the boom was raised the nest was tilted to an angle of 45°.

The nest structure varies depending on the site, but most species usually build a base comprising stones, clods of earth, the lids of trapdoor spiders' nests, sticks or pieces of bark, often forming it into a small rampart in front of the nest. The Familiar

1 

2 

3 

 4

***1** Capped Wheatears nest in rodent burrows or in holes in termite mounds, but may also utilise holes in derelict man-made structures. This bird is about to enter its nest under the floor of a demolished building. **2** The perimeter of a Tractrac Chat's nest under a bush has been built up with small sticks to support the cup. **3** Familiar Chats are very adaptable when nesting. A nest in an old mine excavation is typical in that it has a bulky base. **4** Mocking Chat chicks await food in a Lesser Striped Swallow's nest, the tunnel to which has been broken off.*

1 *A female Mountain Chat tends her chicks in a typical rock-crevice nest. It is believed that the female alone builds the nest.* ***2*** *A female Arnot's Chat brings food to her chicks, which are housed in a hole in a mopane tree.*

Chat may use exceptionally large amounts of such base material, and in one case 361 stones, 15 three-inch nails and approximately 250 pieces of miscellaneous material were counted. Despite the often bulky exterior of chats' nests, the cup is very neatly formed and warmly lined with soft materials such as hair, wool, feathers, fine grasses and plant down.

The Capped Wheatear breeds down the burrows of rodents such as gerbils to a depth of about a metre, or in holes in termite mounds, and it has also been known to nest in a cavity amongst the rubble of a derelict building. In sites such as these it builds a cup-shaped nest of soft materials. The Mocking Chat takes over the bowl-shaped nests of swallows such as the Lesser Striped Swallow, evicting the owner if necessary and pulling out the eggs or chicks within. It usually breaks off the entrance tunnel and pulls out the existing lining, relining the nest with a pad of soft material, especially freshly plucked animal hair. Sometimes it builds its nest in a rock crevice or the stone wall of an outbuilding.

Arnot's Chat, the only southern African chat which inhabits woodland, nests in holes in trees between 2 and 4 metres above the ground. It usually fills the cavity with base material to 30 centimetres below the entrance, then forms a cup of soft plant material on top.

The nesting habits of the Ant-eating Chat differ completely from those of the other species. This chat excavates a tunnel in the side or roof of the burrow of an antbear, hyaena or porcupine, or sometimes in a bank or cutting. The tunnel is usually a little under a metre long, and is excavated by both birds. Digging with their bills and pedalling the earth out with their feet, they complete their work in eight to ten days,

TO CHAT OR NOT TO CHAT – WHERE DOES THE HERERO CHAT BELONG?

The Herero Chat, a localised, uncommon and shy species, was discovered as recently as 1931 and it was only in 1969 that the first nests were found. Its distribution is very restricted, extending from the extreme south-west of Angola to the Naukluft Mountains of central Namibia. Within this area it lives on the escarpment zone fringing the Namib Desert where the annual rainfall is between 75 and 250 millimetres, frequenting arid hillsides with a sparse covering of acacia and commiphora trees.

In appearance the Herero Chat is similar to the Familiar Chat, even having a reddish tail with dark central feathers. The fact that it also resembles the Marico Flycatcher, especially in its feeding behaviour, poses the question: is it a chat-like flycatcher or a flycatcher-like chat? Since it shows affinities to both groups, it is probably intermediate between the two. Taxonomists have placed it in a well-defined genus of its own, the name of which, *Namibornis*, aptly describes this species as a 'bird of the Namib'.

Details of its breeding biology are scant, and the only probable courtship behaviour observed was when one bird was seen making 'wallowing' movements on a branch, as if shaping a nest; its mate replaced it and made similar movements. Then the first bird adopted a begging posture with quivering wings and the second bird fed it an insect. When the birds are breeding, which they do from February to April, they use a triple-noted call *ji-ju-jiiu* to maintain contact, but their song is a series of robin-like warbles and trills described as very similar to that of the Rockrunner.

The first nest found was in a leafy multiple fork on an outer branch of a commiphora tree, at a height of 2 metres. It was bulky and resembled a flycatcher's cup nest. Comprising chaffed fine grasses, vegetable fibres and rootlets, it was crudely woven in such a way that it was attached to the surrounding twigs. The cup was lined with finer soft materials and some animal hair. Other nests found subsequently were similar, except that they had more downy plant lining. They were sited between 1.3 and 4 metres above the ground, one in a knot of mistletoe in an acacia tree.

The only complete clutch recorded comprised two eggs, but another nest contained a single well-grown chick. The eggs were pale greenish white in ground colour with dark red-brown speckles scattered sparingly over the entire surface but concentrated at the large end. Both parents were seen to feed the single chick and they crouched over it and shaded it during the hottest part of the day●

1

2

***1** A shepherd's tree in which a Herero Chat has built its nest. **2** The bulky and crudely woven nest is attached to surrounding twigs.*

or longer if the soil is hard. The nest, cup-shaped like those of other chats, is lined with dry grass and rootlets. Often a previous season's nest may be re-used, but never the same tunnel for two consecutive broods. The chat does not take over the tunnel of another bird species, but may use a swallow's nest such as that of the Greater Striped Swallow, in which case it breaks off the tunnel entrance and evicts the nest occupants if there are any.

Both male and female Mocking Chats have been seen building but in general, as far as is known, the female chat carries out most of the nest construction. However, this conclusion needs to be confirmed by further observations. She works remarkably quickly: a Familiar Chat's nest with a base of 147 stones was completed in three days, and a Mountain Chat's nest was also virtually finished within three days, after the bird had carried an estimated 400 beak-loads of material.

1

2

***1** A female Stonechat tends her chicks at her well-hidden nest. She broods them while they are still small but later helps her mate to feed them. **2** An Ant-eating Chat has excavated its nest tunnel above the entrance to a porcupine's burrow.*

The clutch is usually three eggs but may be 2-4, and the Ant-eating Chat lays 3-5 white eggs. Those of most other species are greenish blue or turquoise in ground colour with rust-red speckles, often in a zone at the large end, and in some cases they are plain. Eggs are laid on consecutive days and incubation usually begins once the clutch is complete. In species where information is available, only the female is known to incubate and she does so for 13 to 15 days. Chats are very alert and wary when nesting, and once they have left the nest they are very cautious about returning to it if an observer is in sight.

The chicks are brooded by the female while they are still small, but both parents feed them and both keep the nest clean, either eating faecal sacs at the nest or removing and dropping them elsewhere. The nestling period for most species is 15 to 18 days, but that of the Mocking Chat is longer at 20 days, and that of the Stonechat may be as short as 13 days. As is the case with many passerines, little is known about how long the young remain dependent once they have left the nest. In some species the dependence period is at least two or three weeks.

Ant-eating Chats may start a new brood a month after the first has left, and the offspring of earlier broods may feed nestlings in subsequent ones. However, their contribution is slight, and two helpers at one nest observed contributed less than 10 per cent of the food brought. Helpers have never been seen at the first brood of the season. When young Ant-eating Chats are alarmed by a predator they use the nest as a bolt-hole, as do the offspring of the Capped Wheatear. Young Mocking Chats have been seen playfully confronting rock hyraxes by running up to them, stopping, and then jumping into the air with a snap of the bill.

Robins and Palm Thrushes

THE 16 SPECIES IN THIS SECTION, collectively referred to as robins, include some of the most attractive birds and some of the finest songsters. They also include the Collared and Rufous-tailed Palm Thrushes – which despite their name are more closely allied to the robins than the thrushes – and the White-breasted Alethe. The habitats of all these species range from evergreen forest to Karoo scrub and acacia thornveld, and although they are often thought to be sedentary, many migrate from the montane forests to the coast in winter.

Some of the forest species, confined to the east of the region, have extremely restricted distributions. The entire southern African population of Swynnerton's Robin, for example, occurs in a few isolated pockets that probably do not exceed a few thousand hectares in eastern Zimbabwe and adjacent Mozambique. The White-breasted Alethe inhabits the Gorongosa and Beira area of Mozambique and no nests have been located in its southern African range. Gunning's Robin, found in coastal Mozambique from Beira northwards, is an elusive species whose nest has not been discovered anywhere in its range.

The amount of information on the breeding biologies of robins varies from very detailed for species such as the Starred, Cape, Natal and Heuglin's species to meagre for others. They are monogamous, territorial and pair for life. The only species for which co-operative breeding has been observed is the Karoo Robin; in two observations a single helper was seen to feed nestlings. Most robins are accomplished songsters, one of the best-known being Heuglin's Robin which also duets antiphonally. A feature of many species is their ability to mimic, not only the calls of other birds, including their alarm calls, but also mammal sounds; one Chorister Robin mimicked the calls of at least 26 other bird species.

Many robins maintain a territory throughout the year by means of calling and by aggressive displays directed at an intruder. As the breeding season approaches the intensity of singing increases. Some species, for example the Natal Robin, use mimicry almost exclusively when competing with other males for a mate, and it appears that the male with the best repertoire is successful. For most species where courtship behaviour has been described, it combines song and posture. In display used by many species the tail is fanned and raised and lowered, an action which has the effect of displaying its rufous colour or the white outer-tail feathers. The male Natal Robin raises his wings in a half-open position, vibrates them, and dances from side to side as he approaches his mate. He also performs a short fluttering flight with audible wing-beats. The Cape Robin fans his tail and assumes various bowing postures in front of the female prior to copulation. The Starred Robin sways slowly from side to side while clinging to the side of the branch on which the female is perched, then he may perform a fluttering flight with fully fanned tail, calling as he does so.

Robins breed in spring and summer, with a peak in November for most forest species. When several consecutive broods are raised the season may be protracted; a pair of Heuglin's Robins successfully raised three broods between early September and early March. The inhabitants of the more arid regions have a less defined season, breeding opportunistically after rain.

1 A male Heuglin's Robin brings food to his incubating mate. 2 Base material has been added to a Natal Robin's nest to support it. 3 Cape Robins have adapted well to breeding in suburban gardens, but conceal the nest well in vegetation such as a creeper. 4 The Starred Robin's nest is unusual in that it is domed, and is even more difficult to find than other robins' nests.

In the species for which details are known, nest construction is usually the task of the female, although the male Natal Robin assists his mate, and male palm thrushes also help to build. Most robins – notable exceptions being the Starred Robin and the palm thrushes – make cup-shaped nests which are neatly lined. Nevertheless, the variety of structures and sites is considerable, even within the same species. For example, a Natal Robin nesting in a hole in a tree stump may merely line the hollow with leaf ribs and rootlets, whereas one using a hollow in the buttress of a tree may need to support the nest with base material. This species also locates its nests in the banks of gullies or in leaf litter on the ground, preferring low situations. In contrast the Chorister Robin, which uses mainly crevices in trees, may nest 12.5 metres above the ground. The majority of other robins, with the exception of the palm thrushes, tend to build their nests below 2 metres.

Although the choice of nest site is sometimes very varied, certain species have preferred situations. The White-throated Robin sets its nest into leaf litter on the ground or among tree roots, placing the nest rim level with the surface; it is always difficult to find because it is concealed so well. Nests near human habitation are sometimes partially hidden beneath discarded household utensils or tins. In the Chirinda Forest in eastern Zimbabwe Swynnerton's Robin shows a marked preference for placing its nest at the base of the broad leaves of *Dracaena fragrans*. Where this plant is not available the nest may be found in the fork of a shrub, in a hole in a tree or among epiphytes on the trunk of a tree, but it is always low down, less than a metre off the ground. The cup is made of skeletonised leaves, leaf ribs and similar material, and lined with fine fibres and rootlets.

The White-browed, Kalahari and Karoo Robins also nest low down, in a bush or on the ground, sometimes in or under a rusty tin in the veld. The Karoo Robin's nest is almost always on the ground under a bush, with a considerable amount of base material built up into a rampart, as in the nests of some chats. The favourite site of the Bearded Robin is in the top a hollow stump and it situates the nest so that the sitting bird is able to look out. The Cape and Heuglin's Robins have adapted well to suburban gardens and often nest in sites such as creepers growing on fences or walls, hanging fern baskets, or in debris caught in a tree. Heuglin's Robins will accept artificial sites put in place for them, for example an old tin hidden in creepers.

Not cup-shaped like those of other robins, the Starred Robin's nest is domed with a side entrance and is positioned on sloping ground, often against a rock or tree trunk. Some nests are situated above ground level, in hollows in banks or on the trunks of fallen trees. They are extremely well hidden, especially when placed among grasses or ferns which may be incorporated into the structure. The female builds the nest, taking about seven days to do so, constructing it of mosses, rootlets, tendrils and dry leaves with a lining of skeletonised leaves.

The palm thrushes differ from all other African robins in that they build nests of mud. As their name indicates, they are always found in association with palm trees, the Collared Palm Thrush mainly along the Zambezi River from Caprivi to Mozambique and the Rufous-tailed Palm Thrush on the Kunene River in northern Namibia. They build their cone-shaped nests out of mud mixed with plant material such as grass, plastering it in place instead of applying it in pellets as swallows do. Fine plant material is used as lining. The nests are placed in palm trees at the base of a frond, sometimes in a crevice in a large tree such as a baobab, or under the eaves of a roof or on a rafter in an outhouse.

Robin clutches may comprise 2-4 eggs, but most species lay three. Only Swynnerton's and Heuglin's Robins regularly lay two eggs and very rarely three. The eggs vary considerably, even within the same species. Both the Chorister and Natal Robins lay dark chocolate or olive-green eggs and it is these, as well as to a lesser extent the eggs of Heuglin's Robin, which best match the chocolate-coloured eggs of the Red-chested Cuckoo. Other species lay eggs that are white, creamy, green or blue in ground colour with red-brown speckles, often concentrated at the large end.

The eggs are laid at daily intervals and incubation usually begins with the completion of the clutch, although sometimes the Cape Robin only starts two days later. In most species only the female incubates, but the male Collared Palm Thrush shares this duty with his mate. The males of some species, for example the Karoo Robin, feed the female while she is sitting. The various species' incubation periods vary from 13 to 18 days, but most are close on 14 days. It is interesting that a small species like the Starred Robin incubates for a relatively long time, from 16 to 18 days.

1 *Like other robins, both male and female White-throated Robins feed the chicks, usually raising a brood of three.*
2 *Swynnerton's Robin, which has a very limited distribution, usually nests in* Dracaena fragrans *plants in the shrub layer of Zimbabwe's Chirinda Forest.*
3 *A White-browed Robin has found shelter for its nest against a rusty tin.*

__1__ Karoo Robins almost always nest on the ground, under a bush or thorn tree. __2__ A Rufous-tailed Palm Thrush's conical nest has been built in a palm tree along the Kunene River. __3__ Palm Thrush nests comprise mud mixed with plant material such as grass and rootlets. A Collared Palm Thrush has plastered its nest to the wall of a building.

Robins are notoriously wary near the nest and are reluctant to return if there is any sign of danger. Incubating White-browed Robins often freeze on the nest and rely on their camouflage until the danger has passed. Brown Robins are silent near the nest and males only sing a considerable distance away. Conversely, the chittering alarm note of Swynnerton's Robin alerts the observer to the approximate locality of a nest.

The skin of newly-hatched young of the various species is dark pink, orange or blackish. They are brooded when small, the male bringing food and passing it to the female. Later both birds feed the chicks and maintain sanitation by removing faecal sacs. When feeding, as well as when building, palm thrushes have the habit of stopping at regular staging posts on the way to the nest. The nestling period for robins ranges from 12 to 17 days, and for most species is close on 14 days. The exception is the Collared Palm Thrush, for which the only record is 20 days.

In those robins for which observations are available, the period of post-nestling dependence may last for 28 to 42 days, although the young can feed themselves long before this. Each Starred Robin parent cares for at least one chick and, like all other robins, both are very alert to the presence of danger, warning the young with alarm calls which include the mimicked alarm calls of other species. The young of this species are unique amongst robins in having two distinct immature plumages: the spotted juvenile plumage typical of robins is moulted after three months and is replaced by a uniformly olive plumage which is retained for at least a year. The immatures' cryptic plumage helps them to avoid detection by predators and, more importantly, spares them from eviction from their parents' territory because they are readily distinguishable from intruding adults. Many immature birds of prey have a distinct plumage for the same reason. An intensive ringing programme monitoring the breeding success of Starred Robins has revealed that half of the eggs in 60 nests kept under observation produced chicks successfully, but only a fifth of these eggs resulted in young that survived to adulthood. The reason for such high mortality is not known.

Warblers

In the context of this book, a wide variety of species in several genera – all members of the warbler family, Sylviidae – have been grouped under the heading 'Warblers'. They include tit-babblers, hyliotas, apalises, crombecs and eremomelas, as well as the Grassbird and the Rockrunner, totalling 33 species which breed in southern Africa. Members of the Sylviidae not included, the cisticolas and prinias, are discussed in the following chapters.

Using the nest type as a criterion, the 33 species have been grouped as follows: Tit-babbler and Layard's Tit-babbler; the Yellow-breasted and Mashona Hyliotas; the African Marsh, Cinnamon Reed, Cape Reed and Greater Swamp Warblers (genus *Acrocephalus*) and the Yellow Warbler; the African Sedge, Barratt's, Knysna and Victorin's Warblers (genus *Bradypterus*) and the Broad-tailed Warbler; the Yellow-throated Warbler; the Bar-throated, Chirinda, Black-headed, Yellow-breasted and Rudd's Apalises; the Red-faced and Long-billed Crombecs; the Yellow-bellied, Karoo, Green-capped and Burnt-necked Eremomelas; the Bleating, Barred and Stierling's Barred Warblers; the Cinnamon-breasted Warbler; the Grassbird and Rockrunner; and the Moustached Warbler. Each group is discussed with emphasis on the type of nest it builds, although other information about breeding is incorporated where it is known or of particular interest.

The warblers are, with few exceptions, an extremely vocal family, and as many species are skulking and elusive, they can often only be identified by their calls. Their songs and calls are used to advertise territories, maintain the pair bond and for courtship. In general the courtship behaviour of warblers is poorly known and there are few characteristic displays, but vocalisations, including duetting in some species, undoubtedly play a part. The breeding season extends from spring to late summer but varies regionally depending on the rainfall; in the south-western Cape a spring and early summer season is clearly defined. In arid regions, as is often the case, breeding may be opportunistic after rain.

Usually two or three eggs are laid, but sometimes four. They vary considerably, and may have a ground colour of white, pale pink or bluish which is lightly freckled or variably spotted and blotched with red-brown, brown or grey. The markings are often concentrated at the large end. Even within a species the eggs may vary, those of the Bleating Warbler, for example, being plain white or pale blue with or without red-brown spotting.

In general the female undertakes incubation duties, although in some species the

1

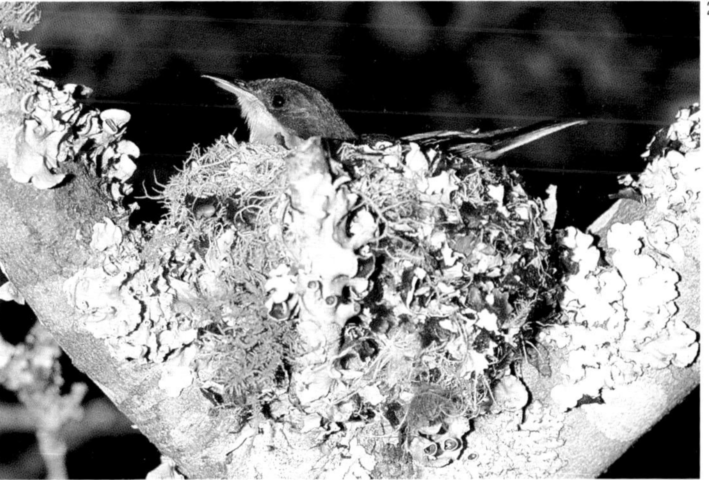

2

3

4

1 *The Tit-babbler builds a small cup of plant material, concealing it well in a small bush or tree.* ***2*** *The neat cup made by the Mashona Hyliota is, like a batis's nest, externally decorated with lichens and is difficult to detect on a high, lichen-covered branch.* ***3*** *Knysna Warblers conceal their nests well in thick shrubbery, and very few have been found.* ***4*** *Victorin's Warbler nests have as rarely been recorded, being located low down in mountain fynbos vegetation.*

2

3

males also do a share. For most warblers the incubation period lasts from 12 to 14 days but, strangely enough, for some it is longer: the Yellow-throated Warbler incubates for 17 days, the Knysna Warbler for 19 days and, longest of all, Victorin's Warbler for 21 days.

Most newly-hatched warbler chicks are naked, although some species have down, and most are characterised by having black spots on the tongue, the size and position of which vary from one species to another. Both parents feed the chicks and attend to the sanitation of the nest. Nestling periods range from 12 to 18 days but most are close on 14 days, and in the case of ground-nesting species the young often leave before they can fly. Not surprisingly for a group which includes so many elusive species, little is known about care of the chicks once they have left the nest.

The nests of the two tit-babbler species are small cups of dry grass and other plant material which includes pieces of the climbing creeper *Galium tomentosum* where it is available. They are placed in a small bush or tree, sometimes in a clump of mistletoe, and are usually well concealed. The hyliotas, like the tit-babblers, have had an uncertain taxonomic history and were once placed with the flycatchers, possibly because their nest is similar to that of the batis, a member of the flycatcher group. They build a cup-shaped nest using fine material such as 'old man's beard' lichen which they bind with cobweb and decorate externally with lichens. It is placed high up in the fork of a branch where it matches the surrounding lichens and is very difficult to detect. Young Mashona Hyliotas have been observed to remain with their parents in their territory until the following breeding season.

4

5

6

1

1 *The Cape Reed Warbler's neatly woven nest has a deep cup.* ***2*** *The African Marsh Warbler's nest is not as neatly finished as the nests of other 'reed' warblers.* ***3*** *The African Sedge Warbler's nest is a loose, untidy cup that is, however, neatly lined.* ***4*** *The Yellow Warbler's nest is similar to nests of other 'reed' warblers and is attached to upright stems of reeds or grass in rank vegetation, but not necessarily over water.* ***5*** *A Yellow-breasted Apalis peers into the entrance to its nest with food for the chicks.* ***6*** *Apalis nests are enclosed, upright and oval, with a side entrance near the top. Rudd's Apalis builds a characteristically flimsy nest and places it in a thorny acacia.*

1

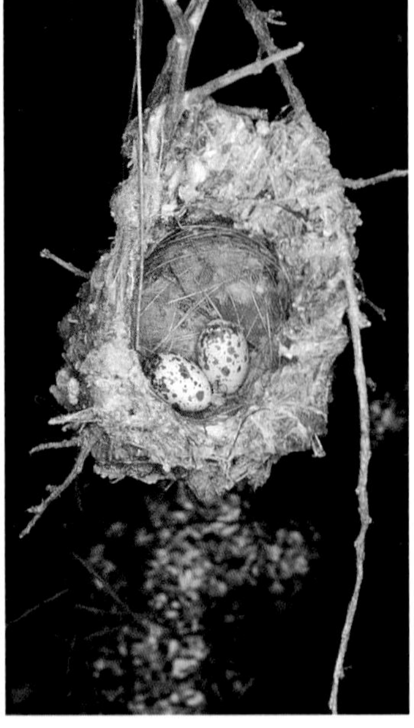
2

3

4

5

__1__, __2__ The finely structured nest of the Long-billed Crombec is, like other crombec nests, purse-shaped. Made of plant material bound with cobwebs, it hangs usually no more than a metre above the ground. __3__ Yellow-bellied Eremomelas sling their cup-shaped nests between two branches. __4__ A Green-capped Eremomela removes a faecal sac. Its tiny nest is slung among leaves high up at the end of a branch. __5__ Karoo Eremomelas, on the other hand, place their deep nests low down inside a bush, lining them thickly with down.

The four *Acrocephalus* or 'reed' warblers and the Yellow Warbler all build nests which are attached between upright stems of reeds, grass or weeds. The Cape Reed Warbler's nest is always sited over water, sometimes beneath a heron's nest, and the shell is made with strips of dry reeds and grasses which the birds often wet to make them more pliable. The cup is very deep, to prevent the eggs falling out when the reeds sway, and it is finely lined with material similar to that used on the exterior, as well as with a few soft feathers. The African Marsh Warbler constructs a nest similar to that of the Cape Reed Warbler but it is not necessarily associated with water and may be in rank vegetation some distance from it. The nest of the little-known Cinnamon Reed Warbler, once considered to be a race of the African Marsh Warbler, is probably also similar. The Greater Swamp Warbler of northern Botswana occurs mainly in the Okavango Delta in papyrus and stands of swamp figs. Extralimitally it makes a typical reed warbler's nest in papyrus, but the only Okavango nest on record was built in the matted outer branches of a swamp fig. The Yellow Warbler's nest is placed in rank vegetation, often away from water, and is a smaller version of the typical reed warbler structure.

The nests of the four *Bradypterus* warblers and the Broad-tailed Warbler are among the most difficult to find in southern Africa, and the author has tried in vain to find nests of Barratt's Warbler in eastern Zimbabwe and the Knysna Warbler at Kirstenbosch near Cape Town. The only compensation for his persistent efforts has been the birds' remarkably loud and melodious song. The nests differ markedly from those of the 'reed' warbler group and are not attached to the stems of plants. The African Sedge Warbler makes an untidy shell of dry flat reed blades and grasses, and lines it with finer material. The nest, so loosely constructed that it falls apart if lifted, is always situated in a swampy area, in a clump of sedge or at the base of reeds. The Broad-tailed Warbler builds a similar structure but not necessarily in swampy ground, sometimes placing it in rank vegetation on dry ground. This species may advertise its territory by making short aerial cruises during which it emits a staccato call and flicks its tail.

Both Barratt's and the Knysna Warblers nest low down in thick shrubbery, and the neat cup-shaped structure has a rather untidy exterior. The nest of Victorin's Warbler is similar, but with a rather more spacious cup which is lined with fine grasses. It is situated in mountain fynbos, either low down in a clump of vegetation or on the ground. Very few nests of these three species have been recorded, and one of the reasons that they are so difficult to locate is that the birds creep to them through the undergrowth like mice. Ironically, once a nest has been found the birds may be extraordinarily tame and return to it if the observer sits quietly nearby.

The Yellow-throated Warbler, an inhabitant of evergreen forest, is another species that was once placed among the flycatchers. Its nest, being dome-shaped with a side entrance, differs from those of all other southern African warblers. It is built of moss and skeletonised leaves with a lining of soft plant material and feathers, and is placed

__1__ Grassbird chicks, like those of most warblers, have black spots on the tongue. __2__ The Rockrunner's bulky grass nest, similar to that of the Grassbird, is well hidden within a clump of grass. __3__ So well hidden are the nests of Cinnamon-breasted Warblers that very few have been found. They are upright oval structures with a side entrance near the top, and are very thickly lined.

on the ground amongst moss, grass or ferns, or on the lip of a bank where it is extremely difficult to find. The young have a covering of down and lack mouth spots, unlike most other warbler nestlings.

The five apalises include three species with very restricted distributions: the little-known Chirinda Apalis is confined to eastern Zimbabwe, the Black-headed Apalis occurs mainly in coastal Mozambique but just reaches eastern Zimbabwe, and Rudd's Apalis is found in Zululand northwards into Mozambique. All the apalises build an enlosed upright oval nest with a side entrance near the top. They use a variety of materials such as moss, pieces of fine creeper, grass and cobweb, and the inside has a soft lining. The Bar-throated and Yellow-breasted species place their nests in a matted bush or creeper, siting them fairly low down, but Rudd's Apalis prefers to breed in thorny acacias. The two forest apalises, the Black-headed and Chirinda, locate their nests as high as 7 metres above the ground and use lichens, moss and, in the case of the latter, also epiphytic liverwort. Their nests are usually on a hanging branch and well concealed amongst 'old man's beard' lichens. Rudd's Apalis differs from the other species in that its nest is loosely woven and flimsy, giving the appearance of lattice work.

The crombecs build characteristic purse-shaped hanging nests which, in the case of the widespread Long-billed Crombec, are usually a metre, sometimes less, above the ground in a bush where there is a hollow among branches. In arid country they are often situated in a bush growing alongside a drainage line. The Red-faced Crombec, on the other hand, places its nest much higher in a tree in its woodland home in Zimbabwe and adjacent Mozambique. Crombec nests are exquisite structures made of plant material bound with cobwebs and decorated externally with items such as pieces of dry leaf or bark, seed pods and the egg-cases of spiders. The nest of the Long-billed Crombec has a tongue of material at the back which attaches it to the branch from which it hangs, and the Red-faced Crombec tends to sling its nest between two twigs, but the distinction is not a constant one. When the parent is incubating it sits low down in the deep nest, facing towards the back with just its head protruding.

Although the nests of the four eremomelas are all cup-shaped, each is distinctive because of either its structure or its situation. The Yellow-bellied Eremomela builds a nest like that of a white-eye, slinging it between two branches, quite often low down in a bare bush in arid country. In such an environment it occurs alongside the Karoo Eremomela which, in the breeding season, emits a monotonous and far-carrying *seep* call for an hour before sunrise. Instead of being slung like that of the Yellow-bellied Eremomela, the Karoo Eremomela's nest is placed among branches inside a low bush and is characterised by its deep cup which is very thickly lined with downy material. It has been suggested that the thick walls insulate the nest by keeping it cool during the day and warm at night.

The Green-capped Eremomela has the distinction of constructing a nest that is one of the hardest of all to find. It is a tiny cup decorated externally with pieces of rust-coloured plant material and is slung among leaves right at the end of a branch of a woodland tree 6 to 7 metres above the ground. Like the Karoo Eremomela it calls monotonously for an hour before sunrise. Although eremomelas often live in family

THE 'TAILOR-BIRDS' OF SOUTHERN AFRICA

Usually when tailor-birds are mentioned it is with reference to south-east Asian species of the genus *Orthotomus* which stitch leaves together to form a cup in such a way that the contents are visible from above. It is not generally known that there are tailor-birds too in southern Africa, including two cisticolas and, on a limited scale, the Tawny-flanked Prinia. The local tailor-birds also stitch leaves together, and most with a skill that probably exceeds that of their Asian counterparts.

The three warblers that qualify as tailor-birds are the Bleating, Barred and Stierling's Barred Warblers. They 'sew' several leaves together by punching short lengths of web through the leaves and teasing them out at either end so that they cannot pull loose, a technique akin to pop-riveting. The leaves form a vertical oval nest with an entrance at the side near thc top, and this is lined with fine grasses, tendrils and downy plant material. The shape of the nest depends on the leaves used, and some are sewn between broad blades of grass.

The Bleating Warbler's nest often appears flimsy whereas the structures built by the barred warblers are far more substantial. The latter also differ in being stitched into a cluster of leaves (as many as 30 leaves may be incorporated) at the end of a hanging branch, as well as in the entrance usually being concealed by overhanging leaves. They are very thickly lined with downy plant material. The Bleating Warbler usually builds low down, usually not more than a metre above the ground, but barred warbler nests are higher, between 1 and 5 metres above the ground, and they are remarkably well concealed, resembling a natural cluster of leaves.

Apart from their exquisite nests, the three warblers have displays that are notable in a group not known for exhibitionism. The Bleating Warbler jumps up and down on a perch as though attached to it with an elastic band, flicks his tail, whirrs his wings and calls at the same time. The barred warblers also perform aerial manoeuvres which incorporate undulating and fluttering flights near the female. All three have penetrating calls which are repeated persistently ●

***1** Bleating Warblers 'pop-rivet' leaves together with short lengths of web to form their nests, which they line with fine grasses and down. **2** This Bleating Warbler nest, unusually, is built among grass stems rather than leaves.*

groups, the Green-capped Eremomela is an especially sociable species that can be found in small groups in a territory that is maintained all year. When two groups meet they challenge each other with chittering calls and flick their tails and snap their bills until the intruders move away. Their sociable behaviour extends to their breeding habits, and all members of the group have been seen to help with nest building, incubation and feeding young. This is the only southern African warbler species which is known to have co-operative breeding habits, although Burnt-necked Eremomelas regularly live in groups and may also breed co-operatively. The latter species occurs in acacia woodland and feeds high up in the canopy. Its small neat cup is well concealed in the outer branches of an acacia and, like the nest of the Green-capped Eremomela, is extremely difficult to find.

The Cinnamon-breasted Warbler inhabits rocky hillsides in arid regions, running over the rocks like a Rockrunner. It is a scarce and elusive species and is usually only located when it calls, but even then it is difficult to see. Its nest is well hidden in thick bushes or clumps of grass, and very few have been found. An upright oval structure with an entrance in the side near the top, it is so thickly lined with downy plant material that the eggs lie just below the entrance.

Another species which conceals its nest well, placing it low down in vegetation, is the Grassbird. Its rather jumbled melodious call is reminiscent of that of Victorin's Warbler, but unlike the latter species the Grassbird often sings from a prominent perch to advertise its presence. Its bulky nest of coarse grass has a neatly lined cup.

The Rockrunner, an inhabitant of rocky hillsides interspersed with grass on the central plateau of northern Namibia, has long been a taxonomic ping-pong ball, having been placed among the babblers, thrushes and warblers at different times. At last it has come to rest among the warblers, its closest affinities being to the Grassbird which it resembles in nesting habits, call and, to a lesser extent, appearance. Its nest is very like that of the Grassbird, with a bulky exterior of grass that is lined with finer grasses, and is extremely well concealed in the centre of a clump of grass. The eggs, two or three in a clutch, are pale buffy pink in ground colour with small red-brown spots concentrated in a zone at the large end. The incubation and nestling periods are unrecorded. At one nest the parents always returned to feed the chick by a set route, using the same perches on the way in. The young leave the nest before they can fly properly and creep around in the undergrowth.

The final species in this diverse range of warblers is the Moustached Warbler which, in contrast to many 'warblers', really does warble. It calls from an elevated perch in much the same way as the Grassbird, but otherwise skulks in rank vegetation. It occurs in a few isolated localities in eastern Zimbabwe and adjacent Mozambique but it is rare and little is known about its habits. The nest is a bulky bowl of dry grass placed low down in vegetation.

Cisticolas

Often discounted as the most frustrating of 'little brown jobs' because they look similar and have three different plumages – breeding, non-breeding and immature – cisticolas can, in fact, be identified on the basis of habitat, behaviour and calls. The 18 species in southern African occur in a variety of habitats such as grasslands, wetlands, bushveld, Karoo scrub and cultivated fields, and some are widely distributed. Others are more restricted, but only the Chirping Cisticola is very limited in its range, mainly to the Okavango Delta in northern Botswana.

Cisticolas are territorial and, with the possible exception of the Fan-tailed Cisticola, also monogamous. Each species has a characteristic call and several have distinctive display flights, from which most species can be identified, especially in the breeding season. Calling and displays serve both territorial and courtship functions. One of the species that displays aerially is the Fan-tailed Cisticola, which cruises at a height of about 15 metres and performs a series of undulations accompanied by a monotonous series of *zit zit zit*... calls at the rate of about one a second. This cisticola has a wide distribution outside Africa from southern Europe to Asia and Australia. In these areas it has been found to be polygynous, but in Africa there are no reports of similar behaviour.

Three other species, the Cloud, Ayres' and Pale-crowned Cisticolas, are known for their high aerial flights – often so high that they cannot be seen without binoculars – which have earned them the name 'cloud scraper' cisticolas. Each has its individual display, some diving down and calling only, others making wing-snapping noises. Some observers maintain that the Cloud Cisticola makes the final part of its dive just above the nest, but even with this clue it is not easy to locate it. Several other species, notably the Black-backed, Chirping, Levaillant's and Croaking Cisticolas, make low, floppy aerial displays during which they jerk their tails. Others such as the Grey-backed, Wailing and Neddicky sing monotonously from a perch; the Neddicky's call is remarkably ventriloquial and difficult to pinpoint.

Cisticolas breed from spring into late summer and, as most depend on grass cover for their nests, breeding is closely linked to rainfall. The typical cisticola nest is an upright oval with a side entrance near the top and it is lined with soft material such as plant down. It is usually situated low down, often at ground level. The Black-backed and Chirping Cisticolas may build their nests in reeds or grass over water, sometimes at a height of about a metre. Without exception, cisticola nests are well concealed, and in many species the birds pull blades of growing grass over the nest or incorporate them into the structure.

The Fan-tailed Cisticola is unique among the cisticolas in building a nest in the shape of a short-necked bottle with the entrance at the top. It binds upright blades of grass together with cobweb, leaving a vertical entrance at the top which is 2 centimetres in diameter. As the incubation period progresses the nest is lined with soft material until it is thickly padded, especially at the bottom. The nest structure of the Pale-crowned Cisticola is similar to that of the Fan-tailed Cisticola, but it does not have a vertical open entrance and grass blades are bent over to form a bower at the top.

The Singing and Red-faced Cisticolas qualify as 'tailor-birds', building their nests within a framework of stitched leaves (see box opposite). The entrance to the nest is fairly large and the interior is fashioned out of broad blades of grass and lined with plant down. Unlike most other cisticola nests, these are placed in rank vegetation about 0.5 to 1.5 metres above the ground.

Little has been recorded about the details of nest construction, except that in the case of the Red-faced Cisticola both sexes build. Building progress may be quite rapid; a Desert Cisticola finished construction in four days and a Neddicky laid its first egg five days after having started the nest.

Three or four eggs are usually laid, but sometimes two or five, and very occasionally more. The colour is very variable, even within the same species, and usually ranges from white to pale greenish blue. The eggs of some species are blue or turquoise and both the Black-backed and Chirping Cisticolas lay eggs that are salmon-pink in ground colour. Some are unmarked whereas others have a freckling of rust-red, or are variously spotted and blotched with brown, grey or lilac.

1

2

3

1 *A Cloud Cisticola's nest is located in short grass at ground level.* ***2*** *The Red-faced Cisticola, like the Singing Cisticola, 'stitches' its nest among leaves.* ***3*** *Rattling Cisticolas build typical cisticola nests – upright ovals with an entrance near the top – and place them low down in grass cover.*

__1__, __2__ The Fan-tailed Cisticola builds a distinctive bottle-shaped nest, with the entrance at the top. A nest cut open shows how cobweb has been used to bind together blades of grass. __3__ The Chirping Cisticola, which occurs mainly in the Okavango Delta, may build its nest in reeds over deep water. __4__ Levaillant's Cisticola is associated with wetland and its nest may be found in thick grass bordering a vlei.

Incubation is thought to be undertaken mainly by the female, but few observations have been made. A characteristic of cisticolas is that they add lining to their nests throughout the incubation period so that it becomes thicker all the time, but they stop once the chicks have hatched. The incubation period lasts for 12 to 15 days in all species for which it has been recorded.

The chicks hatch naked and have characteristic tongue spots. Large young of the Fan-tailed, Pale-crowned, Desert and Croaking Cisticolas have been heard to emit a hissing sound when disturbed in the nest, presumably to deter predators. Both parents feed the chicks and remove and drop the faecal sacs. The nestling period ranges from 12 to 17 days but is usually close on 14 days. Little is known about the care of the chicks once they have left the nest. Tinkling Cisticola young were still being tended by the male 40 days after leaving the nest, although the female had already laid another clutch of eggs which she was incubating. Young Lazy Cisticolas were still under parental care a month after leaving the nest.

Prinias and prinia-like Warblers

As the heading to this account suggests, there are certain problems involving the taxonomic affinities of members of this group, and it is interesting that nest structure has been a criterion for redefining the position of some species. Of the seven species included (which would increase to eight if the separation of the Spotted Prinia into the Spotted and Karoo Prinias is accepted), four are 'true' prinias and the remainder have been classified as warblers, although they bear a number of similarities to prinias. Until recently the Namaqua Warbler was even known as the Namaqua Prinia, but it has been redefined taxonomically as a warbler. A similar situation once pertained to the Rufous-eared Warbler. The third prinia-like warbler is the Red-winged Warbler.

Prinias are characterised by their lively behaviour and calls, and they may occur in small family groups. Roberts's Prinia, for example, is regularly encountered in parties of up to six birds, which attract attention with their noisy calling. All four species are monogamous and territorial, using vocalisations to advertise their territories and intensifying their calling in the breeding season. The Spotted Prinia is the only prinia which has been studied in any detail, and its courtship behaviour involves a dipping display flight to the accompaniment of a staccato ringing call. Both members of a pair have also been seen displaying by drooping their wings and fluttering them. It is interesting to note that there have been two observations of Spotted and Black-chested Prinias interbreeding.

Breeding extends from spring into late summer, and in summer-rainfall areas is associated with the growth of rank grass cover. In the south-western Cape Spotted Prinias may breed as early as July, but in arid regions the nesting of this species and the Black-chested Prinia is often opportunistic in response to rainfall.

A typical prinia nest is an elongated upright oval with a small side entrance at the top and is built in rank vegetation or in low bushes, rarely more than a metre above the ground. It starts off as a hammock of grass slung between two stems, and from this base thin pieces of green grass that dry to brown are woven so closely that the completed nest has the texture of very fine knitting. Sometimes Tawny-flanked Prinias attach their nests to reeds or grasses over water, or to surrounding leaves in which case the nests are loosely stitched to the vegetation. Both this species and Spotted Prinias have been known to build their own nests inside that of a bishop or widow, using the latter as a framework. In the Spotted Prinia both sexes build. The Tawny-flanked Prinia may line its nest with a few pieces of dry grass, but both the Black-chested and Spotted Prinias use plant down such as that of the 'kapokbossie' as lining. A characteristic of prinias is that they add lining to their nests only after incubation has started.

The nest of Roberts's Prinia is an exception. This species, found only in eastern Zimbabwe, also builds an oval nest with a side entrance, but the structure lacks the fine knitted texture of other prinia nests, has a large entrance, and is unlined. It is placed low down in rank vegetation.

Usually three or four eggs are laid, but sometimes two or five, and very occasionally more. Roberts's Prinia lays turquoise-blue eggs with bold chocolate spotting and underlying lilac marks. The eggs of the Spotted Prinia are blue in ground colour with darker blotches and scrolls. The ground colour of the Black-chested and Tawny-flanked Prinias' eggs is extremely variable and may be cream, fawn, pale blue or pinkish, but is always overlaid with blotches and scrolls.

No further details of the breeding biology of Roberts's Prinia are known. In the other prinia species, however, it appears that only the female incubates, and a sitting female may often be identified by her tail feathers which become bent in the narrow confines of the nest. The incubation period ranges from 12 to 14 days for all three species, and once hatched the chicks are tended by both parents. Young Spotted Prinias may be cared for by their parents for at least 21 days after they have left the nest.

The Namaqua Warbler is found in the arid western regions, usually near riverbeds

1

2

***1** When nesting in leafy vegetation, the Tawny-flanked Prinia loosely 'stitches' its nest to the surrounding leaves.*
***2** The finely woven appearance of the Spotted Prinia's nest is characteristic of structures built by 'true' prinias.*

1

2

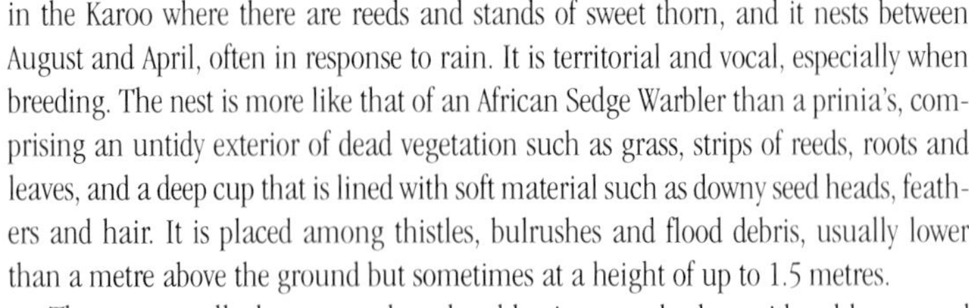

in the Karoo where there are reeds and stands of sweet thorn, and it nests between August and April, often in response to rain. It is territorial and vocal, especially when breeding. The nest is more like that of an African Sedge Warbler than a prinia's, comprising an untidy exterior of dead vegetation such as grass, strips of reeds, roots and leaves, and a deep cup that is lined with soft material such as downy seed heads, feathers and hair. It is placed among thistles, bulrushes and flood debris, usually lower than a metre above the ground but sometimes at a height of up to 1.5 metres.

The eggs, usually three, are pale to deep blue in ground colour with red-brown and grey-brown spots and blotches, but they lack the scrolls that typically decorate prinia eggs. The incubation and nestling periods are recorded as being close on 16 and 15 days respectively. When considered in conjunction with other aspects of this species' biology, the nesting details corroborate the taxonomic redefinition of the Namaqua Warbler.

The Rufous-eared Warbler is a typical species of arid regions such as the Kalahari Desert and the Karoo, and may breed at any time of year in response to rainfall. It creeps through the veld like a mouse and is not easily seen, except when it proclaims its territory by perching on top of a bush and giving vent to a penetrating and rapidly repeated *tee tee tee*... call. The male's courtship display includes body jerks, tail-flicking and wing-quivering as well as singing.

The nest is placed low down in a bush, often on the shady side, and in the Kalahari usually in *Rhigozum trichonotomum* bushes. It is oval, with an entrance in the side near the top, and is made with unwoven dry grass. Unlike prinias, the Rufous-eared Warbler adds the lining of downy plant material before the eggs are laid. The clutch is normally three or four eggs but 2-7 have been recorded. They are very pale blue, appearing almost white, and unmarked. An incubation period of 12 to 13 days is followed by an 11- to 13-day nestling period, during which both parents feed the chicks.

Although it is prinia-like in some of its habits, the Red-winged Warbler also does not make a neatly woven nest and is classified in a genus of its own. It occurs in Mozambique and at low altitudes in adjacent eastern Zimbabwe, inhabiting tall grass in woodlands, rank vegetation and bamboo thickets. It is a cheerful species which, like Roberts's Prinia, is found in small groups of up to six birds that keep in touch by means of melodious tinkling calls.

Little is known about its breeding habits. The nest is placed low down in rank vegetation, and one found in Zimbabwe was a thin-walled, upright oval with a side entrance near the top. It was made of dead grey grass and lined with feathery grass heads and a little cobweb. Despite descriptions to the contrary, it appears that this species does not stitch its nest into leaves. The few available breeding records from Zimbabwe are for November and January, and the two or three eggs laid are pale green with streaks of pale rust-red concentrated at the large end.

3

1 *The Namaqua Warbler's nest is an open, loosely built cup similar to that of the African Sedge Warbler.*
2 *Although prinia-like in its behaviour, the Rufous-eared Warbler does not build a woven nest as do 'true' prinias.*
3 *A Roberts's Prinia nest differs from that of any other 'true' prinia in that it is not finely woven, has a large entrance, and is unlined.*

Flycatchers and Batises

Among the 20 flycatchers, including batises, that breed in southern Africa are some of the region's most attractive birds. In some species, such as the batises, the difference between male and female is striking, whereas in others the sexes are similar. Little is known about certain species, such as the Vanga and Livingstone's Flycatchers, but the biology of the beautiful Paradise Flycatcher, which is widespread and common, has been well documented. Found in a diverse range of habitats ranging from evergreen forest to arid Karoo scrub, most flycatchers are resident. Exceptions are, for example, the Fairy and Paradise Flycatchers which are intra-African migrants.

Flycatchers are territorial and monogamous. Co-operative breeding, in which both adult and immature helpers were involved, has been observed only in the case of the Marico Flycatcher, but it is likely that it will also be recorded in the Pallid Flycatcher, and possibly also in the Wattle-eyed Flycatcher, both species which associate in small groups.

Many flycatchers, especially the batises, are extremely vocal and they call to proclaim a territory, to maintain the pair bond and during courtship. The male batis uses his call in a number of courtship displays, such as when he perches and spreads his tail, lifts his wings and raises his rump feathers, or when he performs a jinking flight, making a 'fripping' sound as he does so and again raising his rump feathers. The Wattle-eyed Flycatcher, which is closely allied to the batises, has a similar zigzag aerial display and the female, like batises, begs for food with a special call and quivering wings. The male feeding the female is an integral part of the courtship behaviour of many flycatchers, and some also exhibit tail-flicking and wing-quivering displays. In the case of the Paradise Flycatcher the male chases after the female in spectacular fashion, flaunting his long tail. On the other hand, no courtship behaviour has been described for the Marico and Pallid Flycatchers, other than that they call from a perch.

The breeding season falls mainly in spring and summer, from September to January, with a peak for many species from September to November. Those that inhabit arid regions, such as the Marico and Chat Flycatchers and the Pririt Batis, often breed in response to rainfall.

Of all southern African birds, flycatchers build some of the most exquisite nests. They vary considerably, however, generally falling into categories according to genus. In the Black and Pallid Flycatchers only the females have been observed building, but in the batises and Wattle-eyed and Paradise Flycatchers both sexes participate. Quite often nests may be found in the same area year after year, sometimes in the same site, and a Paradise Flycatcher has been recorded rearing two consecutive broods in a season in the same nest.

Dusky and Blue-grey Flycatchers nest in a hole in a tree or a bank, or sometimes in a crevice in a building. The former have twice been recorded nesting inside old Cape Weaver nests hanging over water, merely padding the inside with a soft lining. Usually these two species build a padded base of plant material such as moss in

__1__ A Blue-grey Flycatcher has located its nest in a palm tree where a branch has broken off. __2__ Black Flycatchers often nest in tree cavities, especially those which are fire-blackened and thus provide good camouflage for the sitting bird. __3__ Marico and Pallid Flycatchers build similar nests, but the former sometimes lines its nest with feathers. __4__ A Fiscal Flycatcher's nest is well hidden in foliage and is thickly lined with soft plant material. __5__ Dusky Flycatchers nest in any suitable cavity, including a crevice in an embankment. __6__ The bulky nest of a Chat Flycatcher is warmly lined with downy plant material. This species' eggs are characteristically deep blue.

2

3

__1__ Paradise Flycatcher chicks beg vigorously on seeing a parent approaching. __2__ A female Wattle-eyed Flycatcher with her chicks, which she defends boldly if the need arises. __3__ Batises often use lichens to camouflage their nests, but in an area where these are not available a Chin-spot Batis has used pieces of bark instead. __4__ A female Cape Batis sits very tight on her exquisite nest. Incubation falls to her, and while she is carrying out this task she is fed by the male and rarely leaves the nest.

4

which a neat cup is formed, and they line it with rootlets, hair and feathers. The Fan-tailed Flycatcher also nests in holes in trees, either natural cavities or the disused holes of barbets or woodpeckers. Its nest is a small, rudimentary cup of rootlets, grass, bark and hair.

The Black Flycatcher is another cavity-nester, usually in trees but also in a wide variety of other sites, including buildings and abandoned machinery as well as in the nests of other birds such as thrushes, and once in the broken nest of a Red-headed Weaver. A number of records describe nests in holes in trees blackened by fire, a choice that appears to be more than just fortuitous; in these sites the brooding fly-catcher is well camouflaged by its black plumage. Whatever the situation, the cup is always lined with rootlets.

Both the Marico and Pallid Flycatchers make cup-shaped nests of grass, small twigs and roots which they line with finer rootlets and, in the case of the Marico Flycatcher, sometimes with a few feathers. They are placed in trees and are usually 2 to 3 metres above the ground. The Chat Flycatcher's bulky cup is warmly lined with plant down and placed in the centre of a bush, usually at a height of less than 1.5 metres. The Fiscal Flycatcher makes a nest of various plant materials lined with rootlets and plant down. It is concealed in a tree or amongst creepers, and sometimes at the base of the leaves of an aloe.

Two species, the Vanga and Livingstone's Flycatchers, have very restricted distributions in southern Africa, mostly in Mozambique. Very little is known about their breeding behaviour, but the Vanga Flycatcher makes a cup of bark strips bound with cobweb which may be at a height of 30 metres. The only nest of Livingstone's Flycatcher ever recorded was found in Zambia and is described as a dome-shaped nest with a side entrance and constructed of leaves bound with cobweb. It is so different from the nests of other flycatchers that some controversy has arisen about the authenticity of the record.

The Fairy Flycatcher, which has come to rest among the flycatchers after taxonomically alternating between them and the warblers, makes one of the smallest nests in southern Africa, a tiny deep cup with an internal diameter of 3 centimetres.

1

Made with plant material, including pieces of bark bound with cobweb, and given a soft lining, it is well concealed in the centre of a small bush, sometimes amongst debris caught in the branches.

The nests of the Blue-mantled and White-tailed Flycatchers are neat, thin-walled cups made out of plant fibres and bark strips, and bound externally with cobweb. Typically they are placed in the vertical fork of a sapling, often low down. The White-tailed Flycatcher includes in its nest green moss which retains its coloration, and it also binds moss around the bottom of a Y-shaped stem, with the result that the nest has the appearance of a green wine glass.

The batises and the Wattle-eyed Flycatcher build nests that are basically similar: a small neat cup of plant material, bound externally with cobweb and neatly lined with fine fibres. The nest's outside wall and rim are often completely plastered over with lichens so that it blends with surrounding lichen-covered branches. Where lichens are not available, for example in arid regions, pieces of bark are used. The nest is placed in the fork of a tree or sapling, invariably in such a way that it merges with its surroundings.

The neat, lichen-covered cup of the Paradise Flycatcher is one of the most exquisite of the flycatcher nests. It is similar to that of the batises, but is usually placed in a small fork on a thin hanging branch. Quite often these nests are concentrated in a relatively small area, for example in trees around a farmhouse, and pairs may be found breeding in the same area year after year.

Clutches of two or three eggs are most frequently recorded, although sometimes as many as four eggs are laid. In species such as the batises the normal clutch is two. The eggs vary enormously from one species to another, with ground colours ranging from white to greenish blue and, in the case of the Chat Flycatcher, to deep blue. Whatever their ground colour, all the eggs are marked to some degree, varying from a light clouding or freckling to bold spotting, and sometimes these marking are concentrated in a zone at the large end.

Male Wattle-eyed and Paradise Flycatchers assist the female with incubation, but in the other species for which there are records this task is the responsibility of the female. While she is incubating she is often fed on or near the nest by the male. The female batis in particular sits remarkably tight, rarely leaving her post, and when the male approaches to bring food to her she emits a characteristic begging call. The incubation period for most species is close on 14 days, but that of the batises is longer at about 17 days, and the tiny Fairy Flycatcher incubates for 17 or 18 days.

Small chicks are brooded for the first few days and, especially in the batises, the male brings food to the female which she passes to the chicks. Thereafter both birds take part in feeding and either eat faecal sacs at the nest or remove and drop them. Sometimes, if a sac is ejected over the nest edge before the parent is ready, it will deftly swoop down and catch it in mid-air! A Pririt Batis at a particularly exposed nest was seen to stand over the chicks and fan them with its wings. Batises are often extraordinarily tame at the nest and will go about their business despite an observer in full view nearby. The Wattle-eyed Flycatcher, however, defends its nest boldly and will even dive-bomb and strike anyone near it.

Once Paradise Flycatcher chicks are feathered they stand on the rim of the nest and flutter their wings vigorously in preparation for their first flight. This usually occurs when they are between ten and 12 days old. The Chat Flycatcher has a similarly short nestling period, but the young of most species leave after 14 to 15 days, except for the batises which remain in the nest for 16 to 18 days. Young flycatchers may stay with their parents for a few weeks or for very much longer. In the case of the Wattle-eyed Flycatcher they stay for six months, and young Cape Batises stay for up to a year.

__1__ A Blue-mantled Flycatcher's nest is typically placed in a vertical fork and bound into place with cobweb. __2__ The White-tailed Flycatcher incorporates growing moss in its nest, binding it around a Y-shaped stem so that the nest takes on the shape of a wine glass. __3__ The Fairy Flycatcher's tiny nest, only 3 centimetres across, is concealed in the centre of a small bush.

1

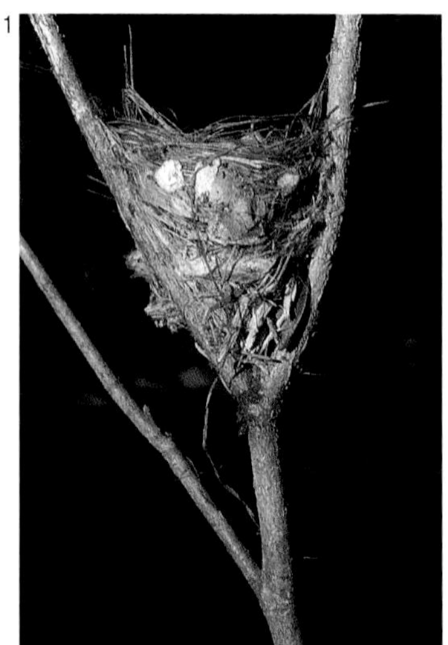

2

3

Wagtails

All three wagtails that breed in southern Africa are monogamous, territorial and apparently pair for life. Two of the three, the Cape and the African Pied Wagtails, are widespread and common and are often gregarious when not breeding. The third, the Long-tailed Wagtail, inhabits perennial rivers and streams in forests, lives in pairs, and is strongly territorial all year. Its territory is linear along a stretch of river, and in a study area in Kwazulu-Natal was measured at 590 metres long on average. The Cape and African Pied Wagtails are aggressively territorial, as demonstrated by their habit of attacking their own reflections in a window or hubcap. In natural circumstances the intruder would give way, making physical contact unnecessary.

No courtship displays have been described for the Cape or African Pied Wagtails, although the male Cape Wagtail may offer twigs or grass to the female at any time of year to maintain the pair bond. Long-tailed Wagtails court by chasing each other in flight, with the pursued bird fanning its tail to display the white outer tail feathers. They also fly up around each other, ascending in a vertical 'spiral dance'. At other times one bird walks towards its mate with its body feathers fluffed up, its wings drooping and its tail raised and fanned to reveal the maximum amount of the white in the plumage. This particularly attractive display is very similar to that used to drive off other members of the same species during territorial encounters.

Although breeding occurs mainly during spring and summer, the nests of Cape and African Pied Wagtails may be found at almost any time of the year in parts of their wide range. In the south-western Cape, however, most Cape Wagtail nests have been recorded in August and September. The Long-tailed Wagtail breeds from August to January, but mainly from September to November. All three species may raise several broods, sometimes as many as four, in a season.

Both sexes construct the nest, using miscellaneous material such as grass, weeds and roots to build up a bulky exterior from a circular base. They then form the neat cup and line it with fine material. The Long-tailed Wagtail also uses moss, and wets the base material before placing it in position, often leaving strands hanging down which serve as camouflage. Unlike the other two species, the Long-tailed Wagtail does not use hair, down or feathers for lining, selecting instead mainly fine rootlets.

Both the African Pied and Cape Wagtails place their nests in a wide variety of sites from ground level upwards. On the ground they build them amongst grass, or they may site them on stumps covered with debris, under bridges or on buildings. Where they live around human habitation they may select any situation, for example a hanging basket of ferns. The Long-tailed Wagtail always nests along streams and rivers, usually on a ledge or in a niche in an embankment, but also occasionally in a tree or under a bridge. It uses the same site year after year, and sometimes the same nest, merely repairing the existing structure.

The wagtails' clutch is usually 2-4 eggs but up to seven have been recorded in the case of the Cape Wagtail. The eggs' ground colour varies from dull greyish white to yellowish with indistinct streaks, spots or cloudings of brown and grey. They are laid on consecutive days, usually in the early morning, and incubation commences when the clutch is complete, with both sexes doing duty in all three species. The incubation period is close on 14 days.

The newly-hatched chicks have pink skin and are covered with tufts of grey down. For the first few days they are brooded during the day by both male and female, and Cape Wagtails brood the chicks at night for at least a week. Both parents feed the chicks and attend to the sanitation of the nest by eating or removing the faecal sacs. In all three species the chicks remain in the nest for close on 15 days, but once they have left the period spent under parental care is variable, depending on whether a subsequent clutch is laid. Long-tailed Wagtails, for example, may care for chicks of a first brood for up to a month, but those of a second for two months. At a Cape Wagtail's nest a parent was seen to lure a boomslang away from its recently fledged chick by feigning injury and keeping just out of reach of the snake.

1

2

3

__1__ Cape Wagtails nest in a variety of sites, including around human habitation. Tufts of grey down cover the newly-hatched chicks. __2__ A Long-tailed Wagtail's nest, situated beside a river, comprises a bulky base and a neat cup lined with rootlets. __3__ Both sexes of the Cape Wagtail brood the chicks and feed them.

Pipits and Longclaws

Ten pipits and three longclaws breed in southern Africa. The latter are easily distinguished from one another by their differently coloured underparts, but pipits are among the most difficult of all birds to identify. The habits, including nesting behaviour, of all members of both groups are very similar so they cannot be used as distinguishing features. Nor can habitat, as although pipits and longclaws occur in a number of different environments such as grassland, woodland and rocky mountain slopes, several species overlap and may be found side by side. To compound the problem some pipits are nomadic, and the Mountain Pipit is an intra-African migrant.

Pipits and longclaws are monogamous and territorial during the breeding season. They have two basic behaviour patterns at this time: calling from a regular perch and calling during an aerial display. Most species perform an aerial display, with variations. The Grassveld (formerly Richard's) Pipit, for example, flies up obliquely and with a dipping motion to a height of at least 30 metres, calling at the top of each undulation, and then drops back to the ground, calling as it descends. The Short-tailed Pipit cruises around in wide circles, calling and making snapping sounds with its wings. Longclaws also call aerially, often with dangling legs, and display their white outer tail feathers in a fluttering flight. They also call from a perch.

Although most species nest in spring and summer between August and January, some, such as the Grassveld Pipit, may breed at any time of the year. Pipit and longclaw nests are all very similar, comprising an exterior of dry grass and a cup neatly lined with fine grass, rootlets and similar material, but no down. They are always well concealed on the ground against a tuft of grass, an earth clod in a ploughed field or a rock, and are usually placed in a hollow so that the rim barely stands out above the level of the surrounding earth. Sometimes grass may be bent over the top of the nest to form a rudimentary bower.

The 2-4 eggs laid are dull white in ground colour with brown and grey speckling over the entire surface, often with markings concentrated at the large end. Some eggs, for example those of the Striped Pipit, have red-brown markings. There is very little information about how the sexes share incubation: in the Grassveld Pipit it is undertaken mainly by the female with some assistance from the male, and in the Yellow-throated and Pink-throated Longclaws incubation is the task of the female. The incubation period for the Grassveld Pipit is 12 to 13 days, and for the three longclaws it is 13 to 14 days.

Like wagtail chicks, the small nestlings have tufts of down and are brooded for the first few days after hatching. Both sexes feed their offspring and remove faecal sacs. The adults are protective of their young and often lure away predators near the nest with injury-feigning displays. They are very wary about returning if there is any sign of danger, and Yellow-throated Longclaws are particularly cautious. Young pipits generally leave the nest after about 14 days, and longclaws after 16 to 17 days. They do not vacate the nest as early as many larks do, but still before they can fly. Because it is so difficult to watch the young once they are out of the nest, there is virtually no information on how long they remain under parental care.

1 

2 

 3

4 

***1** A Grassveld Pipit at its well-concealed nest which it, like all pipits, usually sets into the ground. **2** The deep cup of a Plain-backed Pipit's nest is lined with fine roots. **3** Apparently all gape, Orange-throated Longclaw chicks clamour for food. They will remain in the nest for 16 to 17 days. **4** Both sexes of the Long-billed Pipit feed the chicks and, like other pipit species, are very protective of them.*

Shrikes

It would be true to say that there is virtually no habitat in southern Africa which is not enlivened by at least one shrike species. They are distributed from the evergreen forests of the east to the arid Namib in the west, and without exception they enhance the environment in which they occur. The 24 breeding species range from the ubiquitous Fiscal Shrike to the rare Sousa's Shrike of the extreme north of Botswana and Namibia, and include the confiding helmet shrikes, the brilliantly coloured Crimson-breasted and bush shrikes, the vocal Bokmakierie, and the subtle-hued tchagras with their attractive display flights. Some species fall naturally into subgroups – the boubous, tchagras, bush shrikes and helmet shrikes – and where possible the details of courtship displays, breeding season and egg type are summarised for the subgroup. The helmet shrikes and the White-crowned and Long-tailed Shrikes are sociable and breed co-operatively, and although little is known about the Long-tailed Shrikes, the others have especially interesting habits.

Shrikes are monogamous and territorial, and the pair bond is probably maintained for life. A territory may be occupied year after year, and in one case a Fiscal Shrike's nest was found in the same bush over a period of 14 years. Territorial advertisement may take several forms: the Fiscal Shrike perches prominently, the Puffback Shrike flaunts the white feathers on its back in a display flight, and other species call. Indeed, calling is an essential part of shrike behaviour and as a group the birds are extremely vocal. Duetting, often antiphonally, plays an important role in territorial advertisement, courtship and maintenance of the pair bond. Some displays used for territorial advertisement are also often duplicated during courtship behaviour.

The Fiscal Shrike's courtship behaviour includes a zigzag flight, feeding of the female, and an ecstatic gyrating display at the nest with tail raised and fanned. It is also possible that the habit of impaling food on thorns has courtship significance, advertising the male's fitness as a food provider. Feeding has been recorded as part of the courtship ritual for a number of shrikes, including that of the Grey-headed Bush Shrike. In addition to offering the female food – which she accepts in a submissive wing-quivering posture – the male performs a 'wing-fripping' display flight, flutters his wings and calls. The male Puffback Shrike puffs up his back feathers in display, both perched and in flight, and chases the female. Boubous hop in tandem through the branches while calling, and the male also flies with slow, exaggerated wing-beats. Tchagras perform a characteristic aerial display in which they hold their heads up, spread their tails to reveal the white outer tail feathers, 'frip' their wings and fluff out their grey rump feathers, calling while they do so.

The Bokmakierie provides a good example of the similarity between territorial advertisement and courtship behaviour. Both sexes call from a prominent perch, often in duet but also individually, in a characteristic posture with tail feathers spread, bill pointed upwards and yellow throat exposed. Incorporating elements that are common to the displays of most shrikes, the performance has both a visual and an acoustic effect.

The breeding season falls mainly in spring and summer, with a peak from September to November for many species. In some localities, however, Fiscal Shrikes and Bokmakieries may nest at almost any time of the year. In Namibia the White-tailed Shrike breeds from January to April to coincide with the later rains in the region, and in arid areas breeding occurs opportunistically in response to rainfall. A number of species may raise several broods in a season.

All shrikes make cup- or bowl-shaped nests without a downy lining, but there is considerable variation in the nests' situation and external appearance, as well as in the materials used. In most species for which information is available both sexes build, but in some the male merely accompanies the female as she collects material. The Fiscal Shrike constructs a bulky structure of miscellaneous plant material and may incorporate items such as string and cotton waste. The cup is neatly lined with finer fibres and soft plant material. Quite often this shrike may re-use an old nest or take material from it to build a new one. It places the nest in a bush or tree, preferably a thorn tree, and usually at a height of about 3 metres.

Although Sousa's Shrike has not been recorded breeding in southern Africa, it is resident and must nest within the region. Further north it builds a neat cup of leaf ribs and grass which is bound externally with cobweb and finely lined within. The Long-tailed Shrike makes a bulky bowl of thin thorn twigs and grass which is placed in a thorn tree. Although this species sometimes has helpers at the nest, it appears that only the male and female are involved in construction.

1 

 2

3 

__1__ Fiscal Shrikes build bulky nests of miscellaneous materials, but the cup is always neatly finished. __2__ In contrast, a Sousa's Shrike's nest photographed in Zambia is neatly built and bound externally with cobweb. __3__ The Long-tailed Shrike is a sociable species which normally builds its nest in a thorn tree.

The boubous' nest is a flimsy, bowl-shaped structure of plant material which is finely lined with rootlets, and is sometimes so thin-walled that the eggs may be seen from below. The Crimson-breasted Shrike, despite its resemblance to the boubous, builds a completely different type of nest: a cup made of dry strips of acacia bark. It is usually situated in the thorny acacias which are so much a part of this species' habitat. The Puffback Shrike also incorporates pieces of bark, along with other plant material, in its small but substantial cup, binding it externally with cobweb. The nest is placed in a fork fairly high in a tree.

The nest of the Brubru Shrike is exquisite, resembling the structures of the Spotted Creeper and the cuckoo-shrikes. It is a shallow cup, about 7 centimetres in diameter externally, and is made of plant material such as leaf stalks which are bound with cobweb and decorated on the outside with lichens. Placed about 5 metres above the ground, it is moulded into a fork where there are lichens so that it is extremely difficult to see. An unusual feature of the Brubru Shrike's nest-building behaviour is that it may abandon or destroy completed nests without laying in them. One suggested reason for this is that nest building, which is shared by both birds, may serve a courtship function and synchronise the breeding condition of the pair. If synchrony has not been achieved by the time the nest is complete, then a new nest is started as a continuation of courtship. As each nest incorporates about 600 pieces of

material, the amount of energy that is required to build a new one is considerable.

The tchagras construct a bowl-shaped nest of plant material and line it with fine fibres and rootlets. Both the Three-streaked and Black-crowned Tchagras bind their nests with a little cobweb, but the Southern Tchagra does not. The Marsh Tchagra, which is rare in southern Africa and has a restricted distribution, builds a typical tchagra nest and uses cobweb, but characterises its nest by incorporating a piece of sloughed snake skin into the outside wall. The nests of all species are well concealed and are usually placed fairly low down in a bush. The Bokmakierie's nest is bulky but neatly lined, and is well hidden in the centre of a bush.

The five bush shrikes – the Gorgeous, Orange-breasted, Black-fronted, Olive and Grey-headed – build flimsy platforms of small twigs similar to those of doves, line them with rootlets and other fine material, and conceal them well in bushes or creepers. The Yellow-spotted Nicator's nest is very similar (see page 159) and, if nest structure can be used as a criterion, may justify placing the nicator with the bush shrikes.

The helmet shrike nest is a thin-walled cup of plant material, mainly strips of bark in the case of the White Helmet Shrike, which is profusely bound on the outside with cobweb. Web, collected in the shrike's bill and on the erectile feathers of its forehead, is also used to fasten the structure firmly into a fork or on top of a horizontal branch high in a tree, and is added to the nest throughout incubation and the nesting period until it is thickly plastered with it.

The White-tailed Shrike of Namibia is a delightful species that resembles a terrestrial batis. Its generic name *Lanioturdus*, which translates as 'shrike-thrush', suggests that it is a taxonomic enigma and it has been variously placed with the flycatchers, helmet shrikes (on nest structure) and the bush shrikes. It is now thought to belong with the bush shrikes. Its nest is a bowl of fine grasses bound externally with cobweb like that of helmet shrikes, and it is usually placed 2.5 metres above the ground, but sometimes much higher.

1 *The flimsy nest of the Tropical Boubou is bound externally with a little cobweb.* ***2*** *The Crimson-breasted Shrike characteristically constructs its nest out of dry strips of bark.* ***3*** *A female Puffback Shrike tends her still-blind chick. Placed in a fork high in a tree, the nest is secured to the branches with cobweb.* ***4*** *A Three-streaked Tchagra pushes food into the throat of one of its newly-hatched chicks.* ***5*** *The bowl-shaped nest of the Southern Tchagra is usually placed low down in a bush. Unlike other tchagras, this species does not use cobweb in its nest.*

1 *A male Olive Bush Shrike carries food to its offspring. Both parents share this duty.* **2** *A well-grown Gorgeous Bush Shrike chick cowers on its nest, a flimsy platform of twigs that is typical of bush shrike nests.* **3** *Two fully grown White-tailed Shrike chicks fill their nest. Bound externally with cobweb, it resembles the nests of helmet shrikes.* **4** *Moulded into the fork of a tree, a Brubru Shrike's nest is plastered with lichens and is very difficult to discern.* **5** *Although bulky, the Bokmakierie's nest is always well hidden in a thick bush.*

The White-crowned Shrike is another species which makes abundant use of cobweb, plastering it thickly on the outside of its bulky bowl of dry grass stems. The resulting nest, measuring about 9.5 centimetres across the bowl, is extremely strong and has the consistency of papier mâché. Sometimes nests are repaired and re-used the following breeding season, which is a testimony to their strength. They are placed on a thin outer branch, usually a hanging one, at 4 to 5 metres above the ground.

With the exception of those of White-crowned and helmet shrikes, clutches usually comprise 2-4 eggs, but sometimes larger ones are recorded. In the nests of Long-tailed Shrikes and Bokmakieries, for example, up to six eggs have been counted. White-crowned Shrikes usually lay three or four eggs, but as many as ten eggs have been recorded when two females have laid. Similarly the helmet shrikes' normal clutch of 2-5 has been known to increase to nine, the product of two female White Helmet Shrikes.

The eggs of such a large group of species vary considerably in appearance. The ground colour may be white, cream, pinkish, pale greenish or bluish green, but the eggs of the Bokmakierie are a rich greenish blue and those of the Long-tailed Shrike buffy yellow. The markings may be spots, speckles, scrolls, blotches or cloudings of red-brown, brown and grey, and they are often concentrated at the large end. No species lays plain eggs, although an aberrant clutch of plain white eggs was once found in the nest of a Crimson-breasted Shrike. Several shrikes are parasitised by cuckoos.

Eggs are laid on consecutive days and incubation usually begins on the completion of the clutch. For most species for which there are records both sexes incubate, but the male Fiscal Shrike does not. Brubru Shrikes call while sitting on their well-camouflaged nests, and this is sometimes the only means of locating their position. The incubation period ranges from 13 days in the case of the Puffback Shrike to 19 days for the Brubru Shrike, but for most species it is 15 or 16 days.

The young hatch naked and blind, with skin that varies from blackish to pink in colour. The dark skin of Brubru Shrike nestlings camouflages them well to begin with, and later their speckled plumage renders them almost invisible in the lichen-covered nest. It is usually the female shrike which broods the small chicks, but in some species the male also takes a turn. Both parents feed the young and either eat or remove the faecal sacs. Co-operative behaviour has been observed at Long-tailed Shrike nests where helpers fed the chicks, as well as at a

Crimson-breasted Shrike's nest where two females and a male were in attendance.

The nestling period ranges from 12 to 22 days, and there is no correlation between the size of a species and the length of time its chicks spend in the nest. Chicks of the relatively large Tropical Boubou, for example, leave after about 15 days, whereas those of the smaller Brubru Shrike may stay in the nest for 21 to 22 days. In general young shrikes tend to remain with their parents for a considerable period after having left the nest, often for long after they can fend for themselves. The length of time varies from a few weeks to several months, or even into the following breeding season. The young Puffback Shrike, which remains in its parents' territory until the following breeding season, is of particular interest because it has two plumage stages as an immature: initially it has buff-coloured underparts and then takes on a plumage similar to that of the female. There is a record of an immature feeding a young Puffback Shrike from a subsequent brood.

For variety of appearance, habits and behaviour the shrikes as a group have few equals. Although some have been well studied, much remains to be learnt. The co-operative breeding habits of the Long-tailed Shrike, for example, merit investigation, and the fascinating and delightful White-tailed Shrike, which may yet be found with helpers at the nest, would also be a rewarding species to study in depth.

A SOCIABILITY OF SHRIKES

Of all the birds that breed co-operatively, few have as complex a social system as the helmet shrikes. The White-crowned Shrike has co-operative breeding habits similar to those of the helmet shrikes and is thus included here for discussion. The White and Red-billed Helmet Shrikes are widely distributed in woodland, but the Chestnut-fronted Helmet Shrike is confined mainly to evergreen forest in Mozambique, with small populations in northern Zululand and at the confluence of the Haroni and Lusitu rivers in eastern Zimbabwe. Little is known about its behaviour, but what has been observed indicates that it is similar to that of its better-known and more common relatives.

The three helmet shrike species live year round in flocks averaging seven birds, although they may number up to 30 birds at times. They are permanently sociable in everything they do, including roosting and defending their territory, and when a feeding group moves it does so in unison. A flock always includes a dominant male and female, the core breeding pair. It expands as the young that are produced join it, and dispersal takes place when a group of brothers leaves and teams up with a group of sisters from another flock. A dominant pair emerges in the new flock, and when one dies it is replaced by the next dominant bird of the same sex.

Like other shrikes, the helmet shrikes breed in spring and summer, mostly between September and November. A pair maintains its bond by preening each other, and before copulating they perform a display during which they slowly flap their wings. The male may also initiate courtship by approaching the female with large quantities of nesting material in his bill. The breeding pair selects a site and begins building the nest, but all members of the group assist in construction.

All members also incubate the eggs, which are pale greenish spotted with browns and greys mainly at the large end. The sitting bird does not leave the nest until it is relieved, sometimes calling to the others to indicate that it wants to depart. The whole group flies into the tree, the sitting bird is replaced, and the rest move off again. When the chicks hatch after close on 17 days, all members of the group feed them and attend to nest sanitation, although the chicks also defecate over the edge. They leave the nest after 20 days and remain in the flock until they become helpers during a subsequent breeding season, or depart to form a new group.

The White-crowned Shrike lives in groups of up to 20 birds in winter, but when breeding this number is usually reduced to five or six. They are conspicuous and noisy birds, emitting harsh calls and often wagging their tails up and down in excitement. Breeding takes place mainly from September to December, but in Zimbabwe there is a marked peak in October and November. The dominant pair preen each other, and a flight with exaggerated wing-beats, followed by a glide with wings held in a V, may also be associated with courtship. Copulation has been observed taking place in the nest, and on one occasion a second female arrived and tried to solicit the male's attentions. It appears that more than two birds are involved in constructing the nest, but this has still to be confirmed.

The eggs are creamy in ground colour with grey and brown blotches and spots mainly at the large end, and several birds have been observed helping to incubate them. At one nest it appeared that some members of a group of six merely made 'token sits', often remaining for less than a minute and sometimes hopping on and off the nest in rapid succession. The reason for this restless behaviour could not be established, but it appeared that several birds wanted to participate. The duration of the incubation period has yet to be recorded.

Naked and blind, the newly-hatched chicks have yellowish orange skin and reddish gapes with no mouth markings. The stimulus of the bright gape may be the reason for a remarkable record of two Icterine Warblers feeding White-crowned Shrike chicks in the nest, an observation all the more unusual because these warblers are Palaearctic migrants that do not breed in southern Africa. Several helpers may feed the chicks, but as the birds in the groups studied were not individually marked, it was difficult to know how many were actually feeding. Sanitation is by no means thorough, and the chicks sometimes void their droppings over the nest edge or even in the nest itself. The nestling and post-nestling periods have yet to be recorded, but the young presumably join the group and remain with it as young helmet shrikes do ●

1

2

***1** An incubating White Helmet Shrike will sit tight on the nest until it is relieved by another member of the group.* ***2** Several White-crowned Shrike helpers share the task of feeding the chicks, of which there are usually three or four in a brood.*

STARLINGS

GREGARIOUS, GARRULOUS AND CONSPICUOUS, starlings are a familiar sight in southern Africa. Most of the 14 species in the region are resident – the exceptions being the nomadic Wattled Starling and the Plum-coloured Starling, an intra-African migrant – and half comprise the glossy starlings which share similar nesting habits. Two species, the European Starling and the Indian Myna, were introduced into southern Africa at the turn of the century and have spread widely. Both live near human habitation and are well known, although their breeding biology has not been studied in detail in Africa.

Starlings are territorial and monogamous, although male European Starlings are sometimes promiscuous. Little is known about the courtship behaviour of these species, but singing probably serves for both territorial advertisement and courtship. The ecstatic song of European Starlings, performed with raised head and neck feathers and loosely flapping wings, is undoubtedly an example of this dual function. In the case of the Red-winged Starling a pair stays together all year so there is no need for a courtship ritual, although feeding is sometimes observed.

1

1 The European Starling is versatile in its choice of nest site but seems to prefer man-made structures, including a rusting water tank on Dassen Island.
2 Its eggs are always plain blue.
3 Pied Starlings nest colonially, usually in holes they have excavated in a bank, and helpers contribute to feeding the chicks.

2

3

THE NOMADIC STARLING

Flocks of Wattled Starlings wander from place to place and breed opportunistically when conditions are right. This is usually when good rains in an area are followed by an ample food supply in the form of insects, particularly locust hoppers. Sometimes, presumably when the food supply is insufficient, males come into breeding condition and develop wattles, nests are built, and then the whole colony aborts.

In the south-western Cape, where rainfall is relatively stable from one year to the next, Wattled Starlings may breed annually, whereas in other areas the birds appear almost miraculously from nowhere and start breeding. In one instance in the Zimbabwean lowveld a huge colony was established some 45 years after breeding had last been observed in the area.

Differing substantially from that of other starlings, the Wattled Starling's nest is a large, dome-shaped structure of thorny twigs with a side entrance and a lining of grass and feathers. Sticks are continually added to the structure, so that some attain a considerable size. Both birds build the nest, siting it in a tree – preferably a thorny one – or occasionally on a telephone pole. Stands of eucalypts are favoured in the south-western Cape. Frequently several nests coalesce to form a single mass with a dozen or so compartments in it.

The 2-5 eggs in the clutch hatch after about 11 days. Both parents bring prodigious beakfuls of food to the chicks, although from time to time the male may break off from his duties to perch nearby and sing.

A colony of Wattled Starlings may contain hundreds of birds and when they are breeding the air is filled with their murmuration. Then, almost as suddenly as it started, the activity ceases; the starlings gather into flocks and move on. Only the silent nests in the trees bear witness to the few weeks of frenetic activity●

1

2

3

1 After exceptionally heavy rains in Bushmanland Wattled Starlings began building their nests in the area. 2 Their dome-shaped structures of thorny twigs sometimes merge to form a single unit comprising several compartments. 3 The eggs are laid on a lining of grass and feathers.

1, **2** *Plum-coloured Starlings generally nest in natural tree holes but, like several other starlings, they may also place their bowl of plant material in a hollow fence post, despite the intense heat in such a location.* **3** *Greater Blue-eared Starlings can be persuaded to breed in a nest-box. The lining material it uses includes sloughed snake skin.*

The Plum-coloured Starling is the only species which shows a constant and very marked difference between the sexes, the drabness of the female contrasting with the striking colouring of her mate. When breeding begins the male Wattled Starling develops black wattles, and the plumage of the European Starling becomes iridescent. The season extends through spring and summer, and in arid regions species such as the Pale-winged Starling may breed as late as April or May. The European Starling in the south-western Cape has a clearly defined spring and early summer season and the start of breeding is highly synchronised. Conversely, the Wattled Starling breeds opportunistically – and in large colonies – when conditions are suitable. Several species may raise two broods in a season, and occasionally more.

Generally starlings nest in some form of cavity, the exceptions being the Red-winged, Pale-winged and Wattled Starlings. The European Starling and the Indian Myna nest mainly on man-made structures, often under the roofs of buildings, but they utilise any suitable hole and may nest against the trunks of palm trees where the fronds emerge. On offshore islands, such as Dassen Island off the west coast, the European Starling may nest in a bush or a rusting wreck. The nests of both these alien species vary in size depending on their situation, but the exterior is often bulky and made of a wide range of materials, including grass. A bowl-shaped nest is formed on top of this exterior and lined with grass and feathers.

Pied Starlings, which with Wattled Starlings are the only species to breed colonially, usually nest in banks, excavating a hole 1 to 1.5 metres deep and lining it with vegetation, wool, dung and other miscellaneous material. Their nests have also been found under the eaves of buildings, in girders in a shed, in holes in stone walls, and in stacked lucerne bales. Although they breed co-operatively, only the breeding pair has been seen building the nest. The female incubates, but once the chicks have hatched as many as seven helpers have been seen feeding them. The helpers do not necessarily assist at only one nest and may feed the chicks in several. They are usually sub-adult and juvenile birds (identifiable by their different eye colour), and if a second clutch is laid juveniles from the first brood help to raise the young.

The Plum-coloured Starling breeds in natural holes in trees but also in vertical hollow pipes such as a fence post where the heat is intense. The nest, a bowl of plant material such as grass, is characterised by the addition of green leaves throughout the incubation period. Although some glossy starlings may occasionally use green leaves, it seems that only the Plum-coloured Starling uses them regularly. It would be interesting to analyse the chemical properties of the leaves chosen, because it has been found that those used by European Starlings in North America are effective in controlling the numbers of blood-sucking mites in their nests.

All seven glossy starlings nest in holes in trees, either natural or those excavated by woodpeckers or barbets, and some also nest in hollow pipes and nest-boxes. Competition – from Red-billed Woodhoopoes, rollers, kingfishers and other starlings – for suitable holes is often fierce, and where a suitable site has been secured it is strongly defended and often re-used in subsequent seasons. The nest is lined with grass, feathers and similar material, and both the Glossy and Greater Blue-eared

Starlings include sloughed snake skin. Once, presumably in error, the latter species brought a live blind snake!

Traditionally Red-winged Starlings nested on cliff ledges, but they have adapted to breeding on buildings in towns and the trend seems to be increasing. The nest requires a substantial 'foothold' because it has a bulky circular base of mud mixed with fibres which is built up to a height of 5 to 8 centimetres. The centre is then plugged with grass, hair and fibres to form a neat, deep cup. Red-winged Starlings may breed in the same nest repeatedly, merely refurbishing it each season, and there are two records from Zimbabwe of nests being in continuous use for 33 and 36 years. The Pale-winged Starling, on the other hand, rarely nests on buildings, preferring cliffs where it places the nest in a vertical crevice. It uses no mud in the construction, probably because it breeds in more arid regions.

Most starlings usually lay three or four eggs, but a clutch may range from two to six eggs. They are either blue or turquoise, and those of the European Starling and Indian Myna are always unmarked. The Plum-coloured, Red-winged and Pale-winged Starlings always lay eggs that have small red-brown spots, and those of the other species are either plain or spotted. They are usually, but not invariably, laid at daily intervals, and incubation normally starts when the clutch is complete. Male European and Wattled Starlings and Indian Mynas share incubation, but in other species for which observations are available only the female incubates. The incubation period, where known, ranges from 12 to 18 days, and that of the Red-winged Starling shows remarkable variation, from 12½ to 23 days.

Despite their close association with humans, European Starlings are very wary when breeding, whereas Red-winged Starlings in urban areas are remarkably aggressive at the nest, dive-bombing and sometimes striking people who approach it. Both parents feed the nestlings, which are sparsely covered with down, and eat or remove their faecal sacs. Helpers may assist Long-tailed and Glossy Starlings in their parental duties, and there is a record of two male Plum-coloured Starlings helping the female to feed the chicks. The nestling period, where known, ranges from 20 to 26 days, although that of the Red-winged Starling is again very variable, ranging from 22 to 28 days. The post-nestling period appears to last several weeks, but young European Starlings join flocks of juveniles once they are able to feed themselves.

1

2

3

1 *Red-winged Starling nests are often found on man-made structures, where they need large platforms for their bulky bases.* ***2*** *In its natural habitat of rocky hillsides the Red-winged Starling reverts to its traditional nest site of a cliff ledge, often protected by an overhang.* ***3*** *A Burchell's Starling at its tree-hole nest. A suitable site will be re-used season after season.*

Oxpeckers

Two species of oxpeckers are endemic to Africa and both occur in southern Africa. The Red-billed Oxpecker is widespread and common, but the Yellow-billed Oxpecker was once classified as extinct in South Africa. Recently, however, it re-established itself naturally in the Kruger National Park, and was re-introduced to the Hluhluwe-Umfolozi Game Reserve in Zululand and the Matobo National Park in Zimbabwe.

Oxpeckers have a symbiotic association with the animals on which they live, warning them of danger, feeding on their ectoparasites, especially ticks, collecting hair from them for their nests, and occasionally using them as overnight roosts. Like starlings, to which they are considered to be closely related, they are gregarious and noisy and they roost communally, sometimes with starlings. Although only the Red-billed Oxpecker's breeding biology has been studied in detail, the information available about the Yellow-billed Oxpecker indicates that it has similar nesting habits.

Both species are regular co-operative breeders, with up to five birds in a group. Only the dominant pair is involved in courtship, during which the two birds lower their bodies, open their bills and spread out their wings and tails, then circle each other with quivering wings. The female closes her wings and copulation ensues, taking place on the host animal. Courtship feeding occasionally occurs, also on the host.

The breeding season is in spring and summer, from October to March, and the Red-billed Oxpecker has been recorded raising three broods in this period. Both species place their nests in a natural hole in a tree, although Red-billed Oxpeckers have been seen nesting in a hole in the stone wall of a cattle kraal. In the Kruger National Park leadwood trees were favoured (84 per cent of 43 nests) and the height of the nests ranged from 1.2 to 15 metres, with an average of 8 metres. Prospecting for nest sites takes the form of inspection flights from the back of the host animal. All the birds in a group prospect, but the dominant pair takes the lead and when an acceptable hole is located the male indicates its suitability with quivering wings. Quite often a previous site is re-occupied. Competition for suitable holes is fierce, and Red-billed Oxpeckers have been dispossessed by a Burchell's Starling (at the building stage) and a Striped Kingfisher (after the eggs were laid).

Once the site has been selected, all members of the group help to build the nest. Hair gleaned from host animals constitutes 93 per cent of it, and the balance of the material is grass and rootlets collected from the ground and placed in the hole before the hair is added. Some dung is used to seal any cracks in the wall of the cavity. The hair is formed into a cup 8 centimetres across and 3 centimetres deep, but by the end of the nestling period it becomes a flattened pad.

Usually three eggs are laid, but sometimes two and occasionally more. The Red-billed Oxpecker's eggs are pinkish and spotted with red-brown, and those of the Yellow-billed Oxpecker are white or pale bluish and either plain or spotted. They are laid at daily intervals in the early morning and incubation begins with the first egg. Only the dominant breeding pair incubates and during the day the two birds change over regularly. At night the female incubates alone. The Red-billed Oxpecker incubates for 12 to 13 days, and the Yellow-billed Oxpecker for a similar period.

When the eggs hatch the shells are removed and dropped away from the nest. The chicks are pink with tufts of grey down and their gapes are orange, becoming scarlet as they grow older. They are brooded while small and all members of the group participate in feeding them, carrying food in the mouth cavity, not in the bill. Faecal sacs are removed and the nest is kept scrupulously clean. Throughout the breeding cycle only the immediate nest area is defended. From the age of about 11 days the chicks hiss if disturbed, as do their parents when they are in the nest.

The nestling periods of both species are similar, averaging 30 days, although the smallest chick of a brood may stay in the nest for up to two days after the oldest has left. When they leave the nest the young birds fly straight to a host species, and even though they can feed themselves they may continue to be fed for up to two months. Although a new clutch may be laid only two weeks after the young have left the nest, they may return to it with their parents from time to time during the second cycle. However, there is no clear evidence that they help to feed the chicks. It is assumed that young from a previous nesting season become helpers the following year, but this has not been proved.

Red-billed Oxpeckers use natural holes in trees as nest sites, lining them with hair taken from their mammal hosts.

Sugarbirds

The two sugarbirds of southern Africa are endemic to the region and have ranges that are almost mutually exclusive. The Cape Sugarbird is found in the south-western and southern Cape as far as the King William's Town area, and from there is replaced by Gurney's Sugarbird which occurs from Eastern Cape to the Drakensberg and Transvaal escarpments, with an isolated population in eastern Zimbabwe and adjacent Mozambique. A taxonomic enigma for years, sugarbirds have been placed with the starlings, thrushes and sunbirds. They are now thought to be related to either the thrushes or the sunbirds, but no doubt their taxonomic travels have not yet ended.

Invariably found in association with proteas, sugarbirds rely on these plants for food in the form of nectar and insects, and for nest sites. They are nomadic and their movements are linked to their food supply. Both species, but the Cape Sugarbird in particular, are gregarious, and in some localities they roost communally at night in flocks which may number several hundred birds.

In the breeding season they are monogamous and territorial, and male Cape Sugarbirds guard their mates against other males. They advertise their territory by perching prominently and calling, producing a rather jumbled series of jangling notes which, although difficult to describe, are very characteristic. The male Cape Sugarbird also performs an impressive aerial display, flying up to a height of about 10 metres and executing a series of undulations during which he calls and produces a clacking noise with either his wings or the long tail as it is flicked about. Gurney's Sugarbird is far less demonstrative in his aerial display, and does not produce clacking sounds. The aerial displays also serve a courtship function, and the male Cape Sugarbirds with the longest tails have been found to be most attractive to females. Indeed, a female who mates with a long-tailed male is more likely to produce a second brood in a season. In Gurney's Sugarbird the tails of male and female are the same length and it is not known whether some males have longer tails than others.

The Cape Sugarbird breeds from February to August, and in the south-western Cape there is a marked peak in April and May. The season varies regionally and is closely linked to the flowering of the proteas on which the birds feed. Gurney's Sugarbird breeds from June to February in Kwazulu-Natal and from September to February along the Transvaal escarpment, while in Zimbabwe there are isolated records for July, October, November and April. As is the case with the Cape Sugarbird, breeding is linked to the flowering of proteas.

1 *Cape Sugarbirds breed in winter and select nest sites in the lee of broad-leaved proteas for protection against wind and rain.* ***2*** *The nest has a lining of fine grass which conceals an insulating layer of protea fluff below it.* ***3*** *Unlike the Cape Sugarbird, Gurney's Sugarbird nests mainly during the summer months.*

A sugarbird nest is easily recognised, having an untidy exterior of fine brittle twigs – often heath twigs where these are available – in the centre of which is a very neat cup 6 to 7 centimetres across and 5 to 6 centimetres deep. It is thickly lined – or, more accurately, insulated – with protea fluff which is held in place by an additional lining of fine, reed-like grass stems. The nest is thus ideally designed for inclement weather conditions. In both species only the female builds, completing her task in five to ten days.

Nests are usually placed in protea bushes, either in the middle or on the side opposite to the prevailing wind. Research into the siting of Cape Sugarbird nests has established that bushes with thick foliage and broad leaves are preferred because they provide protection from the windy and cold conditions of winter. In a sample of 91 nests recorded, the average height at which they were placed was 1.2 metres. Gurney's Sugarbird nests tend to be higher, between 1 and 6 metres above the ground.

There are two eggs in a clutch and they vary in colour from pale coffee to dull salmon-pink, sometimes even within the same clutch. The ground colour is overlaid with blotches, spots and scrawls of dark brown and purple, with some underlying lilac marks. The markings are usually erratically concentrated at the large end. The eggs are laid 24 hours apart and incubation begins when the clutch is complete. Only the female incubates, but the male is usually perched prominently somewhere nearby. The incubation period for both species is 17 days.

The shells are removed from the nest when the eggs hatch. The small chicks, pink with a sparse covering of grey down, are brooded by the female during the day until they are five days old and overnight for the first two weeks. They are fed by both parents, although the female is more active in this respect, making about three-quarters of the visits. Small insects make up most of their diet, but observations of a parent carrying no insects and inserting instead its bill into a chick's mouth suggest that nectar is also brought to them. Faeces are removed and deposited in regular 'lavatory' spots in the tops of proteas or silver trees. If alarmed, the parents emit a sharp, ratchet-like call. Although these details refer mainly to the Cape Sugarbird, it is likely that they also apply to Gurney's Sugarbird.

Young Cape Sugarbirds remain in the nest for 17 to 21 days, and young Gurney's Sugarbirds for 21 to 23 days. The post-nestling dependence period of the Cape Sugarbird is about three weeks, but young Gurney's Sugarbirds remain with their parents for two to three months.

Sunbirds

Colourful, vocal and active, sunbirds are conspicuous in the many environments in which they occur. The 20 breeding species found in southern Africa fill diverse habitats from the evergreen forests of the east, inhabited by the Grey, Olive and Collared Sunbirds, to the arid western regions where the Dusky Sunbird is the most representative species. Little is known about the breeding habits of some, such as the Blue-throated and Neergaard's Sunbirds, and few of their nests have been found. In contrast the endemic Orange-breasted Sunbird of the southern and south-western Cape, arguably the most beautiful of all sunbirds, has been intensively studied.

With the exception of the Grey and Olive species, the sexes of our sunbirds are strikingly different: the males are colourful and iridescent, and the females are uniformly drab. Most sunbirds are territorial in the breeding season but often only in the vicinity of the nest, and some, like the Olive Sunbird, defend a food source even when they are not breeding. Sunbirds tend to be aggressive towards members of their own species as well as others, and chasing flights, threat postures and vigorous calling are all aspects of their territorial behaviour. However, their aggression rarely results in physical encounters.

1

2

__1__ The Purple-banded Sunbird characteristically plasters its nest with lichen. __2__ A Neergaard's Sunbird's nest is barely discernible among trails of 'old man's beard' lichen. The fact that very few nests of this species have been found testifies to the effectiveness of the camouflage. __3__ The male Malachite Sunbird with his iridescent colouring contrasts sharply with his drab mate. Where possible, this species likes to suspend its nest over a watercourse.

__1__ The male Lesser Double-collared Sunbird, like most male sunbirds, helps to feed the chicks, although his contribution is variable. __2__ Dry leaves camouflage a White-bellied Sunbird's nest, and the projecting hood breaks up the shape of the entrance. __3__ The Grey Sunbird's nest characteristically hangs from a 'neck'.

Unless birds are marked for identification it is difficult to know whether they are monogamous, but it would appear that most sunbirds do keep a single partner. A pair of ringed Orange-breasted Sunbirds, for example, occupied the same territory for three years. An exception is the Olive Sunbird, the male of which aggressively defends a patch of flowering plants and copulates with several females when they visit them. The females then breed in his territory.

Courtship involves a wide range of behaviour patterns and postures, but does not include the regular flashing of the pectoral tufts, as is widely believed. It seems that the tufts may be exposed at random when a sunbird is excited. Singing is common to all courtship situations. The male often chases the female with his bill almost touching her, and sometimes he hovers over her with a fluttering flight. In perched displays he opens and flutters his drooped wings and raises his tail vertically or wags it up and down. He may throw back his head with bill held vertically, spread his tail, and sing excitedly. Bowing and swaying behaviour, which may be matched by the female, is also part of the courtship repertoire. Sometimes the male becomes so ecstatic that he hangs upside down from his perch and sways from side to side while singing. A characteristic element of courtship behaviour that is common to most sunbirds is vent-pecking, which is thought to provide a pre-copulatory stimulus and occurs particularly while the female is building the nest. The male chases her back and forth as she collects material, following her every movement, and when she alights he pecks at her vent.

It is difficult to define the breeding season of a group with so many species in such a diversity of habitats. As a generalisation it can be said that most sunbirds nest in spring and summer, from August to March, but there is much regional variation, even within the same species. Some sunbirds may lay in most months, but usually there is a clear peak. For example, in Zimbabwe almost three-quarters of 400 nesting records for the Scarlet-chested Sunbird are for September and October. In the more arid regions breeding may be opportunistic after rain, and in the south-western Cape

Orange-breasted Sunbirds nest mainly in June and July, the coldest and wettest months. Sometimes two broods, even occasionally three, may be raised in a season.

The nests of all sunbirds are upright and oval with a side entrance near the top, but they vary considerably both in exterior finish and situation. Most species build a hood or porch which projects downwards over the entrance, partially concealing the round opening. This is such a common feature of sunbird nests that it must have survival value, even though species such as the Orange-breasted Sunbird never include it in their structures. The Grey Sunbird fashions a long neck from which the nest is suspended. Sunbird nests are made of a wide range of dry material such as grass, leaves, moss, lichens, rootlets and fine twigs, and their soft linings may include plant down, protea fluff, cobweb and feathers. Cobweb is an essential component, and the birds gather it by winding it round their bills. They often start a nest by attaching cobweb to a branch and then adding material until enough has accumulated to start shaping the oval.

With very few exceptions, only the female builds the nest, although the male usually accompanies her as she flies back and forth. Construction usually takes about a month, but in some cases it has been finished in seven days and, in one exceptional case of frenetic building, a Collared Sunbird completed her nest in a single day! When the nest itself has been formed the exterior is decorated with such diverse material as lichens, moss, dry leaves, seed pods, spider egg-cases, caterpillar droppings or pieces of paper. Certain species finish their nests in characteristic fashion: for example, the Purple-banded Sunbird plasters its elongated nest with pieces of lichen, and the Violet-backed Sunbird completely covers its structure with dry leaves. Many species add a tail of material which hangs down beneath the nest, disrupting its outline so that it resembles debris caught in a tree. This kind of camouflage is most highly developed in forest species such as the Olive Sunbird.

Nests may be found in a variety of situations and heights, but some species show certain preferences for sites. The Malachite Sunbird, for example, likes to suspend its nest over a stream or donga, and the White-bellied Sunbird places its oval in a prickly pear bush if one is available. Sometimes nests are concealed among the webs of colonial spiders. Quite often sunbirds build in man-made situations: Scarlet-chested Sunbirds have been known to nest in trellises of bougainvillea against house walls, and a Black Sunbird hung its nest from a washing line on a veranda, returning to the same site for three years. Some species nest low down, usually less than a metre above the ground in the case of the Orange-breasted Sunbird, but the Violet-backed Sunbird prefers a height of about 10 metres, and sometimes even 20 metres. Whatever the situation or height, sunbird nests are almost always well concealed either by their camouflage or by foliage.

Sunbirds usually lay two eggs, with the exception of the Bronze Sunbird which lays only one. Clutches of one or three eggs have been infrequently recorded. The eggs' ground colour is usually off-white and it is marked with vertical streaks, spots, cloudings, speckles and scrolls of various shades of grey and brown, depending on the species. In some cases the eggs are so heavily marked that the ground colour is obscured, and in others the markings are concentrated at the large end. The most regular parasite of sunbird nests is Klaas's Cuckoo and its eggs closely match those of the host species (see page 107).

Eggs are laid at daily intervals and incubation usually begins when the clutch is complete, although the White-bellied Sunbird may start sitting after the first egg is laid. Only the female incubates and quite often her kinked tail, bent from the confines of the nest, indicates that her nest is nearby. The male's activities are confined to singing near the nest and chasing off intruders. In those species for which there are records, the incubation period is from 13 to 18 days, and for most it is close on 14 days.

1

2

1 *A female Scarlet-chested Sunbird feeds its chick with a large spider. With insects, spiders form the major part of a sunbird chick's diet.* ***2*** *An untidy exterior and a 'tail' hanging below the Olive Sunbird's nest make it look like debris caught in the tree.*

The newly-hatched chicks are naked and are brooded by the female sporadically during the day for the first few days, and overnight until they are well feathered, even sometimes throughout the nestling period. In most species both sexes feed their offspring, although the contribution of the male is very variable even in the same species. As a generalisation, he probably contributes no more than a third of the meals brought. The chicks are fed on insects and spiders and also probably on nectar. Nests are kept scrupulously clean and faecal sacs are either eaten at the nest or carried away and dropped.

The nestling period, where known, ranges from 13 to 22 days, but for most sunbirds it is close on 15 days. The young remain with their parents in a family group and may probe in flowers for nectar soon after leaving the nest. In a number of species the parents have been observed leading them back to the nest to roost at night, and the female sometimes joins them. The few records available indicate that the young remain with their parents for two to three weeks, but more observations are needed to confirm that this is so.

__1__ Like all sunbirds, the Black Sunbird builds an oval nest which is suspended in a tree. The exterior, however, varies from one species to another, and in the case of the Black Sunbird it comprises lichen. __2__ A female Collared Sunbird at the nest she has built. The 'porch' above the entrance is a feature of many sunbird nests. __3__ A female Violet-backed Sunbird peers out from her nest which, typically, is completely covered with dry leaves. Only female sunbirds incubate the eggs.

THE MID-WINTER SUNBIRD

Orange-breasted Sunbirds nest from February to November in the south-western Cape, but most breeding takes place in the winter months from May to August, with a peak in June and July. Although this is a time of winds, rain and often extreme cold, it is also the flowering season of many ericas which provide a food supply for the sunbirds. Thus they breed at this time, just as the Cape Sugarbird in the same fynbos habitat breeds when the proteas are in bloom (see page 193).

How do the sunbirds cope with adverse weather conditions while breeding? As in the case of the Cape Sugarbird, part of the solution lies in building a well-insulated nest and placing it in a sheltered position. The nest has a particularly thick lining of plant down and is orientated to face east or south-east so that it is protected from the prevailing north-westerly gales. About half of all nests found have been placed in proteas which have thick foliage and the rest were in situations where the surrounding cover provided shelter. Three-quarters of them were less than a metre above the ground and this also had the effect of shielding them from the wind and rain. Thus the female, the eggs and the chicks are protected against the inclement conditions.

The nestling period lasts between 15 and 22 days, and averages 19. It is possible that the relatively large variation results from when there are cold conditions and the female has to brood the young instead of feeding them. It has been observed that when there are cold spells she may enter the nest during the day to brood the chicks, and she broods them overnight for the entire period. After the young have left the nest the adult birds guide them back to it in the evening to roost, and sometimes they even use it for shelter during the day. The female may roost overnight with the young for their first six days out of the nest, and they themselves continue to roost in it for up to 15 days. Thus the Orange-breasted Sunbird's nest is all-important in providing an effective shelter against the elements, and the female's expenditure of energy during numerous trips back and forth to collect lining is an investment well made ●

A female Orange-breasted Sunbird clings to her nest, which is placed low down in the lee of a protea bush whose thick foliage provides shelter in winter. The nest's thick lining helps to keep the eggs and chicks warm.

WHITE-EYES

IN SOUTHERN AFRICA THE LARGE FAMILY of white-eyes is represented by only two species, the Cape and Yellow White-eyes. Their distributions are almost mutually exclusive, the Yellow White-eye replacing the Cape White-eye in the north-east and north of the region. Depending on the locality, the latter varies considerably in coloration, with those on the Orange River being so distinctive that they were once considered a separate species.

White-eyes are well known for their gregarious behaviour, constant activity and cheerful melodious song, especially in the early morning. Although they are widespread and common, it is only recently that their breeding biology has been studied to any extent, and it still merits further research. The information available clearly establishes that the nesting habits of the two species are virtually the same. Although they are gregarious, white-eye pairs roost together and often break away from the group to perch side by side and preen each other. It appears that they are monogamous and pair for life, and they are apparently not territorial, allowing other white-eyes to pass close to an occupied nest without chasing or threatening them in any way. No courtship behaviour has been described, probably because the pair bond is maintained so closely throughout the year.

Both species breed in spring and summer, although in the south-western Cape the Cape White-eye peaks in October and November and in Zimbabwe the Yellow White-eye's peak falls in September and October. The nest is characteristic, a small neat cup which is usually slung between horizontal twigs in a tree or bush, but is sometimes placed among vertical stems. It looks like a hemispherical hammock, almost as deep as it is wide, and the inside diameter of the cup is about 5 centimetres. Built by both birds, it is made from fine dry grass, tendrils, lichens and mosses. Cobweb is used both to bind it externally and to attach it to the surrounding twigs. The interior is sparsely lined with downy material which is held in place with fine fibres or hairs. Although the finished nest appears flimsy, it is securely bound by the cobweb. It is well concealed in foliage between 1 and 5 metres above the ground, usually at a height of about 3 metres. In two remarkable records Cape White-eyes have been found breeding inside Cape Weaver nests suspended over water at the ends

Yellow White-eye chicks are fed by both parents throughout the nestling period.

of trailing willow branches. The nests were not opened in any way, but it was possible to feel that the white-eyes had made a small padded cup inside the bowl of the weaver's nest.

Usually two or three eggs, very occasionally four, are laid, and they are produced at daily intervals in the early morning. The eggs of both species are plain white, pale blue or turquoise, but the variation is not regional as eggs of different colours may be found in the same locality. The Yellow White-eye is parasitised by the Slender-billed Honeyguide and is the only known host of this species in southern Africa (see page 136).

Incubation begins when the clutch is complete and is shared by both birds. Observations of the Cape White-eye established that, on average, the eggs were brooded for 94 per cent of the daylight hours. On one occasion an incubating Yellow White-eye flew off the nest and dropped to the ground where it hopped and fluttered ahead of the observer to lure him away from the eggs. Similar behaviour has been observed at a nest containing chicks. The white-eyes' incubation period is very short, usually no more than 11 or 12 days, and sometimes eggs hatch after only 10½ days.

The newly-hatched chicks have orange skin and are naked except for two small tufts of down just above the eyes. They are brooded constantly for the first three or four days, and the sitting bird is replaced when its mate brings food. Both parents feed the chicks throughout the nestling period, often arriving at the nest at the same time. They keep the nest scrupulously clean, either eating the faecal sacs or removing them. The nestling period is short, lasting only 12 to 14 days. As the young are prone to jumping out of the nest if disturbed, even when they are as young as eight days old, great caution should be exercised when inspecting nests.

When the chicks leave the nest the skin round their eyes is bare, with no sign of the white ring. They do not follow their parents but stay huddled together, waiting for food to be brought to them. It is not known for how long they remain dependent on their parents.

Despite appearances to the contrary, a Cape White-eye's nest is secure enough to accommodate two well-grown chicks, being firmly attached to horizontal twigs with 'old man's beard' lichen and cobweb.

The Red-billed Buffalo Weaver, the White-browed Sparrow-weaver and the Sociable Weaver

Each of these three species is the only representative of its genus in southern Africa, and all are typical of the drier parts of the region. Red-billed Buffalo Weavers occur in dry bushveld, and in some areas are nomadic. Their large, untidy nests are often found on the outer branches of baobabs and camelthorn trees, but also on power pylons and windmills. Each nest is defended by a dominant male, but from time to time nest masses coalesce with the result that different males may 'own' a section of the main structure.

A nest comprises several separate chambers, each of which is connected to the outside by its own tunnel. The whole structure is made of thorny sticks which are broken off trees or collected from the ground. Some sticks may be four times the length of the bird, and may be carried from as far as a kilometre away. Only the male builds, sometimes constructing a new nest in as little as a week, although he continues to maintain it by adding sticks. As he completes a chamber he adds some greenery to line it, but the female is responsible for the final lining of green grass and leaves.

The male courts a prospective mate with drooped, quivering wings and by raising and fanning his tail, calling excitedly all the time. She then inspects a nest chamber and he may bring a sprig of greenery for her. It appears that a dominant male mates with several females who will lay in the chambers in his nest. Copulation takes place away from the nest, preceded by the same wing-quivering display that is used in courtship. The act, which is very brief in most birds, may last for up to two minutes in this species. Buffalo weavers are unique amongst birds in having a permanently erect phalloid organ which grows larger in the breeding season. This appendage is not a penis and there has been considerable speculation about its function.

Breeding, which takes place in summer, is linked to rainfall, and it has been observed that laying may be highly synchronised within a colony. The eggs, 2-4 in a clutch, are laid on consecutive days and are greenish white in ground colour with heavy grey and olive speckling. They are incubated by the female alone, beginning when the first has been laid. After only 11 days the first chick hatches, followed at intervals by its siblings in the same sequence in which the eggs were laid. The shells are not removed completely, but merely dropped to the ground below. The chicks are fed only by the female, and they leave the nest after 20 to 23 days.

The Red-billed Buffalo Weavers' sturdy nests are very durable, sometimes lasting for at least eight years, and in areas where this species is resident they are used for roosting all year. Giant Eagle Owls and White-backed Vultures sometimes breed on top of a buffalo weaver nest, and the latter species and Bateleurs may nest among a colony of the smaller birds, thus effectively camouflaging their own nests.

The White-browed Sparrow-Weaver, an extremely vocal species and a superb songster, is found mainly in acacia and mopane woodland in the drier central and western parts of southern Africa, often near farmhouses. Its habitat usually includes both degraded veld, where it feeds mainly on harvester termites, and an area of good grass cover which supplies material for nest-building. The single tree in which it builds is festooned with what appear to be clumps of dry grass at the ends of the branches, but on closer inspection these turn out to be ball-shaped nests with an entrance at either end. They are always on the leeward side of the tree, which in many regions is its western side. The number of nests in a single tree, perhaps a dozen or more, suggests a substantial breeding colony, but in fact the birds live in small social parties of four or six, and sometimes up to 11. They live co-operatively, with a dominant pair breeding while the others are helpers. The tree is the focal point of their lives and they drive away other birds, including other members of their own species. When two opposing sparrow-weaver groups meet they call loudly and face each other from adjacent trees, holding in their bills grass stems which act as a visual signal and are not subsequently incorporated into the nest. This behaviour is reminiscent of the 'flag-waving' display of Red-billed Woodhoopoes (see page 129).

The nests are used for roosting, a single bird in each one, and the open ends allow for a quick escape should the need arise. However, when the dominant pair breeds one end of one of the nests is sealed and the nest is lined with soft material. In one

1 In their dry bushveld habitat, Red-billed Buffalo Weavers often place their large, untidy nests at the ends of the branches of a baobab. ***2*** *The nest is built of thorny sticks and comprises several chambers, each of which has its own entrance.*

nest 858 feathers were counted. All the members of a group co-operate in building the nests and they maintain them throughout the year. It is not known why the birds should be such compulsive nest-builders, but possibly the nests confuse predators or they may serve as a form of territorial advertisement.

Only the monogamous dominant pair are involved in courtship and they maintain a strong and permanent bond. If the female of the pair dies she is replaced from within the group, but the male is replaced from outside. Prior to copulation they both adopt a crouching position, the male with drooping, quivering wings. He then performs a swaying dance by moving his feet from side to side on the branch, first towards and then away from the female in a series of movements. Such displays can last for up to seven minutes before copulation occurs.

Rainfall is the trigger for breeding, which usually takes place in summer but in the more arid regions is often opportunistic. By using an existing nest for the first clutch of the season the birds can respond quickly to favourable conditions, although for subsequent broods (and there are sometimes a further three) a new nest may be constructed. Two eggs are usually laid, sometimes one or three, and they are pinkish in ground colour with red-brown spots. Only the dominant female incubates and the eggs hatch after 14 days. The chicks are fed by all members of the group, although a greater contribution is made by the females, and they leave the nest after 21 to 23 days. They continue to be fed and still beg when they have been out of the nest for three months, even though they can feed themselves. In the early period after leaving the nest they are guided back to roost by the dominant female.

A Sociable Weaver nest is an impressive structure, and one's impression on seeing one for the first time is complete disbelief, especially when it may measure 7 metres across. There is something incongruous about sparrow-sized birds which weigh a mere 27 grams building a nest the size of a small haystack. Characteristic of the arid regions of the northern Cape and Namibia, some of these large communal structures have been in place for more than 100 years.

Sociable Weavers have two main requirements for their nests: a sturdy support and suitable building material. Camelthorn trees fulfil the first requirement and stiff *Aristida* grasses the second, and both occur throughout the weavers' range. Although camelthorns, and to a lesser extent kokerbooms, are of considerable importance as nest sites, other situations such as telephone poles, windmills and the superstructures of watertanks are utilised. Nests and telephone wires are not compatible, so the

THE WORLD'S LARGEST NEST

The Sociable Weaver's nest is one of the few genuinely communal structures built by any bird. Why should such a remarkable nest have evolved? In the first place it is important to stress that the enormous grass nest could only have evolved in a dry climate, since in conditions of high rainfall and humidity it would become a fermenting compost heap.

The extremes of climate experienced by Sociable Weavers in their arid environment provide the reason for the large nest. In summer, when it is intensely hot, the nest interior is relatively cool and provides shelter for the birds during the middle of the day. At night, however, when the temperature outside drops, the nest chambers remain comfortably warm.

During the winter months, when temperatures drop to around freezing, the nest protects the birds from the cold, and the bigger the nest the better the insulation. By using measuring devices inserted into occupied nest chambers, researchers in the Kalahari Desert have been able to reveal the extent of the insulation against cold. They established that the temperature in an occupied chamber was between 18 and 23 °C above the external air temperature, and even unoccupied chambers were warmer. Also, during winter, the roosting pattern changed from two birds in a nest chamber to as many as five. Not only did the additional birds create more warmth, but from time to time during the night there were bursts of activity which raised the temperature in the chamber.

The benefits of keeping cool in summer, while important, do not match the advantages of keeping warm in winter. Because the birds are insulated from the extreme night cold, the energy expenditure required to keep warm is 40 per cent less than that needed if they roosted in the open. This in turn increases the carrying capacity of the environment because less food is required to regain lost heat, and also less time is needed for foraging. The favourable climate within the nest also enables the weavers to breed during winter if there is an adequate food supply. This has the added advantage of reducing the chance of predation by cobras which are less active in winter. The time and energy required to construct and maintain the huge nest benefit those involved, but at the same time they are passing on the advantages to succeeding generations of Sociable Weavers ●

1

2

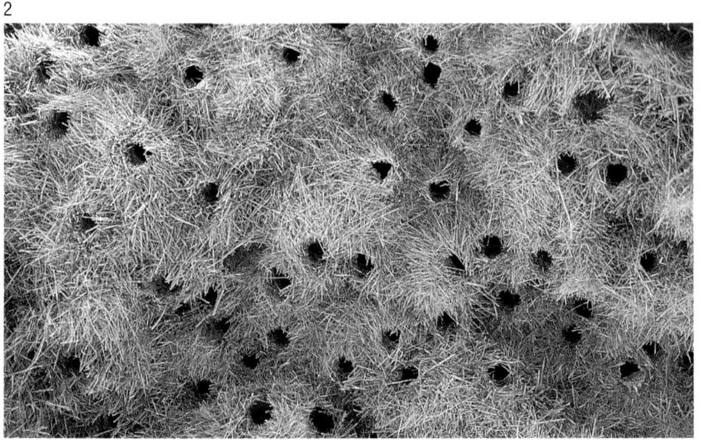

3

4

5

__1__ The Sociable Weaver is a bird of arid habitats, where there is no humidity to transform the mass of grass that is its nest into a compost heap. __2__ Each entrance tunnel on the underside of the communal nest leads directly upward into a chamber. __3__ Nipped-off grasses at the entrance make gaining access to the chamber uncomfortable for anything larger than a Sociable Weaver. __4__ A framework on a telephone pole specifically placed to accommodate a Sociable Weavers' nest is readily accepted by the birds. __5__ Sociable Weavers constantly add to their nests, with the result that some grow to massive proportions.

1

2

1, 2 The White-browed Sparrow-weaver's nest is a characteristically untidy ball of grass which is always placed on the leeward side of a tree.

technicians circumvent the problem by supplying the birds with a framework below the wires, and this they readily accept. Less frequently a nest may be placed on a cliff ledge, or in a tree growing from a cliff, and there is also a record of one supported by beams under the roof of a shed. Whatever the situation, there must be clear access to the nest from beneath.

The birds start building by laying grasses on the chosen base. As they add more and more material the size of the nest gradually increases until it is large enough for chambers to be constructed in the underside. A complete nest may comprise as many as 50 adjacent chambers. Each chamber is retort-shaped, with a lip or threshold at the edge to prevent the eggs or chicks from falling out, and a tunnel that leads directly downwards. It is lined with miscellaneous soft materials. A feature of the tunnel entrance is that sharp straws project downwards, making it painful to insert one's hand into the nest.

To call this species a weaver is something of a misnomer because the birds merely push the grasses into place. The males do most of the building, carrying one grass stalk at a time in a remarkably laborious process, and they cover the roof or superstructure with small thorny twigs. The nest is constantly maintained and added to, and as it grows different levels emerge. Sometimes satellite nests are built on adjacent branches or even in a nearby tree. In a kokerboom the whole top of the tree may be filled with the nest so that it resembles a giant toadstool.

Such a large structure is inevitably attractive to other birds for roosting or breeding. Pygmy Falcons are dependent on Sociable Weavers for a nest site and their relationship with the weavers may be commensal (see page 70). Rosy-faced Lovebirds and Red-headed Finches breed in the nest chambers and Giant Eagle Owls and Barn Owls in hollows on top of the structure. Various other species roost in the compartments. Unwelcome visitors to the nests are boomslangs and Cape cobras; the latter species has been known to devour all the nestlings in a colony. Honey badgers also tear open nests and do considerable damage, but their predation is not as serious as that of the cobras.

Sociable Weavers are gregarious at all times. The colonies are loosely spaced, usually not less than 0.8 kilometres apart, and the birds' foraging range is not more than 1.5 kilometres. They rarely move between colonies. Pairs are monogamous when breeding but do not necessarily stay together from one season to the next. Although this species is not territorial in the accepted sense, a large colony comprising different levels has a social structure which results in birds remaining in their own level.

The male courts the female by following her closely, often with nest material in his bill. In a display on the ground he flicks his tail up and down and walks towards the female who crouches. As he nears her he adopts a stiff upright stance with tail down and bill pointed skywards and calls softly. Such displays sometimes culminate in copulation, but mating is very rarely observed and possibly takes place inside the nest chamber.

Breeding, triggered by rainfall, takes place in any month and during periods of drought none takes place at all. Sometimes a season may extend over nine months and up to four broods are raised. Usually three or four eggs are laid, but the clutch may be 2-6; larger clutches are related to good rainfall. The eggs, dull white and densely spotted with grey, are laid at daily intervals in the early morning. Incubation begins with the first or second egg so that chicks hatch at intervals, and if the food supply is inadequate the smaller chicks die. Both sexes incubate and the eggs hatch after 13 to 14 days.

When the eggs hatch the shells are dropped to the ground below but not removed, as it would be pointless trying to avoid the attention of predators when the nest itself is so conspicuous. The newly-hatched chicks are naked with pink skin and they have well-developed whitish flanges at the corners of the gape which show up well inside the dark nest. They are brooded while small and both parents feed them, mainly on harvester termites which emerge after good rainfall. Droppings, which are not encapsulated in a faecal sac, are merely voided by the chicks out of the chamber entrance with the result that the ground below becomes thickly covered with them. As in the case of the eggshells, no purpose is achieved by carrying them away.

The chicks leave the nest after 21 to 24 days and remain dependent on their parents for at least another 16 days. They still use the nest chamber as a roost even when a new clutch is laid, and may help to feed the chicks of this and subsequent ones. By the time the fourth brood hatches its nestlings may be attended by as many as 11 birds: the parents and nine young. Such help at the end of the breeding season, when food stocks are dwindling and the parents are tired, is probably advantageous.

Sparrows and the Scaly-feathered Finch

Of the five sparrow species which occur in southern Africa, four are 'true' sparrows of the genus *Passer* and the fifth, the Yellow-throated Sparrow, is placed in its own genus, *Petronia*. The sparrow-like Scaly-feathered Finch is also placed in a genus of its own. All the sparrows are widely distributed in the region, and the House Sparrow, introduced into South Africa at the turn of the century, has spread rapidly and is now virtually ubiquitous. It lives in association with humans and is rarely far from habitation. To a lesser extent the Cape and Grey-headed Sparrows also frequent human environments and nest on buildings, but the Great and Yellow-throated Sparrows have remained 'wild'. The delightful little Scaly-feathered Finch is a gregarious species which behaves very much like a sparrow. A bird of arid regions, it can exist without water for long periods; in one experiment captive birds lived on an exclusive diet of grass seeds for 62 days.

Sparrows breed mainly in spring and summer, although the season varies considerably from one region to another, and House and Cape Sparrows may be found breeding at any time of the year. All species may raise several broods in a season. The courtship posture adopted by the male House Sparrow is typical of that of other sparrows: he droops his quivering wings, raises his tail, points his bill upwards, puffs out his chest and calls excitedly. If the female is receptive she adopts a submissive posture and the male mounts her. House Sparrow pairs are known for copulating as many as 30 times in succession.

The House Sparrow makes an untidy nest of dry grass and miscellaneous material with a soft lining, and places it in virtually any hole or crevice that can accommodate it. Almost invariably it nests in association with human habitation, usually under the eaves of a house. Sometimes it usurps the bowl nest of a swallow, breaking the entrance, or it takes over the nest of a Little Swift.

The nests of the Cape Sparrow are untidy domed structures of grass, lined with soft material and with an entrance at one end. They are usually placed in trees, preferably thorny ones, but may be sited on buildings, for example on electrical wires under the eaves. Other situations in which Cape Sparrow nests have been found include swallows' nests, a tin suspended on a pole which had served as a post-box, and a vertical hollow fence post. In the last case the nest was a simple pad with no roof. Although sparrows are monogamous and usually territorial, 23 Cape Sparrow nests have been recorded in a single tree, and five contained eggs.

Whereas the Great Sparrow builds a similar nest, also domed, and locates it in a tree, the Grey-headed Sparrow merely lines a hole in a tree – either natural or excavated by a barbet or woodpecker – with soft material. However, it has also adapted to

The Scaly-feathered Finch's nest is similar to that of sparrows, and is used for roosting as well as breeding.

nesting on buildings, but cannot compete where the more aggressive House Sparrow occurs. The Yellow-throated Sparrow always nests in tree holes, lining it as the Grey-headed Sparrow does.

In all the sparrow species both sexes build the nest, incubate the eggs and feed the chicks. The clutches usually comprise 2-6 eggs, but the Cape Sparrows has been known to lay as many as 12. This is undoubtedly the contribution of two birds, especially as there is convincing circumstantial evidence that females of this species may dump their eggs in the nests of other females. Both the Cape and the Great Sparrows are regularly parasitised by the Diederik Cuckoo (see page 106). The sparrows' dull white or greenish white eggs are heavily marked with grey and brown, often over most of the surface. They are incubated for 12 to 14 days and the nestling period lasts for 15 to 18 days, although Cape Sparrow chicks may remain in the nest for as long as 25 days. Little is known about the duration of parental care once the young leave the nest, but young Cape Sparrows remain with their parents for one or two weeks.

The Scaly-feathered Finch usually lays its clutches between December and April, with a marked March peak in Zimbabwe, but may breed at any time of year in response to rainfall. Its nest, a small version of a Cape Sparrow's, is often placed in a thorn tree and is used for both breeding and roosting. As many as a dozen birds may use a nest overnight, bursting through the thin roof if disturbed. Sometimes the finch roofs over the cup-shaped nest of another species, and there is even a record of its chicks in a weaver's nest.

The clutch usually comprises 3-5 eggs, but up to seven may be laid. They resemble Cape Sparrow eggs but are smaller, and are incubated for close on 11 days. The nestling period is usually 16 days but may vary between 14 and 18 days, and during this time both parents feed the chicks. At one nest watched by the author it was suspected, but not confirmed, that one or two helpers were also feeding. In view of the sociable habits of this species co-operative breeding is likely.

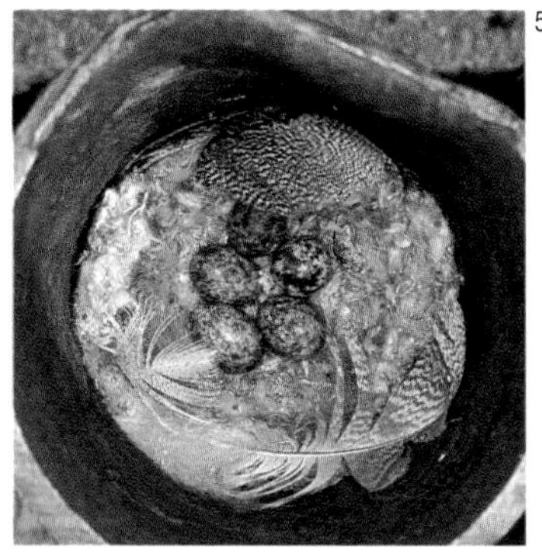

1 *The Grey-headed Sparrow nests in a hole in a tree which may be either natural or, as shown, excavated by a woodpecker or barbet.* ***2, 3*** *House Sparrows place their untidy nests of grass and other materials in any hole or crevice that will accommodate them, such as may be found in the roof of a farm shed. They also sometimes take over the enclosed nests of swallows, including those of the Greater Striped Swallow.* ***4, 5*** *The Cape Sparrow's domed nest is usually placed in a thorn tree, but this species does sometimes select a more unusual situation, such as the inside of a metal fence post.*

Weavers

Among the most fascinating of all birds because of their nesting habits, weavers occur in most habitats in southern Africa. Five of the 13 species in the region – the Forest, Olive-headed, Spectacled, Golden and Red-headed Weavers – nest solitarily, although they are often found in small family groups. The remaining species – the Thick-billed, Spotted-backed, Masked, Lesser Masked, Yellow, Brown-throated, Cape and Chestnut Weavers – nest colonially and are conspicuous and noisy. Their constant activity gives the impression of chaos, but amidst the tumult some of the most intricate of all avian nesting and behaviour patterns are being enacted.

Most weavers, except the Thick-billed and Chestnut Weavers, have yellow in their plumage, and the males are more colourful than the comparatively drab females. Some of the males develop a special plumage feature when breeding, such as the black facial colouring of the 'masked' weavers. In the non-breeding season many weavers are nomadic and roost in large communal flocks.

The breeding seasons of the various species are difficult to define and differ from one region to another. Chestnut Weavers in Namibia, for example, nest opportunistically when it rains and may lay within a week of arriving in an area. Most weavers breed during the wet summer months because they need pliable green material for building their nests. In the winter-rainfall region of the south-western Cape, however, Cape Weavers start breeding as early as July.

In general, species that nest solitarily tend to have no elaborate courtship displays, as they are monogamous and maintain a permanent pair bond. A Forest Weaver pair strengthens its bond by performing a complex duet, and in other solitary species the male may pursue the female, but otherwise he is not demonstrative and does not display at the nest as the colonial weavers do.

The colonial weavers, on the other hand, perform flamboyant displays, notably one in which the male hangs upside down beneath his nest, flapping his wings and swaying from side to side, all to the accompaniment of an excited buzzing call. The male Thick-billed Weaver makes a slow, flapping 'advertisement flight' around the nest area and if a female enters his territory he breaks off to chase her. At the nest, or in reeds nearby, he raises and lowers his wings to reveal white patches that are also prominent in his aerial manoeuvres.

Even within colonies the male defends a small territory, which he may occupy in successive seasons (for eight years in one study). The most aggressive and dominant males are the most attractive to females, as are those with the most persistent and vigorous courtship displays, in which the yellow underside of the wing is an important element. Females are also attracted to colonies which are large, numbering 100 to 200 nests. The colonial weavers practise successive polygyny, which means that once a male has successfully courted a female and she has laid in one of his nests, he then sets out to attract another. Some species may have only two mates, others may have several.

Weavers suspend their nests from hanging branches or palm fronds, or attach them to upright reeds. Some, especially the Red-headed Weaver, even hang them on

1

2

1 *A male Masked Weaver perched above the completed shell of his nest. Once a female has accepted it, she will add a soft lining.* ***2*** *Cape Weaver nests are often seen hanging from the drooping branches of willow trees.*

telegraph wires. With the exception of the Thick-billed Weaver, all the southern African species build a kidney-shaped nest on a horizontal axis, but the nest of each species is distinct because of differences in either the structure or the materials used. Access to the nest is from below, and is direct or via a short spout or a long tubular tunnel. A tunnel probably provides effective protection against snakes, particularly when a new nest is attached at the end of the tunnel of an old one.

Most species build their nests of green grass or shredded strips of palm leaves, and the green colour of fresh nests is important in attracting females to them. Forest and Red-headed Weavers make their nests with pliable tendrils, twigs and long strips of bark which, woven together, give the finished nest the appearance of fine basketwork. There are no southern African breeding records of the Olive-headed Weaver, which is

1 *The Lesser Masked Weaver is one of several weavers which breed colonially.*
2 *Its nest has a spout and may be identified by its 'furry' appearance.*
3 *The Golden Weaver's nest is characterised by an untidy 'porch' of grasses projecting over the entrance.*
4 *The finely woven nest of a Spectacled Weaver is further characterised by its long tunnel.*

1

2

1 The first stage in the construction of a weaver nest, as demonstrated by a Lesser Masked Weaver, is forming a ring. 2 A Cape Weaver perches on his ring, from which he weaves first the brood chamber and then, leaning backwards, the ante-chamber. 3 Having completed the nest, the Cape Weaver has to attract a female to it while it is still fresh and green.

THE SKILL OF THE MASTER-BUILDERS

The nest-building techniques and associated behaviour of weavers have been studied intensively, especially in relation to the Village Weaver in West Africa, which is a race of the Spotted-backed Weaver. The details gleaned from this research apply to most southern African weavers.

Once the male has established his territory he begins to build a nest, first forming a ring. From this focal point he weaves the brood chamber, roof first, the size of it being limited by the distance he can reach out from the ring. Then, still facing in the same direction, he weaves the roof of the ante-chamber by leaning backwards while he works. Finally he forms the entrance below the ring and adds either a spout or a tunnel. The entrance is particularly important and he takes care to finish it off strongly, sometimes even strengthening it after the female has laid. He lines the roof with leaves and grasses to make it waterproof, and when the female has accepted the nest she adds a soft lining to the brood chamber.

The birds' weaving technique is similar to that used by humans and with practice a skilled person can fashion a reasonable weaver's nest. The weavers, too, need to practise to become proficient and young birds have a compulsion to build 'play' nests, and sometimes they even establish 'play' colonies of misshapen structures. By the time they are two years old and ready to breed they can build well and continue to improve with practice. It has been shown with captive birds that young weavers deprived of weaving materials from birth have retarded skills. However, those which are allowed to practise never forget how to weave, even if they are deprived of building material for 2½ years.

A feature of weaver behaviour is that the birds often strip foliage from around the nest, especially when it is suspended from a hanging branch. One explanation is that this makes the nest less accessible to snakes, although they are still able to reach it. Possibly the weaver makes his nest more conspicuous to a female by removing surrounding foliage.

A female weaver, attracted to a dominant male who sings well and builds strong nests, enters his nest and prods about to check its strength. The nest must be fresh and green to arouse her interest, and it has been shown experimentally that faded brown nests are ignored. It is a myth that females destroy sub-standard nests; in species that nest colonially the male himself tears down a nest so that he can build a fresh one to lure a female. Over a period of eight years a single male Masked Weaver constructed a total of 204 nests, most of which were incomplete. Only 82 lasted for more than 10 days, and of these 24 were used for breeding; in only 17 of them were young raised successfully.

3

In a study of Spotted-backed Weavers in the Kruger National Park it was found that a male contributed between 600 and 800 pieces of material to a nest, and a female at least 200 pieces of lining. It was calculated that a male would fly a total of about 20 kilometres to collect material for one nest, and he built on average seven nests per season. The nest is the focal point of a male weaver's life and he exhibits a compulsion to build that far outweighs all other activities. The skill and dedication of these avian master-builders can only arouse our admiration ●

__1__ Red-headed Weavers' nests have a tubular entrance and are made of tendrils and pliable strips of bark and fine twigs. __2__ This species nests solitarily and may re-use a suitable site for many years, sometimes attaching a new nest to the tunnel of an old one. __3__ The Forest Weaver also nests solitarily and constructs a coarsely woven nest with a long tunnel.

found only in Mozambique, but elsewhere in its range its nest is described as being like that of the Forest Weaver except that it is made from 'old man's beard' lichens.

Both the Forest and the Red-headed Weavers build nests with tunnels, but the former's tunnel is longer. The Spectacled Weaver also makes a long tunnel but its nest is immediately distinguishable from that of the Forest Weaver because its fibres are extremely finely woven. The Golden Weaver's nest has an untidy projection of grasses over its entrance, and the Yellow Weaver's nest has no entrance spout at all. In mixed colonies of weavers the structure and finish of the nests usually distinguish which species built them. Where Lesser Masked and Spotted-backed Weavers nest together, for example, the Lesser Masked Weaver's nest has a short spout and a 'furry' appearance because of projecting loose ends, whereas the Spotted-backed Weaver's construction is neatly finished and has no spout.

The nest of the Thick-billed Weaver differs from those of other weavers in the method of its construction and in the fact that it is attached to its supports on a vertical rather than on a horizontal axis. Instead of starting its nest by forming a ring as other weavers do (see box), the Thick-billed Weaver initially builds a bridge between two upright stems, especially those of *Typha* bulrushes or *Phragmites* reeds. The bridge is developed into a cup, the back wall of which is extended upward and then over to form a roof. A rather large entrance is left open, but the incomplete nest is not, as is often thought, used for roosting. The entrance is closed, rather untidily, to a hole just large enough to admit the female when the pair is ready to breed. Despite its heavy bill, the Thick-billed Weaver constructs its nest with extremely fine strips of grasses, rushes and sedges, carrying several pieces at a time. The male clears foliage around the nest as other weavers do, but in his case this entails cutting down nearby reeds instead of stripping branches of their leaves.

In colonial species only the male builds because the nest, intended to attract a female, is an important part of his courtship. The female lines the main structure with soft materials once it is complete and she has accepted it. The nest of the Thick-billed Weaver, however, is not lined because its walls are so thick. Of the solitary species, both sexes or only the male may be involved in construction. Although Spectacled Weaver nests are usually said to be built by the male only, several observations have established that the female may make an important contribution. Female Red-headed Weavers also assist with nest construction. Nests can be completed remarkably quickly, often in a few days and sometimes even in a day.

Clutches of 2-4 eggs are laid in the early morning of consecutive days and within a colony laying may be highly synchronised. The eggs of some species are consistently the same colour, whereas those of other species are highly variable even within the same colony. The Cape Weaver always lays plain turquoise eggs, the Lesser

1 *Having completed the shell of his nest, a male Yellow Weaver will display beneath it in a bid to attract a mate.* ***2*** *A female Thick-billed Weaver inspects an unfinished nest which, unlike other weaver nests, is attached on a vertical axis to supporting reeds.* ***3*** *The entrance to a Thick-billed Weaver's nest has been closed to a small hole, indicating that the occupants are breeding.*

Masked Weaver's eggs are invariably plain white, and the Red-headed Weaver lays blue eggs. The Brown-throated Weaver is unusual in that its eggs are dark brown, although some are blue-green with brown spots. The Thick-billed Weaver usually lays pink eggs with red-brown spots, but sometimes the eggs are white with spots. Those of most other species are variable – usually white, pinkish, greenish or bluish – with spotting of red-brown or grey.

The variability of eggs is particularly noticeable in a species such as the Masked Weaver which, among others, is regularly parasitised by Diederik Cuckoos (see page 105). It has been shown experimentally that individual female Masked Weavers always lay eggs with pigmentation that is unique to themselves. They can recognise their own eggs, and remove a cuckoo's egg if it fails to match theirs.

Incubation usually begins when the clutch is complete, but as Chestnut Weaver chicks hatch at different times, this species presumably starts incubating when the first egg has been laid. As in the case of the Sociable Weaver, this ensures that the size of the brood is reduced if the food supply is insufficient. Only the female incubates in most species for which observations are available, with the exceptions of the Spectacled and Red-headed Weavers. Further study may well reveal that the males of other species that nest solitarily also help with incubation. The incubation period of several common species is still unknown, but for others it lasts from 11 to 16 days.

The newly-hatched chicks are pink and naked, except for a few wisps of down. The task of feeding them, usually on insects, is undertaken mainly by the female, and although the males of some species may assist they do not do so consistently. As with incubation, the males of solitary-nesting species appear to make a greater contribution to parental care than do the males of colonial breeders. In the case of the Chestnut Weaver, the male may desert the colony while his offspring are still in the nest, sometimes even before they hatch. Faeces are removed from the nest initially, but later the chicks defecate out of the entrance. Because of the nature of a Thick-billed Weaver's nest, the droppings tend to accumulate inside it during the latter part of the nestling period. Once the chicks have left the nest the female leans in and cleans it out.

The nestling period of weavers ranges from 14 to 22 days. The young of colonial species accompany the female and join flocks, but those of solitary species such as the Forest Weaver form small family groups. After breeding Thick-billed Weavers split into two groups, the females and immatures moving into open country while the males stay in the forest.

Queleas, Bishops and Widows

Two queleas, three bishops and six widows breed in southern Africa, and they are all gregarious and nomadic in the non-breeding season, often forming flocks made up of several species. When breeding the males assume a striking plumage, and they nest colonially or in loose groups.

Normally thought of as crop-damaging pests which should be eliminated, Red-billed Queleas wander widely and form huge flocks that may number millions of birds. Breeding colonies sometimes cover many hectares, and a single thorn tree can contain 500 nests. The buzz of activity in a colony can be heard from a kilometre away, and the sound of birds flying up is like distant thunder. These queleas breed when conditions are suitable, usually when there is a good food supply after rain, but it is important that they are in good condition before they lay, with substantial reserves of protein and fat. Breeding is highly synchronised and chicks in all nests in a colony may hatch over a period of just a few days.

The male builds the nest and displays near it from a prominent position, his plumage fluffed up and his wings quivering. The nest, usually placed in a thorn tree, is an upright oval made of woven strips of grass with a side entrance near the top and is normally unlined. Sometimes rudimentary practice nests are built, but these are never used for breeding.

Three plain blue eggs are usually laid, a clutch size that apparently relates to the maximum number of chicks that can be raised. Incubation may begin before the clutch is complete, with the result that the chicks hatch at different times and the smaller ones die if there is insufficient food. Both parents incubate, the female at night, and in very hot conditions the eggs may be left unattended for long periods during the day. The incubation period is $9\frac{1}{2}$ to 12 days, and when the chicks hatch they are fed by both parents, initially on insects and later on grass seeds. They leave the nest after 11 to 13 days.

A large colony of Red-billed Queleas is a frequent target of predation. If alarmed the birds become silent, presumably in an attempt to avoid attracting attention, but nothing can prevent the influx of predators which consume the eggs and chicks; birds, mammals and reptiles all feed to satiation. At a colony of 100 000 pairs in Zimbabwe it was estimated that predation depressed the breeding output by 20 per cent. At another colony 32 per cent of the eggs were eaten before they could hatch, and 25 per cent of the surviving chicks were also consumed. How do the queleas withstand these onslaughts? Their survival depends on three factors: the short duration of the breeding cycle, a high level of synchronisation, and the sheer numbers of birds. There are simply too many active nests over a short period of time for the predators to visit all of them.

The Red-headed Quelea is an intra-African migrant which visits coastal Mozambique and sometimes ventures as far south as the Eastern Cape. It does not form large colonies like those of the Red-billed Quelea, and has rarely been recorded breeding in southern Africa. It bred near Durban in 1955, and at Lake St Lucia in 1984/85 a colony of about 750 nests was established in reedbeds at a density of about two nests per square metre. The nest and eggs are similar to those of the Red-billed Quelea but the usual clutch is two eggs. From the information available it appears that the males, unlike Red-billed Quelea males, do not assist with incubation or with feeding the nestlings.

Bishops and widows – often referred to collectively as *Euplectes*, the scientific name for their genus – are characterised by the males' striking breeding plumage in patterns of red and black or yellow and black, and by the fact that several widow species have long tails. The difference between the two groups is that bishops have short tails and do not replace them in the breeding season whereas the widows do. Male bishops and widows are heavier than the drab-plumaged females.

Nomadic in the non-breeding season, the *Euplectes* species frequent moist areas of rank grassland, vleis and marshes when they are breeding. Because of these requirements they nest in summer extending into autumn in the summer-rainfall region, and in spring and early summer in the winter-rainfall region of the southern and south-western Cape.

Bishops and widows are polygynous and in any population there are more females than males. Like colonial weavers they practise successive polygyny, and

1

2

1 Red-billed Queleas build their nests in thorn trees wherever possible, and may fill a single tree with as many as 500 nests. 2 A Red-billed Quelea's nest, built by the male, is loosely woven and usually unlined.

1 *Once she has accepted a nest and mated, a female Red Bishop raises her brood alone while the male sets about attracting another female.* ***2*** *The shell of a newly completed Red Bishop's nest appears flimsy but the female adds lining to it, continuing to do so during the incubation period.* ***3*** *Like other widows, the Yellow-rumped Widow conceals its nest in rank grass.*

once the male has successfully courted a mate and she has accepted the nest frame he has built, he moves on to another female. Red Bishops may have as many as seven mates during a breeding season. All *Euplectes* species are territorial, and a male at the boundary of his territory adopts a stiff upright posture that is clearly aggressive towards intruders. The size of a territory varies considerably from one species to another: the highly colonial Red Bishop, for example, maintains a territory of about 8 square metres in a reedbed. Ringing studies have shown that some species return to the same territory year after year, behaviour which may apply to the *Euplectes* group as a whole.

Courtship comprises both perched and aerial performances, in all of which the plumage is puffed out so that it is displayed to maximum advantage. The bishops and some widows perform a bumblebee-like flight with their back feathers fluffed out, and the Golden Bishop in flight resembles a yellow ball. Sometimes these flights end with an abrupt drop into vegetation, usually beside a female. The bishops, typified by the Red Bishop, indulge in bounding displays from perch to perch and, like weavers, they swivel excitedly from side to side with a buzzing call. The widows with long tails, such as the Red-collared and Long-tailed Widows, perform a slow, flapping flight during which their tail feathers are spread out to appear as broad as possible when viewed from the side. Their flights are so conspicuous that they can be seen from a kilometre away. The importance of tail length was demonstrated in an experiment involving wild Long-tailed Widows, in which some males had their tails artificially lengthened by glueing on an extra piece, others were left as a control group, and yet others had their tails shortened. The results showed conclusively that females preferred males with the longest tails.

The nests of all *Euplectes* are basically similar: an upright oval with an entrance on the side near the top which usually has an untidy porch of grass protruding over it. The male builds the initial framework and the female finishes it off with a soft lining of grass inflorescences. The technique of nest construction, based on a ring or bridge from which the male develops a structure, is like that of weavers, but the latter's habit of demolishing nests has not been observed.

Red Bishops nest colonially, weaving their nests in reedbeds between upright stems. When the eggs are laid they can be seen through the fine weaving of the outer shell, but as incubation progresses the nest lining thickens considerably as the female adds to it. Fire-crowned Bishops, which are localised and relatively rare in southern Africa, nest in a similar manner but their colonies are never large. The Golden Bishop breeds in loose groups in flooded grasslands, placing their nests low down and incorporating growing green grass or sedges into the roof.

The widows do not breed in colonies but their nests may be loosely grouped in an area. They are placed in rank vegetation or grass tufts, and most species do not build higher than a metre above the ground. The nests are always well concealed and often surrounding vegetation is bent over to form a bower that is incorporated into the roof. Frequently an Orange-breasted Waxbill, which rarely builds its own nest, takes over a widow's nest after use.

Three eggs are most often recorded in a clutch, but it may comprise 2-6 eggs. The Red and Fire-crowned Bishops lay attractive plain, glossy, greenish blue eggs, and the Golden Bishop's eggs, unique amongst those of *Euplectes* species, are white with small black spots. The other species lay eggs that have a greenish ground colour with blotches and mottlings of brown, olive and grey. The major parasite of bishops and widows is the Diederik Cuckoo. Several 'tribes' of this cuckoo have evolved, some of which lay plain blue eggs in the nests of Red Bishops, while others lay matching blotched eggs in widows' nests (see page 105).

In all *Euplectes* species incubation and the care of the nestlings is the task of the female. The incubation period is normally 12 to 14 days for all species, and the nestling period is 11 to 17 days. Faecal sacs are removed and nests are kept clean at all times. The young accompany the female after leaving the nest but it is not known for how long they remain dependent on her. Eventually they join the large flocks which are so characteristic during the non-breeding season. By this time the males have shed their resplendent colours or long tails and resemble the females, thus compounding the problem of identifying bishops and widows in their uniformly drab plumage.

Waxbills, Mannikins and other estrildid Finches

This group of 27 species is divided among 12 genera which include twinspots, firefinches, waxbills, finches and mannikins, as well as the Golden-backed Pytilia, Red-faced Crimsonwing and Nyasa Seedcracker. They are all members of the family Estrildidae, and for ease of reference are commonly termed estrildid finches, or simply estrildids. Such an array of species may seem confusing, but they all exhibit remarkably similar habits. When not breeding they are gregarious, sometimes occurring in mixed flocks of several species, and they feed on the ground, mainly on grass seeds. Communal roosting is usual for most species. They are sexually dimorphic, and the fact that the males are often extremely colourful has led to their popularity with aviculturalists, from whom much information on their behaviour and breeding has been derived.

Estrildids are found in a wide range of habitats, from forest to the arid regions of the west. Their breeding season is very variable, and some species may nest in all months. As a generalisation, however, it can be said that most breeding takes place during the latter part of summer, extending into autumn and sometimes even into winter. Although breeding is linked to rainfall, the trigger is not rainfall itself but the resulting availability of food, mainly grass seeds.

The estrildids are monogamous and maintain a lifelong pair bond which is reinforced by bouts of mutual preening and huddling together. In courtship, which is basically similar for most species in the region, the male approaches the female with his head feathers raised and his belly and flank feathers fluffed out. He bows and bobs up and down in front of her while singing, and usually holds a piece of nesting material such as grass or a feather by its extremity in the tip of his bill. Although the carrying of material is not common to all species, for most it is an essential part of courtship. The Quail Finch male 'dances' around the female, and is also said to perform an aerial display in which he calls as he plummets down from a height.

The basic nest type is a horizontal oval which has a side entrance and often an untidy porch. Some nests, notably those of the Black-cheeked Waxbill, have a downward-pointing tunnel about 10 centimetres long. A 'cock nest', or small blind compartment, is found on top of the nests of some species, particularly the Common Waxbill in which it is a regular feature. Despite its name there is no evidence that the male roosts in this compartment, nor is it used as a nursery area if there are too many chicks in the nest, as has been suggested. It is lined with feathers which make it more conspicuous, and it seems likely that its purpose is to deceive predators by drawing attention away from the real entrance.

The males undertake most of the building, but in some species both sexes may be involved, constructing the nest out of grass and lining it with grass inflorescences and feathers, particularly the latter. Forest species such as the Green Twinspot and Red-faced Crimsonwing also incorporate ferns, mosses and skeletonised leaves into the basic structure. Nests are placed between ground level and a height of about 15 metres, and whereas some, especially those on the ground, are remarkably well concealed, others in trees or bushes may be conspicuous. Some species nest either at ground level or at a height of several metres, but Quail and Locust Finches always nest on the ground between grass tufts. The Quail Finch's nest is characterised by having in front of it a patch bare of vegetation.

Some estrildid species deviate from the nesting pattern of the majority, particularly

Typically of estrildids, the Melba Finch builds a nest of grass and lines it mainly with feathers.

1 *A Blue-billed Firefinch emerges from its nest, a horizontal oval which is well concealed in surrounding vegetation.* ***2*** *Blue Waxbill nests are often found near wasp nests, for reasons as yet unconfirmed.* ***3*** *Some estrildids breed in the nests of other birds and merely add their own soft lining of feathers. In this case a Red-headed Finch has taken over the nest of a Masked Weaver.* ***4*** *The Orange-breasted Waxbill regularly makes use of another species' nest, such as that of a Red Bishop.*

in that they may take over the old nests of other birds. The only one which does so regularly (in more than 90 per cent of the records) is the Orange-breasted Waxbill, which uses the old nests of Red Bishops, various widows and weavers, and sometimes prinias and cisticolas. It merely lines the existing nest with feathers and other soft material. Some species occasionally breed in weaver nests: the Green Twinspot and Grey Waxbill, for example, may take over the nests of Forest Weavers.

The Cut-throat and Red-headed Finches breed in a variety of situations, although they also build normal nests, as do the Orange-breasted Waxbill and other estrildids that take over old nests or other sites. They may use a woodpecker hole, a hollow fence post or a hole in a building, or the nests of various weavers. In addition to having been recorded in a Little Swift's nest, the Red-headed Finch regularly breeds in the nests of Red-billed Buffalo Weavers and Chestnut and Sociable Weavers. In the case of the Sociable Weavers, a loose 'colony' of Red-headed Finches may take over part of their nest mass.

Estrildids are not particularly territorial and merely defend a small area immediately around the nest. They often remain in the same area, and sometimes the same tree, year after year, but do not re-use an old nest, preferring to build a new one nearby. In winter Bronze Mannikins build roosting nests but these, comprising only 210 to 400 pieces of grass, are less substantial than breeding nests which contain 540 to 800 pieces of grass.

Certain estrildids, especially the Blue Waxbill, breed in association with social wasps, but their reason for doing so is not known. A common explanation given is that the wasps provide protection for the birds, but it appears that an estrildid nest placed next to a wasp's nest is as vulnerable to predation as any other. A more intriguing suggestion is that trees which contain wasp nests are known to be free of acacia ants which would prey both on the wasps and on the contents of a bird's nest. Blue Waxbills and other estrildids may regard the wasp nests as a sign that the tree is free of ants.

1 The Quail Finch's nest is always on the ground and invariably has a bare patch in front of it. ***2*** *A blind compartment, or 'cock's nest', on top of the Common Waxbill's nest is more visible than the real entrance concealed below. Its function is possibly to mislead predators.*

The estrildids usually lay 3-6 plain white eggs, sometimes 2-7, and occasionally as many as nine. The eggs are laid on consecutive days and incubation does not begin until the clutch is complete or nearing completion. Both sexes share incubation during the day, but at night only the female sits. The birds add lining to the nest throughout the incubation period, and sometimes on leaving the nest deliberately pull a feather over the entrance to conceal it. When the nest is on the ground the parent may hop away ahead of an observer as if to lure him away.

Incubation lasts for between 11 and 14 days, and when the chicks hatch they have pink or blackish skin with a covering of down of varying colours. The nestlings' most striking feature is their mouth markings which are unique to each species. Whydahs and widow-finches parasitise the estrildids and their eggs are also white, but always larger. However, each whydah or widow-finch specialises in a particular species of estrildid and the mouth markings of their chicks match precisely those of the host species (see page 219). Not only do the estrildid chicks have a pattern of black marks within the mouth itself, but there are white tubercles at the corners of the gape. These tubercles have a reflective property and clearly indicate the position of the mouth in the dark nest. Once the chicks are older they feed in a characteristic way by lowering their heads and twisting them upside down so that the parent regurgitates a meal into their upturned gape. They are fed on fresh grass seeds, and termites if these are available. The skin of the crop is translucent so that the composition of a recent meal may easily be seen.

In the early stages of the chicks' development the parents remove their faeces. Later the droppings, which are not encapsulated in a sac, accumulate against the walls of the nest or round the entrance, but they dry hard and do not pose a sanitation problem. Sometimes lining is added to the nest during the nestling period and it covers the droppings that have accumulated.

The nestling period ranges between 15 and 23 days, but for most species it is close on 21 days. It is difficult to make observations on post-nestling behaviour except in captivity, and these may not necessarily apply to wild birds. However, it seems that the young remain with the adults for at least a couple of weeks and the parents are able to recognise their own chicks. Young estrildids may form a pair early in life and then remain together.

Whydahs, Widow-finches and the Cuckoo Finch

In southern Africa the whydahs and widow-finches are each represented by four species, and are referred to as viduines from their family name Viduidae. They and the Cuckoo Finch are all parasitic species, and whereas they are probably most closely related to the bishops and widows, the Cuckoo Finch is more like the weavers and also has different parasitic strategies.

In non-breeding plumage the gregarious viduines, which may form flocks of mixed species, are extremely difficult to identify, and even in breeding plumage the widow-finches pose problems. It now seems that one of the four widow-finches listed in the sixth edition of *Roberts' birds of southern Africa* (1993), the Violet Widow-finch, is no longer considered a valid species and its place has been taken by another which was previously overlooked in our region, the Green Widow-finch.

All the viduines parasitise estrildid finches and each species is specific to one host (see table, page 220), although there is some confusion in the case of the Pin-tailed Whydah. It is known to parasitise the Common Waxbill, and although one authority (*The complete book of southern African birds*, 1989) considers that this is its only host, another (*Roberts' birds of southern Africa*, 1993) lists an additional six hosts. Within an area each viduine parasitises about a third of the host species' nests, but sometimes the figure is as high as 85 to 95 per cent.

The parasitic habits of the whydahs and widow-finches are so complex that they make the cuckoos look like rank amateurs. Painstaking research has revealed that it is possible to deduce the specific host of a viduine merely from the latter's calls, as it mimics its natal host. The previously overlooked Green Widow-finch has been given specific status on the grounds that it mimics the calls of the Red-throated Twinspot, which is undoubtedly its host although a parasitised nest has yet to be found.

When breeding, the male viduines have a plumage that contrasts strikingly with that of their drab mates. The four whydahs sport elongated tails in addition to their striking colour patterns, although it appears that the males with the longest tails are not necessarily more successful in attracting mates (compare Long-tailed Widows, page 214), at least in the case of the Pin-tailed Whydah. However, members of this species which grow their tails early in the breeding season are the first to obtain call sites and establish dominance. The male widow-finches have uniformly black, blue-black or green-black plumage with bills and legs that are red or white. Because of their uniformity even males in breeding plumage are difficult to identify. However, when they call, their identity can be deduced from the host species they are mimicking.

Male viduines are territorial, often aggressively so, and the Pin-tailed Whydah, for example, chases off not only members of his own species, but also any other bird that comes into his domain. In addition to chasing, he shows aggression by adopting an upright posture with sleeked plumage. Research has shown that males with a particular resource in their territory, such as water or a good seed supply, are more likely to attract mates. If a male dies or is removed from a territory his place is soon taken by another.

Each whydah and widow-finch male has several mates, and the Pin-tailed Whydah often protects a harem of six females. Sometimes males in non-breeding plumage which remain subordinate are allowed to join the group. The dominant male has a regular call site which may be a fence post, telephone wire or high dead branch, and sometimes a number of such sites are within sight and hearing of each other. The male incorporates into his repertoire not only his own weaver-like calls, but also perfect imitations of the host species that raised him. A female passing through the area hears the songs of the various males and, also having the call of her host species imprinted on her, responds to the call she recognises. Thus she mates

1

2

1 *A male Pin-tailed Whydah, in breeding plumage and sporting a long tail, defends his territory from a prominent perch.* ***2*** *A female Pin-tailed Whydah is about to enter a Common Waxbill's nest, in which she will lay her egg.*

A Black-chested Prinia feeds a young Cuckoo Finch which has recently left the nest.

with a male of her own species, who speaks the right 'language'. It would appear that widow-finches are confusing not only to the human eye.

In addition to calling from a perch viduines display aerially. The Paradise and Broad-tailed Paradise Whydahs ascend to a height of about 100 metres in a series of steps and then cruise around with their central tail feathers raised like a bustle before plummeting back to a perch. They may spend most of the day displaying in this way and then feed in the evenings. The Pin-tailed and Shaft-tailed Whydahs perform a hovering display over the females, during which they flick their tail feathers. The widow-finches execute a similar bobbing display flight. Females react to courting males with quivering wings and invite copulation, but quite often mating is unsuccessful because another female interferes.

Whydahs and widow-finches lay white eggs which are like those of their hosts but larger. A clutch may comprise 1-4 eggs, but the number deposited in a single nest varies and, as different females may parasitise the same nest, it is not always possible to tell how many one female has laid. During a season a female may lay a series of clutches so that her total productivity could amount to about two dozen eggs. The parasite usually removes an egg of its host for each of its own laid, but it seems that this practice is not invariable. She does not remove any eggs laid by another female viduine, apparently recognising those which are not part of the host's original clutch.

The female whydah or widow-finch, like her cuckoo counterpart, watches the activities of her host carefully and synchronises her laying accordingly. She knows which species she must parasitise by its call, which has been imprinted on her since birth. The victims of viduine parasitism do not evict the larger eggs, but incubate them with their own for a similar period of 11 to 13 days. It is only when the chicks hatch that the most amazing aspects of viduine parasitism emerge. In general appearance they closely resemble the host's chicks but, more remarkably, both the spots inside the mouth and the tubercles at the side of the gape are identical to those of the host's chicks. The matching of mouth markings is comparable to the replication of some cuckoo eggs to the eggs of the host species.

Nor does the deception end with mouth markings. The estrildid finch chick has a unique method of receiving food from its parents: it turns its lowered head upside down and the parent regurgitates into its mouth for a prolonged period without raising its head. The parasite chick feeds in precisely the same way, and its begging call matches that of the host's chicks, except that it is rather more insistent. The parasite does not evict the chicks of its host, and the intensified begging call may help it to be fed preferentially.

HOST SPECIES OF WHYDAHS, WIDOW-FINCHES AND THE CUCKOO FINCH	
PIN-TAILED WHYDAH	Common Waxbill. Also said to parasitise Bronze Mannikin, Orange-breasted and Swee Waxbills, Red-billed Firefinch, Neddicky and Tawny-flanked Prinia.
SHAFT-TAILED WHYDAH	Violet-eared Waxbill
PARADISE WHYDAH	Melba Finch
BROAD-TAILED PARADISE WHYDAH	Golden-backed Pytilia
BLACK WIDOW-FINCH	Blue-billed Firefinch
PURPLE WIDOW-FINCH	Jameson's Firefinch
GREEN WIDOW-FINCH	Red-throated Twinspot (deduced from call)
STEEL-BLUE WIDOW-FINCH	Red-billed Firefinch
CUCKOO FINCH	Fan-tailed, Desert, Cloud, Ayres', Pale-crowned, Croaking, Singing, Levaillant's and Red-faced Cisticolas; Tawny-flanked and Black-chested Prinias

The viduine nestling period of 16 to 21 days is similar to that of the host species. When the young leave the nest their plumage closely matches that of the host's young, the only difference being the lack of colour on the rump. They are cared for as though they were part of a single brood for about a month after leaving the nest, and by this time they have long since assimilated the calls of their foster parents. When their turn to breed eventually arrives the calls imprinted on them from birth become essential, not only for selecting the right mate, but also for choosing the correct host. Thus the breeding cycle of one of the most remarkable of all brood parasites is perpetuated.

The Cuckoo Finch is also a gregarious species and, like a canary in appearance, is easily overlooked. Its courtship behaviour, involving the slow flapping of raised wings and 'swizzling' displays during which the male calls, tends to reinforce the case for it being related to weavers. In terms of its parasitism it is similar to cuckoos in that it usually matches its eggs to those of its host. Grassland cisticolas are its preferred hosts, but it also parasitises prinias.

Cisticolas are discerning hosts, so in their nests the Cuckoo Finch lays matching bluish or white eggs which are either plain or have red-brown speckles. It does not match its eggs to those of prinias but they are accepted. A single egg per nest is laid twice as often as two, and all the host's eggs in a nest are removed when the Cuckoo Finch deposits hers. Thus any found with the parasite's eggs have been laid afterwards. The incubation period is approximately 14 days.

The newly-hatched Cuckoo Finch differs from the naked pink chicks of its cisticola hosts in that it has dark skin and wisps of down, and it lacks mouth spots. Its mouth is purplish and the gape has an orange-yellow rim. The host's young are not evicted but they are usually trampled by the Cuckoo Finch chick and die. However, two Cuckoo Finch chicks in the same nest may survive together. The nestling period is 18 days and the young accompany the host for about two weeks before joining a flock of Cuckoo Finches.

Canaries and Buntings

The canaries, which include the Cape and Drakensberg Siskins, are represented in southern Africa by 14 species, and the buntings by five. The Chaffinch, which was released in Cape Town with the European Starling at the turn of the century, also belongs in this account. The group as a whole is widespread and common in a diverse range of habitats, yet there are prominent gaps in our knowledge about the breeding biology of several of its members: for example, the incubation and nestling periods of the common Cape Bunting have yet to be recorded, the first nest of the Lemon-breasted Canary was discovered only in 1988, and although the Protea Canary – a species endemic to the south-western Cape – is not uncommon in suitable habitat, we still know much more about its feeding habits than about its breeding biology.

The Chaffinch, which remains tenuously established on the Cape Peninsula, lives in plantations of non-indigenous oaks and pines and in suburban gardens. Not having adapted to indigenous habitats, it has not been able to expand its range. Very little is known about its breeding habits in southern Africa and very few nests have ever been found, probably because they are always high up. It breeds from September to November, building a nest similar to Chaffinch nests in Europe. The nest, a cup of rootlets, mosses and lichens, is bound with cobweb, decorated externally with lichens, and lined with various soft materials.

Canaries and buntings are usually gregarious and some, especially in arid regions, are nomadic. Several canaries are fine songsters, and at the onset of the breeding season Cape Canaries fill the air with their incessant melodious trills and twitters. Buntings sing pleasantly too, although their phrases tend to be monotonous. Canaries and buntings use regular song perches, and by singing from them presumably advertise their territory, although as a group canaries and buntings are not aggressively territorial.

Little is known about courtship behaviour, particularly that of the buntings. Some canaries perform a 'butterfly' flight during which they fly around with shallow 'rowing' wing-beats over a distance of about 100 metres. This behaviour is particularly prevalent when the birds are building their nests. Two canaries spiralling upwards, breasts touching, are probably executing another courtship display. Buntings indulge in nuptial chases, and 'butterfly' flights have been seen, although rarely.

Breeding generally occurs in spring and summer, from September to April, although in the southern and eastern Cape some species such as the Cape and Bully Canaries may start breeding in July. In arid regions nesting is often opportunistic after rain has fallen and triggered the production of seeds. Breeding seasons vary not only regionally but also seasonally; a study of the Rock Bunting in Zimbabwe showed

A Streaky-headed Canary's nest is attached to the branch with cobweb.

that its peak laying periods varied from year to year, depending on the rainfall. Most canaries nest solitarily, except the Cape Canary which quite often breeds in loose colonies in which the nests are only a few metres apart. In some cases these aggregations may occur because the birds are attracted to trees around a farmhouse in otherwise treeless country.

A 'typical' canary's nest has an exterior of dry vegetation such as grass, leaf ribs, fine twigs, weed stems and pine needles, and the neat cup is warmly lined with plant down or other soft material. Some species, notably the Streaky-headed Canary, bind their nests with cobweb which is also sometimes used to attach the nest to the surrounding branches. The Black-eared Canary differs in that it makes its nest mostly with 'old man's beard' lichens, binding it with cobweb and lining it with fine rootlets and mosses. Only one Lemon-breasted Canary's nest has been found and it was quite different. Placed in a fold of a palm frond, it had a rather bulky base to keep it in position and was made with creeper stems, leaf pieces and silky pieces of chewed bark from caterpillar tubes. The cup was lined with fine fibres peeled from palm leaves.

Canaries place their nests in low bushes, trees or creepers at heights varying from a metre or less in arid regions to about 18 metres in plantations. The Cape Canary shows a preference for non-indigenous trees such as pines, whereas the Cape Siskin nests in holes in rocks in mountainous terrain, often near water. One siskin nest was in a hole in a tree on the edge of a forest, a site very like that of a Dusky Flycatcher which may account for the nest having been parasitised by a Red-chested Cuckoo. The Drakensberg Siskin also nests in holes in rock, as well as in banks with overhanging vegetation.

The nests of buntings differ from those of canaries mainly in the way they are lined. Except for the Lark-like Bunting, buntings do not use downy plant material,

***1** Newly-hatched canary chicks are sparsely covered with down, and these of the Cape Canary will be brooded at night until they are ten days old. The location of this nest in a vine is unusually low. **2** The eggs of the Golden-breasted Bunting are characteristically marked.*

lining the cup instead with fine fibres, grasses and hairs, especially the tail hairs of cattle and horses if these are available. Cabanis's and the Golden-breasted Buntings make rather flimsy nests of grasses which are placed in trees and bushes. The Cape Bunting's nest is substantially built and usually placed in low bushes, often almost on the ground, although one nest was in a crevice in a cliff 10 metres up. This same nest was found to contain 1400 pieces of material.

The Rock and Lark-like Buntings nest on the ground against a rock or at the base of a bush and sometimes in a hoofprint. Because their nests are on the ground or in a hollow, base material of small earth clods, stones and twigs is often used to build up the exterior. Sometimes when nests are against rocks they are semi-circular, the back half not being completed. Lark-like Bunting nests are often placed on the southern side of a bush or rock so that they are shaded for most of the day. In a Zimbabwe study Rock Buntings were attracted to the numerous fissures on mine dumps and bred in loose colonies.

In both the canaries and buntings usually only the female builds, although the male accompanies her back and forth and sings while she works. In a few canary species the male has also been observed to help with nest construction and at one Streaky-headed Canary's nest the male did most of the work.

Most canaries and buntings usually lay three or four eggs, but the clutch may be 2-5. However, both Cabanis's and the Golden-breasted Buntings lay two or three eggs. Under particularly favourable conditions larger clutches may be laid, as happened when seven eggs were found in a Lark-like Bunting's nest in Bushmanland after exceptionally heavy rainfall. Canary eggs are usually white, greenish or bluish in ground colour with small spots and scrolls of black or various shades of brown. Some species lay eggs that may be either spotted or plain, even sometimes in the same nest. The Cape Siskin always lays plain white eggs. The eggs of Cabanis's and the Golden-breasted Buntings are characteristically scrolled with black in a zone at the large end. The Cape, Rock and Lark-like Buntings lay eggs that are heavily blotched and spotted with various shades of dark brown, red-brown or grey.

Eggs are laid on consecutive days and incubation starts with the completion of the clutch, although the Cape Canary incubates once the third egg is laid, even if the eventual clutch is four. Incubation in both canaries and buntings is the task of the female, while the male often sings from a perch nearby. Male canaries feed the female on or near the nest by regurgitation and she responds with a wing-quivering begging posture like that of a young bird. Male buntings, on the other hand, do not feed incubating females. The incubation period for most species in the group is 12 to 14 days, but that of the Cape Siskin is 16 to 17 days.

The newly-hatched chicks are sparsely covered with down and they are brooded by the female for the first few days. In the case of the Cape Canary they are brooded at night until they are ten days old. The male canary brings food and regurgitates it to the female who passes it on to the chicks. Alternatively, he pushes beneath her to feed the chicks directly. Once the chicks are larger they are fed directly by both birds. Buntings feed their chicks directly, either by regurgitating seeds or by passing them single insects which they carry in the bill. The chicks' translucent crop allows the contents of a recent meal to be seen. The parents may be away for two to three hours in their search for food, often returning to the nest together.

The gelatinous faecal sacs of canary chicks are removed or eaten at the nest until the chicks are about a week old, and then they accumulate on the rim. Cape Siskins have been seen to remove faecal matter throughout the nestling period, and it is likely that buntings do too, as their nests do not become fouled. The nestling period for most species falls within the range of 15 to 21 days, although occasionally longer or shorter periods are recorded. Little is known about the post-nestling period but the young continue to be fed after they have left the nest, and for the first few days they may be located by their insistent begging calls.

1

2

3

1 *A Yellow Canary's nest is warmly lined with downy plant material.* ***2*** *The Cape Bunting's substantial nest is usually situated in a low bush.* ***3*** *Lark-like Buntings nest on the ground, supporting the cup with a ramp of small sticks.* ***4*** *Cape Siskins usually nest in rock hollows, but this pair has chosen a tree hole instead.*

4

THE SOUTHERN AFRICAN REGION

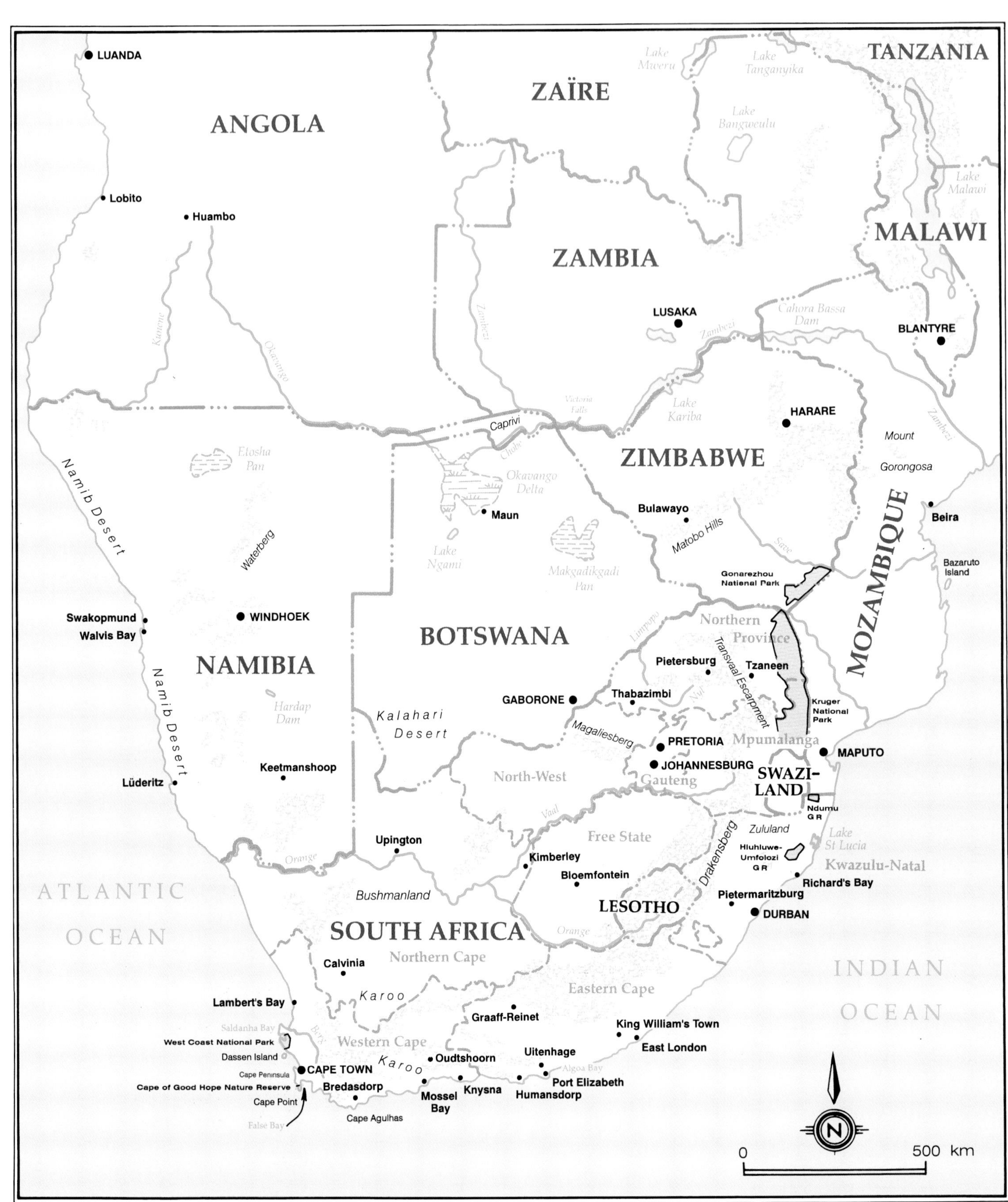

Acknowledgements

This book has had a protracted incubation period lasting some 45 years. The embryo developed slowly to begin with, but finally it grew to such proportions that hatching could no longer be delayed. Its gestation combined two elements: the acquisition of photographs and the accumulation of information on the breeding habits of birds.

Over such a long time span it is impossible to recall all those co-operative people who rendered assistance in so many different ways, and it is inevitable that some may have been overlooked.

The 17 years I lived in Zimbabwe, as well as subsequent visits, have undoubtedly been the most productive, and I was fortunate to have had the stimulation there of some of the finest ornithologists in southern Africa: Richard Brooke, Ken Cackett, Val Gargett, Peter Ginn, Dr Hans Grobler, Ron Hartley, Michael Irwin, Des Jackson, Alex Masterson, Peter Mwadziwana, Dr Colin Saunders, the late Dr Reay Smithers and Carl Vernon (who introduced me to a simple but extremely effective card index system for recording observations and references).

In Zimbabwe I was also indebted to many friends for hospitality and logistical support: Dave and Meg Barbour, Rolf and Maia Cheneaux-Repond, Cecilia Manson, Mike and Thelma Mylne, Douglas, Olive-Mary, Willy and Claire Robinson, John and Jo Scott, Bob and Jenny Thomson, Dave and Margie Tredgold, and Brian and Shelagh Worsley. Over the years successive members of the Falcon College Natural History Society were my companions away from the classrooms as we studied birds in the Matabeleland bushveld near Bulawayo. The late Douglas Ncube was my friend and field assistant for 11 years, and my book *Birds of prey of southern Africa* was dedicated as a tribute to his memory.

In South Africa my study and photography of the birds of the arid regions were undertaken mainly on the farms of Neil and Neva MacGregor (at Nieuwoudtville in Namaqualand) and Willouw and Annette Cilliers (at Brandvlei in Bushmanland) and I am grateful to them for their hospitality and many kindnesses over the years. Others have assisted in many ways: Pieter and Ann Albertyn, Colin Bell and the staff of Wilderness Safaris, Dr Chris Brown, Dr Mike Cherry, Derek and Vernon Coley, John Cooper, Dr Rob Crawford, Ronnie and Linda Crowther, Dr Rob Davies, Richard Dean, Bruce Dyer, John Fisher, Mike Fraser, Doug Galpin, Ian Garland, Dalton Gibbs, Ben Kakebeeke, Dr Alan and Meg Kemp, Andries and Stephanie Krügel, Howard Langley, Gavin Lawrie, Wicus and Hanlie Leeuwner, Tim and June Liversedge, Professor Gordon Maclean, Walter Mangold, Liz McMahon, Dr Peter Mundy, Nico and Ella Myburgh, Ken Newman, Carl Nortier, Dr Terry Oatley, Tony Pooley, Lance and Renate Scoble, Dr Rob Simmons, Claire Spottiswoode, Warwick Tarboton and François van der Merwe.

Professor Roy Siegfried is thanked for allowing me to use the research facilities at the Niven Library of the Percy FitzPatrick Institute of African Ornithology, and Vivienne Stiemens, the librarian, was generous with her help in many ways. David Allan of the Avian Demography Unit at the University of Cape Town, in addition to other support, read the introduction and made valuable suggestions for its improvement. Gill Wheeler not only generously lent me my first word processor but she also instructed me in the mysteries of its operation.

The photographs are an essential ingredient of this book and it was impossible to obtain all the necessary ones myself. Listed below are all those who have contributed pictures, and whether it was merely one, or several, each submission has proved invaluable. Warwick Tarboton and Peter Ginn have made substantial contributions, and Hugh Chittenden and Nico Myburgh went to special lengths to obtain pictures of difficult or elusive species; both Hugh and Nico have been a constant source of encouragement through their enthusiasm and support.

In his capacity as consultant, Morné du Plessis patiently assessed the text of a non-scientist and constructively and tactfully suggested where revision would be beneficial. As important, however, was his friendship and steady encouragement which kept me on track through some difficult periods.

The production of the book was in the capable hands of Pieter Struik and Leni Martin of Fernwood Press, ably assisted by designer Neville Poulter, and Bunny Gallie and Pam Struik. Leni's task was the most onerous, but calmly and skilfully she used her prodigious editorial talents to keep the book on an even keel over a period of nearly three years.

The book is dedicated to my children for their patience, their assistance and their understanding over the years. Finally, it would be no exaggeration to say that, without the encouragement and love of my wife Jenny, this book would not have come to fruition.

Photographic credits

The following people have kindly contributed photographs to this book. Copyright for these photographs remains with the relevant individuals.

David Allan 16 (Lesser Jacana), 53/2, 70/2, 79/1, 93/4, 173/3
Hugh Chittenden 19/2, 84/6, 100/2, 105/4, 120/2, 140/1, 148/2, 150/1, 150/2, 152/3, 158/3, 159/2, 171/6, 182/1, 194/1, 194/2, 196/3
Derek Coley 162/1, 187/2
Alex Cremer 72/box
Ronnie Crowther 168/2
Richard Dean 140 (Angola Pitta)
Pat Donaldson 38/2, 211/2
Bruce Dyer 94/4
Rudi Erasmus 112/1, 112/2
Peter Ginn 77/6, 78 (African Finfoot), 80 (Red-crested Korhaan), 115/4, 159/1, 172/4, 185/3, 188/1, 191/2, 198/3, 208/3, 216/2, 221/2
Ron Hartley 70/4
Andrew Jenkins 69/1
Alan Kemp 89/box, 133/3
Gavin Lawrie 100/1
John Ledger 52/1
Wicus Leeuwner 151/1, 151/2
Tim and June Liversedge 45/1, 45/2, 46/1, 46/2, 46/3
Derek Longrigg 148/3, 171/2
Gordon Maclean 84/3
Peter Mundy 53/3
Nico Myburgh 12/1, 13/1, 15/2, 17/2, 21 (Namaqua Dove), 22 (Golden-tailed Woodpecker), 23 (Blue-billed Firefinch), 35/2, 47/2, 47/3, 49, 51/3, 62/1, 62/4, 70/box, 79/2, 81/box, 99, 102/box, 104/4, 111/2, 113/2, 117/2, 121/2, 122, 124/1, 132, 135/3, 142/1, 142/2, 149/2, 153, 154/1, 154/2, 154/3, 154/4, 156/2, 169/1, 169/3, 175/1, 179/2, 182/3, 184/2, 209, 212/3, 216/1
Ken Newman 105/5, 123/3, 138/2, 192, 216/4, 217/1, 218/2, 219
W. Nichol 47/1, 76
Charles Ratcliffe 72/1
Noel Snyder 51/2
Warwick Tarboton 35/3, 37/2, 37/box, 48/5, 53/1, 69/2, 72/3, 5/3, 80/1, 81 (African Jacana with chicks), 98/3, 104/3, 144/3, 158/2, 173/2, 179/3
Barry Taylor 77/3
C.J. Uys 45/3, 178/1
Alan Weaving 75/2, 162/4, 169/2
Brian Worsley 31/1, 165/box (both), 178/3

Glossary

Altricial. Describing a chick which is helpless when it hatches, having closed eyes and little or no down, and is fed by its parents. Cf. Precocial.

Antiphonal duetting. The male and female of a pair producing syllables alternately, often with such good synchronisation that the duet sounds like a single call.

Brood parasitism. The exploitation by one species of the parental behaviour of another, such as a cuckoo laying its egg in the nest of a host which then rears the cuckoo's chick.

Brood patch. A bare ventral area which is well supplied with blood vessels and is used to cover the eggs during incubation.

Cainism. *See* Siblicide.

Cloaca. The single terminal opening of a bird which serves for reproduction and excretion.

Commensal. Describes an association in which one species benefits from the presence of another, but the second species is indifferent to the presence of the first.

Co-operative breeding. The phenomenon of non-breeding birds attending the nests of breeding members of their own species and assisting with some or all of the functions of breeding behaviour.

Covert feathers. The contour feathers of a bird that cover the bases of the main flight feathers. Hence underwing coverts, undertail coverts, etc.

Crèche. An assembly of the young of colonial species that are still dependent on their parents.

Crepuscular. Mainly active in the dim light of dusk and dawn.

Cursorial. Describing birds which are adapted for running, such as coursers.

Ectoparasite. An external parasite such as a tick.

Endemic. Occurring in one geographic area and nowhere else.

Estrildid finches. A general term for members of the family Estrildidae, which includes waxbills, mannikins, twinspots and firefinches.

Faecal sac. A white, gelatinous 'envelope' which contains the faeces of the young of many species.

'Funk hole'. A place where young birds, and sometimes the adults, may hide, such as the chimney situated above the entrance of a hornbill's nest.

Ground colour. The main colour of an eggshell which may have underlying and overlying markings on it.

Gular air-sac. A bare inflatable sac on the throat of some species.

Gular patch. A bare, often colourful patch on the throat of some species.

Juvenile. A young bird in its first plumage of true feathers.

Lores. The area between the base of the bill and the eyes of a bird.

Mandible. Technically, the lower part of a bird's bill, but generally used to refer to both upper and lower parts.

Melanistic. A black or mainly blackish plumage form resulting from an excess of the pigment melanin.

Migrant. A species which makes regular movements, usually long-distance, at predictable times of the year between a breeding and a non-breeding area. Cf. Nomadic, Resident.

Monogamy. A breeding system whereby a bird has only one mate at a time, usually on a long term basis but sometimes only for a single breeding season. Cf. Polygamy.

Natal flock. The group in which young birds were born.

Nomadic. Describing a species which makes irregular movements, usually to an area of good food supply. Cf. Migrant, Resident.

Oviduct. The section of the reproductive tract in which the egg is formed.

Pair bond. The relationship between two birds of the opposite sex.

Palaearctic. The region that includes Europe, northern Asia and North Africa.

Passerine. Birds of the order Passeriformes (*passer* = sparrow). Characterised by their perching and singing functions, they are often divided into 'perching birds' and 'song birds', but in practice the two are synonymous.

Polyandry. The mating of one female with two or more males. Cf. Polygyny.

Polygamy. A breeding system whereby a bird (usually a male) has more than one mate. Cf. Monogamy.

Polygyny. A mating system in which individual males mate with two or more females during one breeding season. In successive polygyny he mates with two or more females in succession, and in simultaneous polygyny he mates with any available females who offer themselves to him. Cf. Polyandry.

Precocial. Describing a nestling that is covered with down when it hatches and is active almost immediately, or within a few hours of hatching. It is often able to feed by itself. Cf. Altricial.

Preen gland. A gland at the base of the tail which exudes a fatty secretion (preen oil) when squeezed between the mandibles. The oil is then applied to the feathers.

Pyriform. Describing an egg that is pointed at one end, such as that of a plover.

Red data book for birds. A book that lists bird species which are threatened or potentially threatened by the activities of man. The major categories under which the species are listed are extinct, endangered, rare and vulnerable.

Regurgitated pellets. Casts of undigested food. Owls, for example, regurgitate hair and bones, and bee-eaters regurgitate chitinous exoskeletons of insects.

'Releaser'. A device whereby a specific response is elicited, such as the red bill tip of a Kelp Gull which elicits a begging reaction from the chick.

Resident. A species which remains in the same general area throughout the year. Cf. Migrant, Nomadic.

Scapulars. The contour feathers above the 'shoulder' of a bird.

Sexual dimorphism. Differences in appearance between the male and female of the same species. They may be in size, shape, plumage or colour, or a combination of some of these.

Siblicide. The death of smaller and weaker nestlings in a brood. In obligate siblicide, which usually occurs in species that typically lay two eggs, the older chick almost always kills its weaker sibling by attacking and relentlessly pecking it. In facultative siblicide, which is the more common phenomenon, the attacks are less persistent and do not necessarily result in death. In these cases the death of the younger chick(s) is often linked to food deprivation.

Speculum. A distinctively coloured patch on the bird's wing which is visible in flight or when the wing is stretched. It is commonly found in ducks.

Territorial. Describing behaviour whereby an area is defended, usually for courtship and breeding purposes, but also to protect food resources.

Viduine. A member of the family Viduidae, represented in southern Africa by the whydahs and widow-finches.

Viscera. A bird's soft internal parts (entrails).

Useful addresses

BirdLife South Africa (formerly known as the Southern African Ornithological Society) aims to promote the enjoyment, understanding, study and conservation of southern African birds and their environments. It has a wide membership and branches throughout the country. Black Eagle Publishing, in association with BirdLife South Africa, produces a quality bi-monthly magazine, *Africa – Birds & Birding*, which supports these aims and offers highly illustrated and comprehensive coverage of the natural history of the continent's birdlife.

Unless stated otherwise, the addresses given are in South Africa.

BirdLife South Africa
P O Box 84394
2034 Greenside

Cape Bird Club
P O Box 5022
8000 Cape Town

Eastern Cape Wild Bird Society
P O Box 27454
6057 Greenacres

Lowveld Bird Club
P O Box 4113
1200 Nelspruit

Natal Bird Club
P O Box 1218
4000 Durban

Natal Midlands Bird Club
P O Box 2772
3200 Pietermaritzburg

Northern Transvaal Bird Club
P O Box 4158
0001 Pretoria

OFS Ornithological Society
P O Box 6614
9300 Bloemfontein

Rand Barbets Bird Club
P O Box 130355
2021 Bryanston

Sandton Bird Club
P O Box 650890
2010 Benmore

Wesvaal Bird Club
P O Box 2413
2520 Potchefstroom

Witwatersrand Bird Club
P O Box 72091
2122 Parkview

Namibia Bird Club
P O Box 67
Windhoek, Namibia

The Ornithological Association of Zimbabwe
P O Box CY161
Causeway, Zimbabwe

The Botswana Bird Club
P O Box 71
Gaborone, Botswana

Black Eagle Publishing
(Subscriptions: *Africa – Birds & Birding*)
P O Box 44223
7735 Claremont

Avian Demography Unit
Department of Statistical Sciences
University of Cape Town
7700 Rondebosch

Bibliography

General references

The following were used extensively throughout the text or for substantial parts of it.

Brooke, M. & Birkhead, T. 1991. *The Cambridge encyclopaedia of ornithology*. Cambridge University Press, Cambridge.

Brooke, R.K. 1984. South African red data book – birds. *South African National Scientific Programmes Report* No. 97. CSIR, Pretoria.

Brown, L.H., Urban, E.K. & Newman, K. 1982. *The birds of Africa* (Vol. 1). Academic Press, London.

Campbell, B. & Lack, E. 1985. *A dictionary of birds*. T. & A.D. Poyser, Calton.

Fry, C.H., Keith, S. & Urban, E.K. 1988. *The birds of Africa* (Vol. 3). Academic Press, London.

Ginn, P.J., McIlleron, W.G. & Milstein, P. le S. 1989. *The complete book of southern African birds*. Struik Winchester, Cape Town.

Irwin, M.P.S. 1981. *The birds of Zimbabwe*. Quest, Harare.

Keith, S., Urban, E.K. & Fry, C.H. 1992. *The birds of Africa* (Vol. 4). Academic Press, London.

Maclean, G.L. 1993. *Roberts' birds of southern Africa* (sixth edition). Trustees of the John Voelcker Bird Book Fund, Cape Town.

Newman, K. 1971. *Birdlife in southern Africa*. Purnell, Johannesburg.

Rowan, M.K. 1983. *The doves, louries and cuckoos of southern Africa*. David Philip, Cape Town.

Skead, C.J. 1967. *The sunbirds of southern Africa; also the sugarbirds, the white-eyes and the Spotted Creeper*. Balkema, Cape Town.

Steyn, P. 1982. *Birds of prey of southern Africa*. David Philip, Cape Town.

Urban, E.K., Fry, C.H. & Keith, S. 1986. *The birds of Africa* (Vol. 2). Academic Press, London.

Introduction

Büttiker, W. 1960. Artificial nesting devices in southern Africa. *Ostrich* 31:39-48.

Dean, W.R.J. & Milton, S.J. 1993. The use of *Galium tomentosum* (Rubiaceae) as nest material for birds in the southern Karoo. *Ostrich* 64:187-189.

Dean, W.R.J. & Siegfried, W.R. 1990. The use of wool as nest material by birds in the Karoo, South Africa: bane or bonus? *South African Journal of Ecology* 1:31-32.

Dean, W.R.J., Milton, S.J. & Siegfried, W.R. 1990. Dispersal of seeds as nest material by birds in semiarid Karoo shrubland. *Ecology* 71:1229-1306.

Gargett, V. 1990. *The Black Eagle – a study*. Acorn Books & Russel Friedman Books, Randburg & Halfway House.

Grimes, L.G. 1976. The occurrence of cooperative breeding behaviour in African birds. *Ostrich* 47:1-15.

Johnson, D. & Johnson, S. 1993. *Gardening with indigenous trees and shrubs*. Southern Books, Halfway House.

Kiff, L. 1991. The egg came first. *Terra* 30:5-20.

Maclean, G.L. 1971. The breeding seasons of birds in the south-western Kalahari. *Ostrich* Suppl. 8:179-192.

Maclean, G.L. 1974. Factors governing breeding seasons in African birds in non-arid habitats. *Proceedings of the 16th International Ornithological Congress*, pp. 258-271.

Maclean, G.L. 1990. *Ornithology for Africa*. University of Natal Press, Pietermaritzburg.

Mock, D.W., Drummond, H. & Stinson, C.H. 1990. Avian siblicide. *American Scientist* 78:438-449.

Moreau, R.E. 1950. The breeding seasons of African birds. 1. Land birds. *Ibis* 92:223-267.

Priest, C.D. 1948. *Eggs of birds breeding in southern Africa*. Robert Maclehose, Glasgow.

Simmons, R. 1988. Offspring quality and the evolution of cainism. *Ibis* 130:339-357.

Trendler, R. & Hes, L. 1995. *Attracting birds to your garden in southern Africa*. Struik Publishers, Cape Town.

Vernon, C.J. 1978. Breeding seasons of birds in deciduous woodland at Zimbabwe, Rhodesia, 1970-1974. *Ostrich* 49:102-115.

Wimberger, P.H. 1984. The use of green plant material in bird nests to avoid ectoparasites. *Auk* 101:615-618.

Winterbottom, J. 1971. *Priest's eggs of southern African birds*. Winchester Press, Johannesburg.

Winterbottom, J.M. & Rowan, M.K. 1962. The effect of rainfall on breeding of birds in arid regions. *Ostrich* 33(2):77-78.

The Ostrich

Bertram, B.C.R. 1979. Ostriches recognise their own eggs and discard others. *Nature* 279:233-234.

Bertram, B.C.R. 1980. Breeding systems and strategies of Ostriches. *Proceedings of the 17th International Ornithological Congress*, pp. 890-894.

Bertram, B.C.R. & Burger, A.E. 1981. Are Ostrich eggs the wrong colour? *Ibis* 123:207-210.

Brooke, R.K. 1979. Tool using by the Egyptian Vulture to the detriment of the Ostrich. *Ostrich* 50:119-120.

Sauer, E.G.F. & Sauer, E.M. 1966. The behaviour and ecology of the South African Ostrich. *Living Bird* 5:45-75.

Smit, D.J.V.Z. 1963. *Ostrich farming in the Little Karoo*. Department of Agricultural and Technical Services, Pretoria.

Wannenburgh, A., Johnson, P. & Bannister, A. 1984. *The Bushmen*. Struik Publishers, Cape Town.

The Jackass Penguin

Eggleton, P. & Siegfried, W.R. 1979. Displays of the Jackass Penguin. *Ostrich* 50:139-167.

Grebes

Broekhuysen, G.J. 1962. Does Black-necked Grebe cover its eggs? *Bokmakierie* 14:2-4.

Broekhuysen, G.J. 1973. Behavioural responses of Dabchicks *Podiceps ruficollis* to disturbance while incubating. *Ostrich* 44:111-117.

Broekhuysen, G.J., & Frost, P.G.H. 1968. Nesting behaviour of the Black-necked Grebe *Podiceps nigricollis* (Brehm) in southern Africa. 1. The reaction of disturbed incubating birds. *Bonner Zoologischen Beiträge* 19:350-361.

Broekhuysen, G.J., & Frost, P.G.H. 1968. Nesting behaviour of the Black-necked Grebe *Podiceps nigricollis* in southern Africa. 2. Laying, clutch size, incubation and nesting success. *Ostrich* 39:242-252.

Dean, W.R.J. 1977. Breeding of the Great Crested Grebe at Barberspan. *Ostrich* Suppl. 12:43-48.

Albatrosses, Petrels, Shearwaters, Prions and Skuas

Harrison, P. 1983. *Seabirds – an identification guide*. Croom Helm, Beckenham.

Jameson, W. 1958. *The Wandering Albatross*. Rupert Hart-Davis, London.

Pelicans

Brown, L.H. & Urban, E.K. 1969. The breeding biology of the Great White Pelican *Pelecanus onocrotalus roseus* at Lake Shala, Ethiopia. *Ibis* 111:199-237.

Burke, V.E.M. & Brown, L.H. 1970. Observations on the breeding of the Pink-backed Pelican *Pelecanus rufescens*. *Ibis* 112:499-512.

Cooper, J. 1980. Fatal sibling aggression in pelicans – a review. *Ostrich* 51:183-186.

Din, N.A. & Eltringham, S.K. 1974. Breeding of the Pink-backed Pelican *Pelecanus rufescens* in Rwenzori National Park, Uganda. *Ibis* 116:477-493.

Feely, J.M. 1962. Observations on the breeding of the White Pelican *Pelecanus onocrotalus*, at Lake St. Lucia, Zululand, during 1957 and 1958. *Lammergeyer* 2:10-20.

Gannets

Broekhuysen, G.J. & Rudebeck, G. 1951. Some notes on the Cape Gannet. *Ostrich* 22:132-138.

Crawford, R.J.M. & Cochrane, K.L. 1990. Onset of breeding by Cape Gannets *Morus capensis* influenced by availability of nesting material. *Ostrich* 61:147-149.

Gibson-Hill, C.A. 1948. Display and posturing in the Cape Gannet *Morus capensis*. *Ibis* 90:568-572.

Jarvis, M.J.F. 1970. Interactions between man and the South African Gannet *Sula capensis*. *Ostrich* Suppl. 8:497-513.

Jarvis, M.J.F. 1972. The systematic position of the South African Gannet. *Ostrich* 43:211-216.

McGill, L. 1972. Bird Island. *African Wild Life* 26:66-68.

Randall, R. & Ross, G.J.B. 1979. Increasing population of Cape Gannets on Bird Island, Algoa Bay, and observations on breeding success. *Ostrich* 50:168-175.

Cormorants and the Darter

Berry, H.H. 1976. The wind that makes the birds breed. *African Wildlife* 30:17-21.

Cooper, J. 1986. Biology of the Bank Cormorant. 4. Nest construction and characteristics. *Ostrich* 57:170-179.

Cooper, J. 1987. Biology of the Bank Cormorant. 5. Clutch size, eggs and incubation. *Ostrich* 58:1-8.

Olver, M.D. 1984. Breeding biology of the Reed Cormorant. *Ostrich* 55:33-140.

Olver, M.D. & Kuyper, M.A. 1978. Breeding biology of the White-breasted Cormorant in Natal. *Ostrich* 49:25-30.

Williams, A.J. & Cooper, J. 1983. The Crowned Cormorant: breeding biology, diet and offspring-reduction strategy. *Ostrich* 54:213-219.

Herons, Egrets and Bitterns

Blaker, D. 1969. Behaviour of the Cattle Egret *Ardeola ibis*. *Ostrich* 40:75-129.

Dean, W.R.J. 1988. Breeding of Goliath Heron at Barberspan, Transvaal. *Ostrich* 59:75-76.

Editor 1982. Little Egrets breeding in Kelp Gull colony. *Bee-eater* 33(1):1.

Fry, C.H. & Hosken, J.H. 1986. Further observations on the breeding of Slaty Egrets *Egretta vinaceigula* and Rufous-bellied Herons *Ardeola rufiventris*. *Ostrich* 57:61-64.

Hustler, K. & Williamson, C. 1985. The Rail Heron in the Hwange National Park. *Honeyguide* 31:145-147.

Langley, C.H. 1983. Biology of the Little Bittern in the southwestern Cape. *Ostrich* 54:83-94.

Siegfried, W.R. 1971. The nest of the Cattle Egret. *Ostrich* 42:193-197.

Skead, C.J. 1966. A study of the Cattle Egret *Ardeola ibis*, Linnaeus. *Ostrich* Suppl. 6:109-139.

Tarboton, W.R. 1967. Rufous Heron *Ardeola rufiventris* breeding in the Transvaal. *Ostrich* 38:207.

Tarboton, W.R. 1980. Notes on the Dwarf Bittern. *Witwatersrand Bird Club News* 108:3-5.

Tomlinson, D. 1974. Colour changes in the Great White Egret. *Wild Rhodesia* 5:19.

Tomlinson, D.N.S. 1974. Studies of the Purple Heron. 1. Heronry structure, nesting habits and reproductive success. *Ostrich* 45:175-181.

Tomlinson, D.N.S. 1974. Studies of the Purple Heron. 2. Behaviour patterns. *Ostrich* 45:209-223.

Tomlinson, D.N.S. 1975. Studies of the Purple Heron. 3. Egg and chick development. *Ostrich* 46:157-165.

Tomlinson, D.N.S. 1976. Breeding behaviour of the Great White Egret. *Ostrich* 47:161-178.

Tomlinson, D.N.S. 1979. Interspecific relations in a mixed heronry. *Ostrich* 50:193-198.

The Hamerkop

Cheke, A.S. 1968. Copulation in the Hamerkop *Scopus umbretta*. *Ibis* 110:201-203.

Gentis, S. 1976. Co-operative nest building by Hamerkops. *Honeyguide* 88:48.

Godfrey, R. 1941. *Bird-lore of the eastern Cape Province*. Bantu Studies, Monograph Series No. 2. Witwatersrand University Press, Johannesburg.

Goodfellow, C.F. 1958. Display in the Hamerkop *Scopus umbretta*. *Ostrich* 29:1-4.

Kahl, M.P. 1967. Observations on the behaviour of the Hamerkop *Scopus umbretta* in Uganda. *Ibis* 109:25-32.

Liversidge, R. 1963. The nesting of the Hamerkop *Scopus umbretta*. *Ostrich* 34:55-62.

Lorber, P. 1985. What makes a Hamerkop's nest. *Honeyguide* 31:49.

Nel, J.E. 1966. Nesting Hamerkops. *Bokmakierie* 18:70-72.

Siegfried, W.R. 1975. On the nest of the Hamerkop. *Ostrich* 46:267.

Uys, C.J. 1967. Hamerkop *Scopus umbretta* nesting on a house. *Ostrich* 38:199-200.

Wilson, R.T. & Wilson, M.P. 1984. Breeding biology of the Hamerkop in central Mali. *Proceedings of the Fifth Pan-African Ornithological Congress* pp. 855-865.

Wilson, R.T. & Wilson, M.P. 1986. Nest building by the Hamerkop *Scopus umbretta*. *Ostrich* 57:224-232.

Wislon, R.T., Wilson, M.P. & Durkin, J.W. 1987. Aspects of the reproductive ecology of the Hamerkop *Scopus umbretta* in central Mali. *Ibis* 129:382-388.

Storks

Anthony, A.J. & Sherry, B.Y. 1980. Openbill Storks breeding in the southeastern lowveld of Zimbabwe Rhodesia. *Ostrich* 51:1-6.

Broekhuysen, G. 1965. Nesting of the White Stork (*Ciconia ciconia* (L)) in South Africa. *Die Vogelwarte* 23:5-11.

Broekhuysen, G.J. 1971. White Storks breeding in the Bredasdorp district, most southern part of the wintering quarters. *Die Vogelwarte* 26:164-169.

Hancock, J.A., Kushlan, J.A. & Kahl, M.P. 1992. *Storks, ibises and spoonbills of the world*. Academic Press, London.

Kahl, M.P. 1971. Behaviour and taxonomic relationships of the storks. *Living Bird* 10:151-170.

Lorber, P. 1982. Further notes on the Black Stork in Zimbabwe. *Honeyguide* 110:8-14.

Roberts, A. 1941. Notes on some birds of the Cape Province. *Ostrich* 11:124.

Ryder, J.H. & Ryder, B.A. 1978. First breeding records of Black Storks in Malawi. *Ostrich* 49:51.

Schüz, E. 1973. White Stork colonization – a social factor also? *Bokmakierie* 25:69-70.

Scott, J.A. 1975. Observations on the breeding of the Woollynecked Stork. *Ostrich* 46:201-207.

Siegfried, W.R. 1967. The distribution and status of the Black Stork in southern Africa. *Ostrich* 38:179-185.

Tarboton, W. 1982. Breeding status of the Black Stork in the Transvaal. *Ostrich* 53:151-156.

Uys, C.J. 1968. Breeding of the White Stork *Ciconia ciconia* L. at Mossel Bay, Cape. *Ostrich* 39:30-32.

Ibises and the African Spoonbill

Cooper, K.H. & Edwards, K.Z. 1969. A survey of the Bald Ibis in Natal. *Bokmakierie* 21:4-9.

Hancock, J.A., Kushlan, J.A. & Kahl, M.P. 1992. *Storks, ibises and spoonbills of the world*. Academic Press, London.

Kahl, M.P. 1983. Breeding displays of the African Spoonbill *Platalea alba*. *Ibis* 125:324-338.

Martin, J., Martin, E. & Martin, R. 1975. In search of Booted eagles. *Bokmakierie* 27:4-38.

Milstein, P. le S. 1973. Buttons and Bald Ibises. *Bokmakierie* 25:57-60.

Milstein, P. le S. 1974. More Bald Ibis buttons. *Bokmakierie* 26:88.

Milstein, P. le S. & Siegfried, W.R. 1970. Transvaal status of the Bald Ibis. *Bokmakierie* 22:36-39.

Raseroka, B.H. 1975. Breeding of the Hadeda Ibis. *Ostrich* 46:208-212.

Siegfried, W.R. 1966. The present and past distribution of the Bald Ibis in the Cape Province of the Cape of Good Hope. *Ostrich* 37:216-218.

Siegfried, W.R. 1971. The status of the Bald Ibis of southern Africa. *Biological Conservation* 3:88-91.

Skead, C.J. 1951. A study of the Hadedah Ibis *Hagedashia h. hagedash*. *Ibis* 93:360-382.

Urban, E.K. 1974. Breeding of the Sacred Ibis *Threskiornis aethiopica* at Lake Shala, Ethiopia. *Ibis* 116:263-277.

Uys, C.J. & Broekhuysen, G.J. 1966. Hadeda *Hagedashia hagedash* nesting on telegraph pole. *Ostrich* 37:239-240.

Whitelaw, D. 1968. Notes on the breeding biology of the African Spoonbill *Platalea alba*. *Ostrich* 39:236-241.

Flamingos

Berry, H.H. 1972. Flamingo breeding on Etosha Pan, South West Africa, during 1971. *Madoqua* 5 (Series 1):5-31.

Boshoff, A.F. 1979. A breeding record of the Greater Flamingo in the Cape Province. *Ostrich* 50:124.

Brown, L.H. 1958. The breeding of the Greater Flamingo *Phoenicopterus ruber* at Lake Elementeita, Kenya colony. *Ibis* 100:388-420.

Brown, L. 1973. *The mystery of the flamingos*. East African Publishing House, Nairobi.

Brown, L.H. & Root, A. 1971. The breeding behaviour of the Lesser Flamingo *Phoenicopterus minor*. *Ibis* 113:147-172.

Brown, L.H., Powell-Cotton, D. & Hopcraft, J.B.D. 1973. The breeding of the Greater Flamingo and the Great White Pelican in East Africa. *Ibis* 115:352-374.

Kahl, M.P. 1970. East Africa's majestic flamingos. *National Geographic* 137:276-292.

Middlemiss, E. 1961. Notes on the Greater Flamingo. *Bokmakierie* 13:9-14.

Porter, R.N. & Forrest, G.W. 1974. First successful breeding of Greater Flamingos in Natal, South Africa. *Lammergeyer* 21:26-33.

Uys, C.J., Broekhuysen, G.J., Martin, J. & Macleod, J.G. 1963. Observations on the breeding of the Greater Flamingo *Phoenicopterus ruber* Linnaeus in the Bredasdorp district, South Africa. *Ostrich* 34:129-154.

Ducks and Geese

Boulton, R. & Woodall, P. 1974. The breeding season of waterfowl in Rhodesia. *Honeyguide* 36:38.

Clark, A. 1964. The Maccoa Duck (*Oxyura maccoa* (Eyton)). *Ostrich* 35:264-276.

Clark, A. 1969. The breeding of the Hottentot Teal. *Ostrich* 40:33-36.

Clark, A. 1974. The breeding of Hottentot Teal. *Bokmakierie* 26:31-32.

Clark, A. 1976. Observations on the breeding of whistling ducks in southern Africa. *Ostrich* 47:59-64.

Clark, A. 1979. The breeding of the Whitebacked Duck on the Witwatersrand. *Ostrich* 50:59-60.

Clark, A. 1980. Notes on the breeding biology of the Spurwinged Goose. *Ostrich* 51:179-182.

Edelsten, G. 1932. Notes on the nest of the South African Shelduck. *Ostrich* 3:61.

Ford, G. 1978. Botletle birdwatching. *Honeyguide* 95:31-32.

Johnsgard, P.A. 1968. Some observations on Maccoa Duck behaviour. *Ostrich* 39:219-222.

Johnsgard, P.A. & Kear, J. 1968. A review of parental carrying by waterfowl. *Living Bird* 7:89-102.

Jones, M.A. 1978. White-faced Whistling Duck *Dendrocygna viduata* (Linnaeus, 1766) carrying their young. *Honeyguide* 94:19-21.

Kear, J. 1967. Notes on the eggs and downy young of *Thalassornis leuconotus*. *Ostrich* 38:227-229.

MacNae, W. 1959. Notes on the biology of the Maccoa Duck. *Bokmakierie* 11:49-52.

McKinney, F., Siegfried, W.R., Ball, I.J. & Frost, P.G.H. 1978. Behavioural specializations for river life in the African Black Duck (*Anas sparsa*). *Zeitschrift für Tierpsychologie* 48:349-400.

Middlemiss, E. 1958. The Southern Pochard. *Ostrich* Suppl. 3:1-34.

Milstein, P. le S. 1973. Maccoa Duck *Oxyura punctata* parasitising Fulvous Duck *Dendrocygna bicolor* nest. *Bokmakierie* 25:74.

Milstein, P. le S. 1975. How baby Egyptian Geese leave a high nest. *Bokmakierie* 27:49-51.

Rowan, M.K. 1963. The Yellowbill Duck *Anas undulata* Dubois in South Africa. *Ostrich* Suppl. 5:1-56.

Siegfried, W.R. 1965. The Cape Shoveller *Anas smithii* (Hartert) in southern Africa. *Ostrich* 36:155-198.

Siegfried, W.R. 1968. The Black Duck in the south-western Cape. *Ostrich* 39:61-75.

Siegfried, W.R. 1969. The proportion of yolk in the egg of the Maccoa Duck. *Wildfowl* 20:78.

Siegfried, W.R. 1974. Brood care, pair bonds and

plumage in the southern African Anatini. *Wildfowl* 25:33-40.
Siegfried, W.R. 1976. Sex ratio in the Cape Shelduck. *Ostrich* 47:113-116.
Siegfried, W.R. 1978. Social behaviour of the African Comb Duck. *Living Bird* 17:85-104.
Siegfried, W.R. & Heyl, C.W. 1976. Long-standing pair-bonds in Cape Teal. *Ostrich* 47:130-131.
Siegfried, W.R., Burger, A.E. & Caldwell, P.J. 1976. Incubation of Ruddy and Maccoa Ducks. *Condor* 78:512-517.
Siegfried, W.R., Burger, A.E. & Frost, P.G.H. 1976. Energy requirements for breeding in the Maccoa Duck. *Ardea* 64:171-191.
Skead, D.M. 1977. Pair-forming and breeding behaviour of the Cape Shoveller at Barberspan. *Ostrich* Suppl. 12:75-81.
Swennen, C. 1968. Nest protection of Eiderducks and Shovelers by means of faeces. *Ardea* 56:248-258.
Winterbottom, J.M. 1974. The Cape Teal. *Ostrich* 45:110-132.
Wintle, C.C. 1981. Notes on the breeding of the White-backed Duck. *Honeyguide* 105:13-20.
Zaloumis, E.A. 1976. Incubation period of the African Pygmy Goose. *Ostrich* 47:231.

The Secretary Bird

Steyn, P. & Myburgh, N. 1991. Notes made at two Secretarybird nests. *Birding in Southern Africa* 44:19-21.

Vultures

Mundy, P., Butchart, D., Ledger, J. & Piper, S. 1992. *The vultures of Africa*. Acorn/Russel Friedman Books, Randburg & Halfway House.

Kites

Mendelsohn, J. 1984. The timing of breeding in Blackshouldered Kites in southern Africa. *Proceedings of the Fifth Pan-African Ornithological Congress* pp. 799-808.
Slotow, R. & Perrin, M.R. 1992. The importance of large prey for Blackshouldered Kite reproduction. *Ostrich* 63:180-182.

The Cuckoo Hawk and the Bat Hawk

Carlyon, J. 1987. Two breeding records of the Cuckoo Hawk in the Transvaal, RSA. *Gabar* 2:45-46.
Chittenden, H. 1984. Aspects of Cuckoo Hawk *Aviceda cuculoides* breeding biology. *Proceedings of the Second Symposium on African Predatory Birds* pp. 47-56.
Dewhurst, C.F., Cunningham-Van Someren, G.R., Elliott, C.C.H., Thomsett, S. & Wilson, A.C. 1989. Some observations on the nesting habits and food of the Cuckoo Falcon *Aviceda cuculoides* in Kenya. *Gabar* 4(2):11-15.
Hall, D.G., Tarboton, W.R. & Theron, F. 1991. More on the enigmatic Cuckoo Hawk. *Gabar* 6:47-50.
Harris, T., Dunning, J. & Hoets, D. 1990. The darker side of Bat Hawks. *Birding in Southern Africa* 42:86-90.
Jones, J.M.B. 1985. The breeding cycle of the African Cuckoo Hawk. *Honeyguide* 31:196-202.

Eagles

Allan, D. 1984. Crowned Eagle nesting on a cliff. *Witwatersrand Bird Club News* 124:3.
Allan, D. 1988. Raptors nesting on transmission pylons. *African Wildlife* 42:325-327.
Brown, C.J. 1985. Booted Eagles breeding in Namibia. *Madoqua* 14:189-191.
Brown, C.J. & Lawson, J.L. 1989. Birds and electricity transmission lines in South West Africa/Namibia. *Madoqua* 16:59-67.
Boshoff, A.F. 1993. Density, breeding performance and stability of Martial Eagles *Polemaetus bellicosus* on electricity pylons in the Nama-Karoo, South Africa. *Proceedings of the Eighth Pan-African Ornithological Congress* pp. 95-104.
Boshoff, A. & Fabricus, C. 1980. Black Eagles nesting on man-made structures. *Bokmakierie* 38:67-70.
Davies, R.A.G. & Vogel, J.C. 1992. Radiocarbon dating of sticks from a Black Eagle's nest in the Karoo, South Africa. *Gabar* 7:24-25.
Edwards, S.F.C. 1985. Breeding of the Western Banded Snake Eagle in Zimbabwe. *Honeyguide* 31:213-216.
Gargett, V. 1990. *The Black Eagle – a study*. Acorn Books and Russel Friedman Books, Randburg and Halfway House.
Hustler, K. 1988. Some interesting records from Lake Kariba. *Honeyguide* 34:25-26.
Ledger, J., Hobbs, J. & Van Rensburg, D. 1987. First record of a Black Eagle nesting on an electricity transmission tower. *African Wildlife* 41:60-66.
Martin, R., Martin, J. & Martin, E. 1991. Booted Eagles: an estimate of breeding numbers in the Cape. *Promerops* 198:8-9.
Mundy, P.J. 1973. Green leaves for a tablecloth. *Honeyguide* 74:34.
Simmons, R.E. & Mendelsohn, J.M. 1993. A critical review of cartwheeling flights of raptors. *Ostrich* 64:13-24.
Steyn, P. & Grobler, J.H. 1981. Breeding biology of the Booted Eagle in South Africa. *Ostrich* 52:108-118.
Steyn, P. & Grobler, J.H. 1985. Supplementary observations on the breeding biology of Booted Eagles in South Africa. *Ostrich* 56:151-156.
Upton, W.L. 1988. Tawny Eagle breeding. *Mirafra* 5:83-84.
Watson, R.T. 1988. The influence of nestling predation on nest site selection and behaviour of the Bateleur. *South African Journal of Zoology* 23:143-149.
Watson, R.T. 1990. Breeding biology of the Bateleur. *Ostrich* 61:13-23.
Wimberger, P.H. 1984. The use of green plant material in bird nests to avoid ectoparasites. *Auk* 101:615-618.

Buzzards

Allan, D.G. 1992. Long distance movements of Forest Buzzards in South Africa. *Gabar* 7:26-27.
Palmer, N.G., Norton, P.M. & Robertson, A.S. 1985. Aspects of the biology of the Forest Buzzard. *Ostrich* 56:67-73.
Stinton, J.C. 1990. Augur Buzzards rearing two young. *Honeyguide* 36:37.

Sparrowhawks and Goshawks

Colebrook-Robjent, J.F.R. 1986. On the validity of the genus *Micronisus*. *Gabar* 1:7-8.
Dewhurst, C.F. 1986. The breeding biology of the African Goshawk at Karen, Nairobi, Kenya. *Ostrich* 57:1-8.
Henschel, J.R., Mendelsohn, J.M. & Simmons, R.E. 1991. Is the association between Gabar Goshawks and social spiders *Stegodyphus* mutualism or theft? *Gabar* 6:57-60.
Kemp, A.C. 1986. The Gabar Goshawk: taxonomy, ecology and further research. *Gabar* 1:4-6.
Kemp, A.C. 1988. Flight and nest display by Gabar Goshawk. *Gabar* 3:53.
MacDonald, I.A.W. 1986. Do Redbreasted Sparrowhawks belong in the Karoo? *Bokmakierie* 38:3-4.
Malan, G. 1992. Nest-lining used by Pale Chanting Goshawks in the Little Karoo, South Africa. *Gabar* 7:56-61.
Simmons, R. 1986. In defence of Karoo Redbreasted Sparrowhawks. *Bokmakierie* 38:82-83.
Simmons, R. 1990. African Goshawk encounter. *African Wildlife* 44:91.
Steyn, P. 1988. Observations on the Redbreasted Sparrowhawk. *Bokmakierie* 40:66-73.
Steyn, P. 1992. Gabar Goshawks and colonial spiders. *Gabar* 7:21.
Steyn, P. & Myburgh, N. 1992. Observations on the breeding of Pale Chanting Goshawks. *Birding in Southern Africa* 44:80-83.

Harriers

Roberts, E.L. 1987. Proximity of Black Harrier nests. *Promerops* 181:16.
Simmons, R. 1991. Comparisons and functions of sky-dancing displays of *Circus* harriers: untangling the marsh harrier complex. *Ostrich* 62:45-51.
Simmons, R. 1991. The efficiency and evolution of aerial food passing in harriers. *Gabar* 6:51-56.
Van Jaarsveld, J. 1986. Notes on a Black Harrier's nest. *Bokmakierie* 38:75.

The Gymnogene and the Osprey

Dean, W.R.J. 1983. Osprey breeding records in South Africa. *Ostrich* 54:241-242.
Jackson, P.M.M. 1985. Martial Eagle breeding in the Gutu district. *Honeyguide* 31:213.
Smeenk, C. & Smeenk-Enserink, N. 1983. Observations on the Harrier Hawk *Polyboroides typus* in Nigeria, with comparative notes on the neotropical Crane Hawk *Geranospiza caerulescens*. *Ardea* 71:133-143.

Falcons and Kestrels

Braby, R., Paterson, J. & Brown, C.J. 1987. Peregrine Falcon breeding in the Namib desert. *Gabar* 2:43-44.
Brown, C.J., Paxton, M.W. & Henrichsen, I. 1987. Aspects of the biology of the Greater Kestrel in SWA/Namibia. *Madoqua* 15:147-156.
De Swardt, D.H. 1990. Pygmy Falcon catches Sociable Weaver in flight. *Gabar* 5:27.
Hartley, R. 1993. The Batonka gorges, haven for birds of prey. *African Wildlife* 47:74-78.
Hartley, R.R., Bodington, G., Dunkley, A.S. & Groenewald, A. 1993. Notes on the breeding biology, hunting behavior, and ecology of the Taita Falcon in Zimbabwe. *The Journal of Raptor Research* 27:133-141.
Hustler, K. 1983. Breeding biology of the Greater Kestrel. *Ostrich* 54:129-140.
Hustler, K. 1983. Breeding biology of the Peregrine Falcon in Zimbabwe. *Ostrich* 54:161-171.
Jenkins, A. 1994. Caught in the act – mamba attack on Peregrine brood. *African Wildlife* 48:13-15.
Jenkins, A.R. & Wagner, S.T. 1991. First breeding of the Taita Falcon in South Africa. *Ostrich* 62:78.
Kemp, A.C. 1993. Breeding biology of Lanner Falcons near Pretoria, South Africa. *Ostrich* 64:26-31.
Paxton, M. & Brown, C. 1987. Rednecked Falcons nesting in palm trees in Namibia. *Gabar* 2:13-13.
Pepler, D. & Martin, R. 1991. Peregrines nesting on man-made structures. *Promerops* 200:21-22.
Pepler, D., Van Hensbergen, H.J. & Martin, R. 1991. Breeding density and nest site characteristics of the Peregrine Falcon *Falco peregrinus minor* in the southwestern Cape, South Africa. *Ostrich* 62:23-28.
Saunders, K. 1989. High rise accommodation. *Bokmakierie* 41:61.
Thomsett, S. 1991. Polyandrous Pygmy Falcon? *Gabar* 6:73.
Tree, A.J. 1989. Recent reports. *Honeyguide* 35:30.

Francolins, Quails, Guineafowls and Buttonquails

Crowe, T.M. & Siegfried, W.R. 1978. It's raining guineafowl in the northern Cape. *South African Journal of Science* 74:261-262.
Dudley, E.P.V. 1971. Development of chicks of Harlequin Quail. *Ostrich* 42:79-80.
Hartley, R.R. & Mundy, P.J. 1992. Management of terrestrial gamebird hunting in Zimbabwe in relation to breeding seasons. *Gibier Faune Sauvage* 9:837-846.
Jarvis, M.J.F. 1977. *Gamebird management and utilisation*. Natural Resources Board of Rhodesia, Causeway.
Little, R. & Crowe, T. 1993. Greywing Francolin hunting: a key to conservation in the 'new' South Africa. *Birding in Southern Africa* 45:85-91.
Masterson, A.N.B. 1973. Notes on the Hottentot Buttonquail. *Honeyguide* 74:12-16.
Skead, C.J. 1962. A study of the Crowned Guinea Fowl *Numida meleagris coronata* Gurney. *Ostrich* 33:51-65.
Van Niekerk, J.H. 1980. Some socio-biological features of Crowned Guineafowl in the Krugersdorp Game Reserve. *Bokmakierie* 32:102-108.
Van Niekerk, J.H. 1983. Observations on courtship in Swainson's Francolin. *Bokmakierie* 35:90-92.
Wintle, C.C. 1975. Notes on the breeding habits of the Kurrichane Buttonquail. *Honeyguide* 82:27-30.

Cranes

Lees, S.G. 1977. Crowned Crane nesting in a tree. *Honeyguide* 92:49.
Maozeka, F. 1993. Observations on three pairs of Wattled Crane breeding on the Great Dyke, Mazoe district. *Honeyguide* 39:182-188.
Paterson-Jones, C. 1994. Down on the farm – Blue Crane conservation in the Overberg. *Africa – Environment & Wildlife* 2:65-68.
Porter, D.J., Craven, H.S., Johnson, D.N. & Porter, M.J. 1992. *Proceedings of the First Southern African Crane Conference*. The Southern African Crane Foundation, Durban.
Steyn, P. & Ellman-Brown, P. 1974. Crowned Crane nesting on a tree. *Ostrich* 45:40-42.
Steyn, P. & Tredgold, D. 1977. Crowned Cranes covering eggshells. *Bokmakierie* 29:82-83.
Tarboton, W.R. 1984. The status and conservation of the Wattled Crane in the Transvaal. *Proceedings of the Fifth Pan-African Ornithological Congress* pp. 665-678.
Tyson, N. 1988. Blue Crane poisoning. *African Wildlife* 42:4-5.
Van Ee, C.A. 1966. Notes on the breeding behaviour of the Blue Crane *Tetrapteryx paradisea*. *Ostrich* 37:23-29.
Walkinshaw, L.H. 1965. The Wattled Crane *Bugeranus carunculatus* (Gmelin). *Ostrich* 36:73-81.
West, O. 1963. Notes on the Wattled Crane *Bugeranus carunculatus* (Gmelin). *Ostrich* 34:67-77.

Rails, Crakes, Flufftails, Gallinules, Moorhens and Coots

Broekhuysen, G.J., Lestrange, G.K. & Myburgh, N. 1964. The nest of the Red-chested Flufftail (*Sarothrura rufa* (Vieillot)). *Ostrich* 35:17-120.
Brooke, R.K. 1975. Cooperative breeding, duetting, allopreening and swimming in the Black Crake. *Ostrich* 46:190-191.
Dean, W.R.J. 1980. Brood division by the Red-knobbed Coot. *Ostrich* 51:125-126.
Hopkinson, G. & Masterson, G.N.B. 1975. Notes on the Striped Crake. *Honeyguide* 84:12-21.
Kakebeeke, B. 1993. Striped Flufftail found breeding in Somerset West. *Birding in Southern Africa* 45:9-11.
Keith, S., Benson, C.W. & Irwin, M.P.S. 1970. The genus *Sarothrura* (Aves, Rallidae). *Bulletin of the American Museum of Natural History* 143:1-184.
Manson, A.J. 1986. Notes on the breeding of the Buff-spotted Flufftail at Seldomseen, Vumba. *Honeyguide* 32:137-142.
Schmitt, M.B. 1976. Observations on the Cape Rail in the southern Transvaal. *Ostrich* 47:16-26.
Siegfried, W.R. & Frost, P.G.H. 1975. Continuous breeding and associated behaviour in the Moorhen *Gallinula chloropus*. *Ibis* 117:102-109.
Steyn, P. & Myburgh, N. 1986. A tale of two flufftails. *African Wildlife* 40:22-27.
Steyn, P. & Myburgh, N. 1986. A tale of two flufftails – a sequel. *African Wildlife* 40:126.
Taylor, P.B. 1985. Field studies of the African Crake *Crex egregia* in Zambia and Kenya. *Ostrich* 56:170-185.
Wintle, C.C. & Taylor, P.B. 1993. Sequential polyandry, behaviour and moult in captive Striped Crakes *Aenigmatolimnas marginalis*. *Ostrich* 64:115-122.

The African Finfoot

Del Toro, M.A. 1971. On the breeding biology of the American Finfoot in southern Mexico. *Living Bird* 10:79-88.
Skead, C.J. 1962. Peter's Finfoot *Podica senegalensis* (Vieillot) at the nest. *Ostrich* 33:31-33.

Bustards and Korhaans

Boobyer, M.G. 1988. Denizens of the desert. *Bokmakierie* 40:47-50.
Crowe, T.M., Essop, M.F., Allan, D.G., Brooke, R.K. & Komen, J. 1994. 'Overlooked' units of comparative and conservation biology: a case study of a small African bustard, the Black Korhaan *Eupodotis afra*. *Ibis* 136:166-175.
Kemp, A. & Tarboton, W. 1976. Small South African bustards. *Bokmakierie* 28:40-43.

Jacanas

Steyn, P. 1973. African Jacana at last. *Bokmakierie* 25:34-37.
Tarboton, W.R. 1991. Polyandry in the African Jacana. Ph.D. thesis, Witwatersrand University.
Tarboton, W.R. 1992. Aspects of the breeding biology of the African Jacana. *Ostrich* 63:141-157.
Tarboton, W.R. 1992. Variability in African Jacana eggs and clutches. *Ostrich* 63:158-164.
Tarboton, W.R. & Fry, C.H. 1986. Breeding and other behaviour of the Lesser Jacana. *Ostrich* 57:233-243.
Wilson, G. 1974. Incubating behaviour of the African Jacana. *Ostrich* 45:185-188.

Snipes

Elgood, J.H. & Donald, R.G. 1962. Breeding of the Painted Snipe *Rostratula benghalensis* in southwest Nigeria. *Ibis* 104:253-256.
Masterson, A.N.B. 1971. Snipe in Rhodesia. *Honeyguide* 65:30-35.
Schmidt, R.K. 1961. Incubation period of the Painted Snipe *Rostratula benghalensis*. *Ostrich* 33:183-184.
Steyn, P. 1957. Photographing the Ethiopian Snipe. *African Wild Life* 11:124-126.

The African Black Oystercatcher

Brown, A. 1984. Belly-soaking in the Black Oystercatcher. *Promerops* 162:12.
Hall, K.R.L. 1959. Observations on the nest-sites and nesting behaviour of the Black Oystercatcher *Haematopus moquini* in the Cape Peninsula. *Ostrich* 30:143-154.
Hockey, P.A.R. 1983. Aspects of the breeding biology of the African Black Oystercatcher. *Ostrich* 54:26-35.
Hockey, P.A.R. & Cooper, J. 1980. Paralytic shellfish poisoning – a controlling factor in Black Oystercatcher populations? *Ostrich* 51:188-190.
Jeffery, R.G. 1987. Influence of human disturbance on the nesting success of African Black Oystercatchers. *South African Journal of Wild life Research* 17:71-72.
Summers, R.W. & Cooper, J. 1977. The population, ecology and conservation of the Black Oystercatcher. *Ostrich* 48:28-40.

Plovers

Ade, B. 1979. Some observations on the breeding of Crowned Plovers. *Bokmakierie* 31:9-16.
Begg, G.W. & Maclean, G.L. 1976. Belly-soaking in the Whitecrowned Plover. *Ostrich* 47:65.
Broekhuysen, G.J. & Macleod, J.R.G. 1948. Breeding record of the Blacksmith Plover *Hoplopterus armatus* for the neighbourhood of Cape Town. *Ostrich* 19:237-239.
Brown, A. & Gottschalk, H. 1986. Unusual clutch for Blacksmith Plover. *Promerops* 175:10.
Brown, L.H. 1972. Partial burying of eggs by Blacksmith Plover. *Ostrich* 43:130.
Conway, W.G. & Bell, J. 1968. Observations on the behavior of Kittlitz's Sandplovers at the New York Zoological Park. *Living Bird* 7:57-70.
Crowe, A.A. & Crowe, T.M. 1984. Variation in the breeding season of the White-fronted Sandplover in southern Africa. *Proceedings of the Fifth Pan-African Ornithological Congress* pp. 787-798.
Hall, K.R.L. 1958. Observations on the nesting sites and nesting behaviour of the Kittlitz's Sandplover *Charadrius pecuarius*. *Ostrich* 29:113-125.
Hall, K.R.L. 1959. Nest records and additional behaviour notes for Kittlitz's Sandplover *Charadrius pecuarius* in the S.W. Cape Province. *Ostrich* 30:33-38.
Hall, K.R.L. 1959. A study of the Blacksmith Plover *Hoplopterus armatus* in the Cape Town area. 1. Distribution and breeding data. *Ostrich* 30:117-126.
Hall, K.R.L. 1960. Egg-covering by the White-fronted Sandplover *Charadrius marginatus*. *Ibis* 102:545-553.
Hall, K.R.L. 1964. A study of the Blacksmith Plover *Hoplopterus armatus* in the Cape Town area: 2. Behaviour. *Ostrich* 35:3-16.
Jeffery, R.G. & Liversidge, R. 1951. Notes on the Chestnut-banded Sandplover. *Ostrich* 22:68-76.
Little, J. de V. 1967. Some aspects of the behaviour of the Wattled Plover *Afribyx senegalensis* (Linnaeus). *Ostrich* 38:259-280.
Liversidge, R. 1965. Egg covering in *Charadrius marginatus*. *Ostrich* 36:59-61.
Maclean, G.L. 1974. Egg covering by the Charadrii. *Ostrich* 45:167-174.
Maclean, G.L. & Moran, V.C. 1965. The choice of nest site in the White-fronted Sandplover *Charadrius marginatus*. *Ostrich* 36:63-72.

Roberts, M.G. 1977. Belly-soaking in the Whitefronted Plover. *Ostrich* 48:111.
Saunders, C.R. 1970. Observations on breeding of the Long-toed or White-winged Plover *Hemiparra crassirostris leucoptera* (Reichenow). *Honeyguide* 62:27-29.
Skead, C.J. 1955. A study of the Crowned Plover. *Ostrich* 26:88-98.
Van der Merwe, F.J. 1973. Essay on Crowned Plovers. *Bokmakierie* 25:81-86.
Ward, D. 1989. Behaviour associated with breeding of Crowned, Blackwinged and Lesser Blackwinged Plovers. *Ostrich* 60:141-150.

The Avocet and the Black-winged Stilt
Every, B. 1974. Abnormal clutch size for the Blackwinged Stilt. *Ostrich* 45:260.

Dikkops
Bigalke, R. 1933. Observations on the breeding habits of the Cape Thickknee, *Burhinops capensis* (Licht.), in captivity. *Ostrich* 4:41-48.
Maclean, G.L. 1966. Studies of the behaviour of a young Cape Dikkop *Burhinus capensis* (Lichtenstein). *Ostrich* Suppl. 6:155-170.
Uys, C.J. 1961. Notes on our Thick-knees. *Bokmakierie* 13:2-3.
Uys, C.J. 1983. The urban dikkop. *Promerops* 159:14-16.

Coursers
Kemp, A.C. & Maclean, G.L. 1973. Neonatal plumage patterns of Three-banded and Temminck's Coursers and their bearing on courser genera. *Ostrich* 44:80-81.
Kemp, A.C. & Maclean, G.L. 1973. Nesting of the Three-banded Courser. *Ostrich* 44:82-83.
Maclean, G.L. 1967. The breeding biology and behaviour of the Double-banded Courser *Rhinoptilus africanus* (Temminck). *Ibis* 109:556-569.
Maclean, G.L. 1970. The neonatal plumage of the Double-banded Courser. *Ostrich* 41:215-216.
Steyn, P. 1965. Temminck's Courser. *African Wild Life* 19:29-32.

Pratincoles
Maclean, G. 1973. Red-winged Pratincoles nesting in Natal. *Bokmakierie* 25:61-63.
Pollard, C.J.W. 1982. Status of the Rock Pratincole in the Victoria Falls area. *Honeyguide* 109:29-30.
Williams, G.D., Coppinger, M.P. & Maclean, G.L. 1989. Distribution and breeding of the Rock Pratincole on the upper and middle Zambezi River. *Ostrich* 60:55-64.

Gulls
Broekhuysen, G.J. & Elliott, C.C.H. 1974. Hartlaub's Gulls breeding on the roof of a building. *Bokmakierie* 26:66-67.
Burger, J. & Gochfield, M. 1981. Colony and habitat selection of six Kelp Gull *Larus dominicanus* colonies in South Africa. *Ibis* 123:298-310.
Crawford, R.J.M., Cooper, J. & Shelton, P.A. 1982. Distribution, population size, breeding and conservation of the Kelp Gull in southern Africa. *Ostrich* 53:164-177.
Sinclair, J.C. 1977. Interbreeding of Grey-headed and Hartlaub's Gulls. *Bokmakierie* 29:70-71.
Tinbergen, N. & Broekhuysen, G.J. 1954. On the threat and courtship behaviour of Hartlaub's Gull, *Hydrocoloeus novae-hollandiae hartlaubi* (Bruch). *Ostrich* 25:50-61.
Williams, A.J., Cooper, J. & Hockey, P.A.R. 1984. Aspects of the breeding biology of the Kelp Gull at Marion Island and in South Africa. *Ostrich* 55:147-157.

Terns and the African Skimmer
Braby, R., Braby, S.J. & Simmons, R.E. 1992. 5000 Damara Terns in the northern Namib Desert: a reassessment of world population numbers. *Ostrich* 63:133-135.
Butchart, D., Stannard, K. & Bell, C. 1989. Skimming on thin ice. *Quagga* 26:6-8.
Clinning, C.F. 1978. The biology and conservation of the Damara Tern in South West Africa. *Madoqua* 11:31-39.
Cooper, J., Brooke, R.K., Cyrus, D.P., Martin, A.P., Taylor, R.H. & Williams, A.J. 1992. Distribution, population size and conservation of the Caspian Tern *Sterna caspia* in southern Africa. *Ostrich* 63:58-67.
Coppinger, M.P., Williams, G.D. & Maclean, G.L. 1988. Distribution and breeding biology of the African Skimmer on the upper and middle Zambezi River. *Ostrich* 59:85-96.
Couto, J.T. & Couto, F.M. 1994. African Skimmers on the highveld. *Honeyguide* 40:95-96.
De Villiers, D. & Simmons, R. (In press.) The high incidence of two-egg clutches in a Damara Tern colony in southwestern Namibia. *Madoqua*.
Pollard, C.J.W. 1989. Observations at an African Skimmer colony at Victoria Falls. *Honeyguide* 35:21-22.
Randall, R.M. & Randall, B.M. 1981. Roseate Tern breeding biology and factors responsible for low chick production in Algoa Bay, South Africa. *Ostrich* 52:17-24.
Randall, R. & Randall, B. 1986. The Roseate Tern – South Africa's rarest breeding seabird. *Quagga* 14:22-24.
Randall, R.M. & McLachlan, A. 1982. Damara Terns breeding in the eastern Cape, South Africa. *Ostrich* 53:50-51.
Roberts, M.G. 1976. Belly-soaking and chick transport in the African Skimmer. *Ostrich* 47:126.
Simmons, R. 1993. Catching and counting the elusive Damara Tern. *African Wildlife* 47:150-152.
Simmons, R. & Braine, S. 1994. Breeding, foraging, trapping and sexing of Damara Terns in the Skeleton Coast Park, Namibia. *Ostrich* 65:264-273.
Steyn, P. 1960. Nesting of the Whiskered Tern in southern Cape. *Bokmakierie* 12:35-36.
Steyn, P. 1966. Whiskered Terns. *Bokmakierie* 18:83-85.
Tarboton, W.R., Clinning, C.F. & Grond, M. 1975. Whiskered Terns breeding in the Transvaal. *Ostrich* 46:188.
Uys, C.J. 1978. Swift Terns breeding along the western Cape coast. *Bokmakierie* 30:64-66.
Uys, C.J. 1988. Nesting Damara and Caspian Terns along the southern Cape coast. *Promerops* 182:14-15.

Sandgrouse
Blaine, S. & Tarboton, W. 1990. A year with the Yellow-throated Sandgrouse. *African Wildlife* 44:272-274.
Dixon, J. & Louw, G. 1978. Seasonal effects of nutrition, reproduction and aspects of thermoregulation in the Namaqua Sandgrouse (*Pterocles namaqua*). *Madoqua* 11:19-29.
Joubert, C.S.W. & Maclean, G.L. 1973. Structure of the water-holding feathers of the Namaqua Sandgrouse. *Zoologica Africana* 8:141-152.
Lovegrove, B. 1993. *The living deserts of southern Africa*. Fernwood Press, Cape Town.
Maclean, G.L. 1968. Field studies on the sandgrouse of the Kalahari Desert. *Living Bird* 7:209-235.
Maclean, G.L. 1976. Adaptations of sandgrouse for life in arid lands. *Proceedings of the 16th International Ornithological Congress* pp. 502-616.

Pigeons and Doves
Elliott, C.C.H. & Cooper, J. 1980. The breeding biology of an urban population of Rock Pigeons *Columba guinea*. *Ostrich* 51:198-203.
Skead, D.M. 1971. A study of the Rock Pigeon *Columba guinea*. *Ostrich* 42:65-69.
Tarboton, W.R. & Vernon, C.J. 1971. Notes on the breeding of the Green Pigeon *Treron australis*. *Ostrich* 42:190-192.

Parrots, the Rose-ringed Parakeet and Lovebirds
Dilger, W.C. 1964. Evolution in the African parrot genus *Agapornis*. *Living Bird* 3:135-147.
Tarboton, W. 1976. Meyer's parrots at the nest. *Bokmakierie* 28:44.

Louries
Jarvis, M.J.F. & Currie, M.H. 1979. Breeding captive Knysna and Purplecrested Louries. *Ostrich* 50:38-44.
Steyn, P. 1965. Grey Loerie. *Bokmakierie* 17:40.

Cuckoos
Colebrook-Robjent, J.F.R. & Greenberg, D.A. 1976. Great Spotted Cuckoo *Clamator glandarius*: first breeding record for Zambia and a new host species. *Ostrich* 47:229-230.
Dean, W.R.J., MacDonald, I.A.W. & Vernon, C.J. 1974. Possible breeding of *Cercococcyx montanus*. *Ostrich* 45:188.
Grobler, J.H. & Steyn, P. 1980. Breeding habits of the Boulder Chat and its parasitism by the Redchested Cuckoo. *Ostrich* 51:253-254.
Jensen, R.A.C. & Clinning, C.F. 1974. Breeding biology of two cuckoos and their hosts in South West Africa. *Living Bird* 13:5-50.
Jensen, R.A.C. & Jensen, M.K. 1969. On the breeding biology of the southern African cuckoos. *Ostrich* 40:163-181.
Jensen, R.A.C. & Vernon, C.J. 1970. On the biology of the Didric Cuckoo in southern Africa. *Ostrich* 41:237-246.
Liversidge, R. 1970. The biology of the Jacobin Cuckoo *Clamator jacobinus*. *Ostrich* Suppl. 8:117-137.
Maclean, G.L. & Maclean, C. 1993. A new host for the Redchested Cuckoo in south Africa. *Ostrich* 64:136.
Mundy, P.J. 1973. Vocal mimicry of their hosts by nestlings of the Great Spotted Cuckoo and Striped Crested Cuckoo. *Ibis* 115:602-604.
Mundy, P.J. & Cook, A.W. 1977. Observations on the breeding of the Pied Crow and Great Spotted Cuckoo in northern Nigeria. *Ostrich* 48:72-84.
Oatley, T.B. 1970. Robin hosts of the Redchested Cuckoo in Natal. *Ostrich* 41:232-236.
Oatley, T.B. 1980. Eggs of two cuckoo genera in one nest, and a new host for the Emerald Cuckoo. *Ostrich* 51:126-127.
Onderstall, J. 1987. Letter to the editor. *Hornbill* 8:7.
Pryce, E. 1989. Black Cuckoo raised by Swamp Boubous. *Babbler* 18:38.
Reed, R.A. 1968. Studies of the Diederik Cuckoo *Chrysococcyx caprius* in the Transvaal. *Ibis* 110:321-331.
Siegfried, W.R. 1985. Nest-site fidelity of Malachite Sunbird and parasitism by Klaas's Cuckoo. *Ostrich* 56:277.
Steyn, P. 1973. Some notes on the breeding biology of the Striped Cuckoo. *Ostrich* 44:163-169.
Steyn, P. & Howells, W.W. 1975. Supplementary notes on the breeding biology of the Striped Cuckoo. *Ostrich* 46:258-260.
Tarboton, W. 1975. African Cuckoo parasitising Forktailed Drongo. *Ostrich* 46:186-188.
Tarboton, W. 1986. African Cuckoo: the agony and ecstasy of being a parasite. *Bokmakierie* 38:109-111.
Vernon, C.J. 1970. Pre-incubation embryonic development and egg 'dumping' by the Jacobin Cuckoo. *Ostrich* 41:259-260.
Vernon, C.J. 1982. Notes on the breeding of the Striped Crested Cuckoo. *Honeyguide* 111/112: 10-11.
Vernon, C.J. 1984. The breeding biology of the Thickbilled Cuckoo. *Proceedings of the Fifth Pan-African Ornithological Congress* pp. 825-840.
Vincent, J. 1934. The birds of northern Portuguese East Africa. *Ibis* 4 (Series 13):757-799.

Coucals
Brooke, R.K., Crowe, T.M. & Chambal, M.S. 1990. Copulatory behaviour in Burchell's Coucal *Centropus burchelli*. *Ostrich* 61:87.
Steyn, P. 1972. The development of Senegal Coucal nestlings. *Ostrich* 43:56-59.
Vernon, C.J. 1971. Notes on the biology of the Black Coucal. *Ostrich* 42:242-258.

Owls
Armstrong, A.J. 1991. On the biology of the Marsh Owl, and some comparisons with the Grass Owl. *Honeyguide* 37:148-159.
Brown, C.J., Riekert, B.R. & Morsbach, R.J. 1987. The breeding biology of the African Scops Owl. *Ostrich* 58:58-64.
Carlyon, J. 1988. Nest spacing and territorial fidelity in Marsh Owls. *Gabar* 3:32-34.
Carlyon, J. & Van Zyl, M. 1987. Six egg clutch of Spotted Eagle Owl. *Gabar* 2:25.
Erasmus, R. 1988. Observations on the Grass Owl. *African Wildlife* 42:13-17.
Erasmus, R. 1989. In the nest of a Pearl-spotted Owl. *African Wildlife* 43:128-133.
Lambour, K. & Steyn, P. 1986. Prey and behaviour at a Wood Owl's nest. *African Wildlife* 40:187.
Riekert, R.R. 1986. Some observations on the breeding biology of the African Scops Owl. *Madoqua* 14:425-428.
Steyn, P. 1984. *A delight of owls*. David Philip, Cape Town.
Steyn, P. & Myburgh, N. 1992. Notes taken at a Wood Owl's nest. *Birding in Southern Africa* 43:115-118.
Vernon, C.J. 1984. Episode with Spotted Eagle Owls. *Diaz Diary* 129:5-6.
Wilson, R.T. & Wilson, M.P. 1981. Notes on the Giant Eagle Owl *Bubo lacteus* in central Mali. *Ardea* 69:205-208.

Nightjars
Clancey, P.A. 1972. The Freckled Nightjar in a built-up urban area. *Ostrich* 43:63.
Colebrook-Robjent, J.F.R. 1984. Nests and eggs of some African nightjars. *Ostrich* 55:5-11.
Jackson, H.D. 1985. Aspects of the breeding biology of the Fierynecked Nightjar. *Ostrich* 56:263-276.
Langley, C.H. 1984. Observations on two nests of the Fierynecked Nightjar. *Ostrich* 55:1-4.
Shaw, J.R. 1993. Are Pennant-winged Nightjars polygamous? *Honeyguide* 39:56-59.
Steyn, P. 1971. Notes on the breeding biology of the Freckled Nightjar. *Ostrich* Suppl. 9:179-188.
Steyn, P. & Myburgh, N.J. 1985. Notes at a Fierynecked Nightjar's nest. *Ostrich* 46:265-266.

Swifts and Spinetails
Bromhall, D. 1980. *Devil birds*. Hutchinson, London.
Brooke, R.K. 1957. On the breeding of the White-rumped Swift (*Apus cafer cafer*) in Gatooma. *Ostrich* 23:164-169.
Brooke, R.K. 1971. Breeding of swifts in Ethiopian Africa and adjacent islands. *Ostrich* 42:5-36.
Brooke, R.K. 1973. Distributional and biological notes on the Mottled Swift in Rhodesia. *Ostrich* 44:106-110.
Brooke, R.K. & Avery, G. 1992. Notes on some swifts in the northern Cape. *Mirafra* 9:2-3.
Brown, C.J. 1989. Some breeding sites of Horus and Bradfield's Swifts in South West Africa/Namibia. *Madoqua* 16:69-70.
Carr, B.A. 1984. Nest eviction of Rock Martins by Little Swifts. *Ostrich* 55:223-224.
Colahan, B.D., Earlé, R.A. & Herholdt, J.J. 1991. Alpine Swifts breeding on man-made structure in the Orange Free State. *Birding in Southern Africa* 43:125.
Dean, W.R.J. & Jensen, R.A.C. 1974. The nest and eggs of Bradfield's Swifts. *Ostrich* 45:44.
Donnelly, B.G. & Howells, W.W. 1982. On the breeding of Alpine Swifts near Beit Bridge, Zimbabwe. *Honeyguide* 111/112:24-25.
Harwin, R.M. 1960. Notes on the Horus Swift. *Ostrich* 31:20-24.
Lack, D. 1956. *Swifts in a tower*. Methuen, London.
Lockley, R.M. 1970. Non-stop flight and migration in the Common Swift *Apus apus*. *Ostrich* Suppl. 8:265-269.
Martin, R., Martin, J. & Martin, E. 1990. Alpine Swifts suspected of breeding on man-made structures. *Promerops* 192:11.
Ryan, P.G. & Rose, B. 1985. Bradfield's Swifts using cliffs and palm trees in Namibia. *Ostrich* 56:218.
Saunders, C. 1988. Violence around the nest. *Hartebeest* 20:14-15.
Schmidt, R.K. 1965. Brutbiologie des Weißbürzeglers – *Apus cafer cafer* auf der Kaphalbinsel. *Journal für Ornithologie* 106:295-306.
Schmidt, R.K. 1986. 25 years of ringing White-rumped Swifts. *Safring News* 15:3-10.
Steyn, P. 1952. The nesting of the African Great Swift. *Ostrich* 23:221-222.
Steyn P. 1966. On the breeding of the Mottled Swift *Apus aequatorialis* (Von Müller). *Ostrich* 37:193-194.
Steyn, P. 1990. Alpine Swift – long use of nest site. *Promerops* 196:3.
Tree, A.J. & Cary, R.C. 1989. Unusual nesting site of Little Swift. *Honeyguide* 35:22-23.

Mousebirds
James, G.L. 1948. Display of the Red-faced Coly or Mousebird. *Ostrich* 19:170.
Newman, K.B. 1968. Courtship display and copulation of Red-faced Mousebirds *Colius indicus*. *Ostrich* 39:203.
Newman, K.B. 1971. Precopulatory behaviour in the Whitebacked Mousebird. *Ostrich* 42:154.
Rowan, M.K. 1967. A study of the colies of southern Africa. *Ostrich* 38:63-115.
Steyn, P. 1971. Life upon death. *Bokmakierie* 23:50.
Van Nierop, S.F. 1972. Precopulatory behaviour in the Whitebacked Mousebird. *Ostrich* 43:180.
Vernon, C.J. 1970. Courtship display of the Speckled Coly. *Ostrich* 41:218.
Woodall, P.F. 1974. Precopulatory behaviour in mousebirds. *Ostrich* 45:42.

The Narina Trogon
Harcus, J.L. 1976. Presumed anti-predator behaviour of Narina Trogon. *Ostrich* 47:129-130.
Nuttall, R. 1984. Narina Trogon and others. *Diaz Diary* 127:4-5.

Kingfishers
Argyle, D. 1971. Spotted eggs of Brown-hooded Kingfisher. *Ostrich* 42:230.
Arkell, G.B.F. 1979. Aspects of the feeding and breeding biology of the Giant Kingfisher. *Ostrich* 50:176-181.
Greig-Smith, P.W. 1978. Behaviour of Woodland Kingfishers in Ghana. *Ostrich* 49:67-75.
Hanmer, D.B. & Godwin, D.F. 1987. Striped Kingfisher breeding in an old swallow nest. *Bokmakierie* 39:7.
Jones, J.M.B. 1983. Breeding observations on the Giant Kingfisher. *Honeyguide* 116:16-18.
Meintjies, E.P.H. 1943. The Angola Kingfisher. *Ostrich* 14:104-106.
Milstein, P. le S. 1962. The Angola Kingfisher *Halcyon senegalensis*. *Ostrich* 33(3):3-12.
Moreau, R.E. 1944. The Half-collared Kingfisher (*Alcedo semitorquata*) at the nest. *Ostrich* 15:161-177.
Pike, E.O. 1966. The Mangrove Kingfisher. *Bokmakierie* 18:58.
Reyer, H.U. 1980. Sexual dimorphism and co-operative breeding of the Striped Kingfisher. *Ostrich* 51:117-118.
Reyer, H.U. 1984. Investment and relatedness: a cost/benefit analysis of breeding and helping in the Pied Kingfisher (*Ceryle rudis*). *Animal Behaviour* 32:1163-1178.
Steyn, P. 1970. Notes on the nesting of the Striped Kingfisher. *Bokmakierie* 22:64-65.

Bee-eaters
Dean, W.R.J. & Brooke, R.K. 1991. Review of the Olive Bee-eater *Merops superciliosus* breeding in eastern Africa south of 14°S. *Ostrich* 62:86-88.
Emlen, S.T. & Demong, N.J. 1984. Bee-eaters of Baharini. *Natural History* 10:50-59.
Fry, C.H. 1972. The biology of African bee-eaters. *Living Bird* 11:75-112.
Fry, C.H. 1984. *The bee-eaters*. Russel Friedman Books, Halfway House.
Harwin, R.M. & Rockingham-Gill, D.V. 1981. Aspects of the biology of the southern races of the Swallow-tailed Bee-eater. *Honeyguide* 106:4-10.

Rollers
Cackett, K. 1976. Some observations on nest holes and hole-nesters. *Honeyguide* 87:17-22.
Stander, S.G. 1958. Observations on the Lilac-breasted Roller. *Bokmakierie* 10:29.

The Hoopoe and Woodhoopoes
Du Plessis, M.A. 1989. The influence of roost-cavity availability on flock size in the Redbilled Woodhoopoe *Phoeniculus purpureus*. *Ostrich* Suppl. 14:97-104.
Du Plessis, M.A. 1989. Behavioural ecology of the Redbilled Woodhoopoe *Phoeniculus purpureus* in South Africa. Ph.D. thesis, University of Cape Town.
Hoesch, W. 1933. Brutbiologische Beobachtungen am Sichelhopf. *Ornithologische Monatsberichte* 41:33-37.
Ligon, D.J. & Ligon, S. 1978. The communal

social system of the Green Woodhoopoe in Kenya. *Living Bird* 17:159-197.

Rowan, M.K. 1970. Communal nesting in Redbilled Woodhoopoes. *Ostrich* 41:257-258.

Skead, C.J. 1945. Display of the S.A. Hoopoe. *Ostrich* 16:77-78.

Skead, C.J. 1945. Chick feeding by the S.A. Hoopoe. *Ostrich* 16:184-188.

Skead, C.J. 1950. A study of the African Hoopoe. *Ibis* 92:434-463.

Hornbills

Courtnay-Latimer, M. 1942. The Ground Hornbill. *Ostrich* 13:121-136.

Du Plessis, M.A. 1994. Cooperative breeding in the Trumpeter Hornbill *Bycanistes bucinator*. *Ostrich* 65:45-47.

Kemp, A.C. 1970. Some observations on the sealed-in nesting method of hornbills (Family: Bucerotidae). *Ostrich* Suppl. 8:149-155.

Kemp, A.C. 1973. Environmental factors affecting the onset of breeding in some southern African hornbills, *Tockus* spp. *Journal of Reproductive Fertility* Suppl. 19:319-331.

Kemp, A.C. 1976. A study of the ecology, behaviour and systematics of *Tockus* hornbills (Aves: Bucerotidae). *Transvaal Museum Memoir* 20:1-125.

Kemp, A.C. 1978. A review of the hornbills: biology and radiation. *Living Bird* 17:105-136.

Kemp, A.C. & Kemp, M.I. 1972. A study of the biology of Monteiro's Hornbill. *Annals of the Transvaal Museum* 27:255-268.

Kemp, A.C. & Kemp, M.I. 1980. The biology of the Southern Ground Hornbill *Bucorvus leadbeateri* (Vigors) (Aves:Bucerotidae). *Annals of the Transvaal Museum* 32:65-100.

Moreau, R.E. & Moreau, W.M. 1941. Breeding biology of Silvery-cheeked Hornbills. *Auk* 58:13-27.

Ranger, G. 1949-1952. Life of the Crowned Hornbill. *Ostrich* 20:54-65 & 152-167; 21:2-13; 22:77-93; 23:26-36.

Barbets

Büttiker, W. 1960. Artificial nesting devices in southern Africa. *Ostrich* 31:39-48.

Holliday, C.S. & Tait, I.C. 1953. Notes on the nidification of *Buccanodon olivacea woodwardi* (Shelly). *Ostrich* 24:115-117.

MacDonald, I.A.W. 1986. Range expansion in the Pied Barbet and the spread of alien trees in southern Africa. *Ostrich* 57:75-94.

Oatley, T.B. 1968. Observations by W.M. Austen on the breeding biology of the White-eared Barbet *Buccanodon leucotis* (Sundevall). *Lammergeyer* 8:7-14.

Prozesky, O.P.M. 1986. A study of the behaviour of the Crested Barbet *Tachyphonus vaillantii*. *Ostrich* Suppl. 6:171-182.

Roberts, A. 1939. Swifts and others birds nesting in buildings. *Ostrich* 10:85-99.

Skead, C.J. 1950. A study of the Black-collared Barbet *Lybius torquatus* with notes on its parasitism by the Lesser Honeyguide *Indicator minor*. *Ostrich* 21:84-96.

Honeyguides

Cyrus, D. 1988. Observations on the parasitism of Blackcollared Barbets *Lybius torquatus* by the Lesser Honeyguide *Indicator minor* at St Lucia Forest Station. *Ostrich* 59:138-139.

Dean, W.R.J., Siegfried, W.R. & MacDonald, I.A.W. 1990. The fallacy, fact, and fate of guiding behaviour in the Greater Honeyguide. *Conservation Biology* 4:99-101.

Diamond, A.W. & Place, A.R. 1988. Wax digestion by Black-throated Honey-guides *Indicator indicator*. *Ibis* 130:558-561.

Friedman, H. 1955. The honey-guides. *United States National Museum Bulletin* 208.

Friedman, H. 1958. Advances in our knowledge of honey-guides. *Proceedings of the United States National Museum* 108:309-320.

Friedman, H. 1968. Additional data on brood parasitism in honey-guides. *Proceedings of the United States National Museum* 124:1-8.

Friedman, H. 1970. Further information on the breeding biology of the honey-guides. *Los Angeles County Museum Contributions in Science* 205:1-5.

Fry, C.H. 1974. Vocal mimesis in nestling Greater Honeyguides. *Bulletin of the British Ornithologists' Club* 94:58-59.

Hosken, J.H. 1966. Sharp-billed Honeyguide *Prodotiscus regulus* being fed by a pair of cisticolas. *Ostrich* 37:235.

Maclean, G.L. 1971. Sharp-billed Honeyguide parasitising Neddicky in Natal. *Ostrich* 42:75-77.

Parkes, D.A. 1994. Slender-billed Honeyguide parasitism of Yellow White-eye. *Honeyguide* 40:97-98.

Ranger, G.A. 1955. On three species of honeyguide; the Greater (*Indicator indicator*), the Lesser (*Indicator minor*) and the Scaly-throated (*Indicator variegatus*). *Ostrich* 26:70-87.

Steyn, P. & Myburgh, N. 1986. Lesser Honeyguide breeding at Somerset West. *Promerops* 173:16.

Steyn, P. & Scott, J. 1974. Blackcollared Barbets evicting a Lesser Honeyguide. *Ostrich* 45:143.

Vernon, C.J. 1974. *Prodotiscus regulus* parasitising *Camaroptera brevicaudata*. *Ostrich* 45:261.

Vernon, C.J. 1987. On the Eastern Green-backed Honeyguide. *Honeyguide* 33:6-12.

Woodpeckers and the Red-throated Wryneck

Attwell, G.D. 1952. The breeding of the Cardinal Woodpecker at Gatooma, Southern Rhodesia. *Ostrich* 23:88-91.

Short, L.L. 1971. Notes on South African woodpeckers. *Ostrich* 42:89-98.

Tarboton, W. 1970. Notes on the Beared Woodpecker. *Bokmakierie* 22:81-84.

Tarboton, W. 1976. Aspects of the biology of *Jynx ruficollis*. *Ostrich* 47:99-112.

The African Broadbill

Lawson, W.J. 1961. Probable courtship display of the Broadbill *Smithornis capensis*. *Ibis* 103a:289-290.

The Angola Pitta

Masterson, A. 1987. Pitta chatter. *Bokmakierie* 39:23-25.

Larks and Finch-larks

Allan, D.G., Batchelor, G.R. & Tarboton, W.R. 1983. Breeding of Botha's Lark. *Ostrich* 54:55-57.

Boyer, H.J. 1988. Breeding biology of the Dune Lark. *Ostrich* 59:30-37.

Dean, W.R.J. 1989. The nest, egg and nestlings of the Red Lark *Certhilauda barra*. *Ostrich* 60:158.

Hustler, K. 1985. First breeding record of the Short-clawed Lark. *Honeyguide* 31:109-110.

Irwin, M.P.S. & Lorber, P. 1983. The breeding season of the Chestnut-backed Finch Lark in Zimbabwe. *Honeyguide* 114/115:22-24.

Maclean, G.L. 1970. Breeding behaviour of the larks in the Kalahari sandveld. *Annals of the Natal Museum* 20:381-401.

Maclean, G.L. 1970. The biology of the larks (Alaudidae) of the Kalahari sandveld. *Zoologica Africana* 5:7-39.

Masterson, A.N.B. & Parkes, D.A. 1993. Melodious Larks breeding near Felixburg. *Honeyguide* 39:189-192.

Myburgh, N. & Steyn, P. 1989. Notes on a Red Lark's nest. *Birding in Southern Africa* 41:114-115.

Steyn, P. 1964. A note on the Chestnut-backed Finch-Lark. *Bokmakierie* 16:2-4.

Steyn, P. 1988. Cooperative breeding in the Spikeheeled Lark. *Ostrich* 59:182.

Steyn, P. & Myburgh, N. 1989. Notes on Sclater's Lark. *Birding in Southern Africa* 41:67-69.

Steyn, P. & Myburgh, N. 1991. Further notes on Sclater's Lark. *Birding in Southern Africa* 43:73-76.

Vernon, C.J. 1983. Notes on the Monotonous or White-tailed Bush Lark in Zimbabwe. *Honeyguide* 113:19-21.

Winterbottom, J.M. & Wilson, A.H. 1959. Notes on the breeding of Red-capped Lark *Calandrella cinerea* (Gmel.) at Cape Town. *Ostrich* Suppl. 3:289-299.

Swallows and Martins

Allan, D. 1986. The Blue Swallow – next bird for extinction? *Quagga* 15:27-29.

Allan, D. 1988. The Blue Swallow in with a chance. *Quagga* 22:5-7.

Brooke, R.K. & Vernon, C.J. 1961. Aspects of the breeding biology of the Rock Martin. *Ostrich* 32:51-52.

Earlé, R.A. 1986. Breeding biology of the South African Cliff Swallow. *Ostrich* 57:138-156.

Earlé, R.A. 1987. Moult and breeding seasons of the Greyrumped Swallow. *Ostrich* 58:181-182.

Earlé, R.A. 1989. Breeding biology of the Redbreasted Swallow *Hirundo semirufa*. *Ostrich* 60:13-21.

Earlé, R.A. & Underhill, L.G. 1991. The effect of brood size on the growth of South African Cliff Swallow *Hirundo spilodera* chicks. *Ostrich* 62:13-22.

Lockhart, P.S. 1970. House Martins nesting at Somerset West. *Ostrich* 41:254-255.

Myburgh, N. & Steyn, P. 1979. Unusual nest of a Rock Martin. *Bokmakierie* 31:64.

Roberts, E.L. 1986. Banded Sand Martins in the southwestern Cape. *Promerops* 176:13.

Schmidt, R.K. 1959. Notes on the Pearl-breasted Swallow *Hirundo dimidiata* in the south-western Cape. *Ostrich* 30:155-158.

Schmidt, R.K. 1962. Breeding of the Larger Striped Swallow *Cecropis cucullata* in the South West Cape. *Ostrich* 33:3-8.

Snell, M.L. 1969. Notes on the breeding of the Blue Swallow. *Ostrich* 40:65-74.

Snell, M.L. 1979. The vulnerable Blue Swallow. *Bokmakierie* 31:74-78.

Van der Merwe, F. 1986. Banded Sand Martins in the South-western Cape. *Promerops* 175:9.

Cuckoo-shrikes

Skead, C.J. 1966. A study of the Black Cuckoo-Shrike *Campephaga phoenicia* (Latham). *Ostrich* 37:71-75.

Whittingham, A.P. 1964. Notes on the nesting habits of the White-breasted Cuckoo-Shrike (*Coracina pectoralis*). *Ostrich* 35:63-64.

Drongos

Manson, C. & Manson, A.J. 1983. Unusual behaviour of Square-tailed Drongo. *Honeyguide* 114/115:54.

Scott, J.A. 1983. A further note on nest destruction by Square-tailed Drongos. *Honeyguide* 116:19.

Tarboton, W. 1986. African Cuckoo: the agony and ecstasy of being a parasite. *Bokmakierie* 38:109-111.

Orioles

Clancey, P.A. 1970. On an oriole – new to the South African list. *Bokmakierie* 22:53-54.

Langham, K. 1976. Fledging age of Black-headed Orioles. *Honeyguide* 87:27-28.

Crows and Ravens

Allan, D. 1981. Growth rates of nestling Black Crows. *Ostrich* 52:189-190.

Goodwin, D. 1976. *Crows of the world*. British Museum (Natural History), London.

Jones, M.A. 1983. The Pied Crow in Harare, Zimbabwe. *Honeyguide* 116:6-13.

Masterson, A.N.B. 1993. Changes in the behaviour of urban crows. *Honeyguide* 39:138-141.

Mundy, P.J. & Cook, A.W. 1977. Observations on the breeding of the Pied Crow and Great Spotted Cuckoo in northern Nigeria. *Ostrich* 48:72-84.

Skead, C.J. 1952. A study of the Black Crow. *Ibis* 94:434-450.

Uys, C.J. 1966. At the nest of the Cape Raven. *Bokmakierie* 18:38-41.

Winterbottom, J.M. 1975. Notes on the South African species of *Corvus*. *Ostrich* 46:236-250.

Tits

Tarboton, W.R. 1981. Cooperative breeding and group territoriality in the Black Tit. *Ostrich* 52:212-225.

Penduline Tits

Cole, D. 1949. Notes on the Cape Penduline Tit *Anthroscopus minutus*. *Ostrich* 20:31-32.

Milstein, P. le S. 1975. Observations on Penduline Tit nest structure. *Bokmakierie* 27:8-9.

Skead, C.J. 1959. A study of the Cape Penduline Tit *Anthroscopus minutus* (Shaw & Nodder). *Ostrich* Suppl. 3:274-288.

Uys, C.J. 1966. Impressions of the Cape Penduline Tit and its nest in the south-west Cape. *Bokmakierie* 18:80-82.

The Spotted Creeper

Masterson, A.N.B. 1970. Notes on the Spotted Creeper. *Honeyguide* 61:35-36.

Steyn, P. 1974. A confiding creeper. *Bokmakierie* 26:80-82.

Babblers

Barbour, D. 1972. Sociability at a Pied Babbler's nest. *Honeyguide* 70:34-35.

Lindeque, M. & Kapner, J. 1993. Cooperative group defence by Pied Babblers *Turdoides bicolor* results in death of avian predator. *Ostrich* 64:189.

Vernon, C.J. 1976. Communal feeding of nestlings by the Arrow-marked Babbler. *Ostrich* 47:134-136.

The Boulder Chat and Rockjumpers

Grobler, J.H. & Steyn, P. 1980. Breeding habits of the Boulder Chat and its parasitism by the Redchested Cuckoo. *Ostrich* 51:253-254.

Irwin, M.P.S. 1985. *Chaetops* and the Afrotropical Timaliidae (Babblers). *Honeyguide* 31:99-100.

Martin, J. 1964. Nestlings of the Rufous Rockjumper (*Chaetops frenatus*) being fed by two males and one female. *Ostrich* 35:62.

Bulbuls, the Bush Blackcap and the Yellow-spotted Nicator

Frost, S. 1979. Pre-copulatory display of Yellow-bellied Bulbul. *Ostrich* 50:185.

Liversidge, R. 1970. The ecological life history of the Cape Bulbul. Ph.D. thesis, University of Cape Town.

Longrigg, T.D. 1978. Pre-copulatory display of Sombre Bulbul. *Ostrich* 49:202-203.

Marshall, B.E. 1969. Unusual nesting site of Black-eyed Bulbul. *Ostrich* 40:135.

Steyn P. 1959. Distraction display of Cape Bulbul. *Bokmakierie* 11:16.

Van der Merwe, F. 1987. Incubation period of Cape Bulbul. *Promerops* 178:15.

Van der Merwe, F. 1988. Nestling period of the Cape Bulbul. *Promerops* 182:12.

Van der Merwe, F. 1990. Distraction displays by Rameron Pigeon, Cape Bulbul and Dabchick. *Promerops* 193:10-12.

Thrushes

Chittenden, H. 1982. Kurrichane Thrush builds seven nests in one season. *Bokmakierie* 34:67-68.

Earlé, R.A. & Oatley, T.B. 1983. Populations, ecology and breeding of the Orange Thrush at two sites in eastern South Africa. *Ostrich* 54:205-212.

Hezekia, G. 1987. Helpers at nest of Ground-scraper Thrush. *Honeyguide* 33:18.

Kirkpatrick, C.W.M. 1993. Miombo Rock Thrush incubation period. *Honeyguide* 39:25-26.

Lorber, P. 1973. Multiple brooding in the Kurrichane Thrush *Turdus libonyanus*. *Ostrich* 44:84.

Parnell, G.W. 1974. Treble-brooding of Kurrichane Thrush. *Honeyguide* 79:37.

Vernon, C.J. 1968. A year's census of Marandellas, Rhodesia. *Ostrich* 39:12-24.

Winterbottom, M.G. 1966. A study of the Cape Thrush *Turdus olivaceus* L. *Ostrich* 37:17-22.

Chats

Barbour, D. 1972. Some notes on Arnot's Chat. *Bokmakierie* 24:16-17.

Earlé, R.A. & Herholdt, J.J. 1986. Co-operative breeding in the Anteating Chat. *Ostrich* 57:188-189.

Earlé, R.A. & Herholdt, J.J. 1988. Breeding and moult of the Anteating Chat *Myrmecocichla formicivora*. *Ostrich* 59:155-161.

Farkas, T. 1961. Notes on the behaviour of the Mocking Chat *Thamnolaea cinnamomeiventris* in western Transvaal. *Ostrich* 32:122-127.

Farkas, T. 1966. Notes on the breeding activities and post-embryonic developments in the Mocking Chat *Thamnolaea cinnamomeiventris cinnamomeiventris*. *Ostrich* Suppl. 6:95-107.

Jensen, R.A.C. & Jensen, M.K. 1971. First breeding records of the Herero Chat *Namibornis herero*, and taxonomic implications. *Ostrich* Suppl. 8:105-116.

Plowes, D.C.H. 1948. The Mountain Chat at the nest. *Ostrich* 19:80-88.

Steyn, P. 1966. Observations on the breeding biology of the Familiar Chat *Cercomela familiaris* (Stephens). *Ostrich* 37:176-183.

Steyn, P. 1967. Mobile Familiar Chat's nest. *Bokmakierie* 19:67.

Taylor, J.S. 1946. Notes on the Mountain Chat. *Ostrich* 17:248-253.

Robins and Palm Thrushes

Beasley, A. 1988. White-throated Robin in Botanic Garden, Harare. *Honeyguide* 34:127-128.

Beasley, A.J. 1994. Some observations of the breeding activities of the White-throated Robin. *Honeyguide* 40:75-77.

Donnelly, B.G. 1967. Some observations on the behaviour and call in the Palm Thrush *Cichladusa arquata*. *Ostrich* 38:230-232.

Farkas, T. 1969. Notes on the biology and ethology of the Natal Robin *Cossypha natalensis*. *Ibis* 111:281-292.

Farkas, T. 1973. Notes on the biology and ethology of Heuglin's Robin *Cossypha heuglini*. *Ostrich* 44:95-105.

Lamberts, A.E. 1985. Duetting and antiphonal singing observed in Heuglin's Robin. *Bokmakierie* 37:40-41.

Manson, A.J. 1990. The biology of Swynnerton's Robin. *Honeyguide* 36:5-13.

Oatley, T. 1959. Notes on the genus *Cossypha*, with particular reference to *C. natalensis* (Smith) and *C. dichroa* (Gmelin). *Ostrich* Suppl. 3:426-434.

Oatley, T. 1971. The function of vocal imitation by African *Cossyphas*. *Ostrich* Suppl. 8:85-89.

Oatley, T.B. 1982. The Starred Robin in Natal. 3. Breeding, populations and plumages. *Ostrich* 53:206-221.

Rowan, M.K. 1969. A study of the Cape Robin in southern Africa. *Living Bird* 8:5-32.

Steyn, P. 1983. Karoo Scrub Robin – cooperative breeding. *Promerops* 157:12.

Warblers

Allan, D.G. & Bassi, J. 1988. Breeding of the Broadtailed Warbler in South Africa. *Ostrich* 59:137.

Beven, G. 1944. Rudd's Bush-Warbler (*Apalis ruddi*). *Ostrich* 15:178-187.

Clinning, C.F. & Tarboton, W.R. 1972. Notes on the Damara Rockjumper *Achaetops pycnopygius*. *Madoqua* 1 (5): 57-61.

Earlé, R.A. 1980. Notes on the breeding of the Bleating Bush Warbler. *Ostrich* 51:128.

Frost, P.G.H. & Vernon, C.J. 1978. Notes on the Green Eremomela. *Ostrich* 49:87-89.

Hofmeyr, J.H., Hofmeyr, P.K., Broekhuysen, G.J. & Stanford, W. 1961. The nest of the Knysna Scrub Warbler (*Bradypterus sylvaticus*) and some notes on parental behaviour. *Ostrich* 32:177-180.

Lorber, P., Manson, A.J. & Sharp, S. 1983. The undescribed nest of Chirinda Apalis. *Honeyguide* 114/115:69-71.

Martin, J. & Martin, R. 1965. The nest of the Cinnamon-breasted Warbler *Euryptila subcinnamomea*. *Ostrich* 36:136-137.

Masterson, A.N.B. 1972. The nests and eggs of Rhodesian 'Tailor-birds'. *Honeyguide* 69:21-25.

Parkes, D.A. 1993. Nests of the Red-faced and Long-billed Crombecs. *Honeyguide* 39:91-93.

Plowes, D.C.H. 1972. The nest of the Red-winged Warbler (*Heliolais erythroptera rhodoptera* (Shelly)). *Honeyguide* 69:34-35.

Pringle, J.S. 1977. Breeding of the Knysna Scrub Warbler. *Ostrich* 48:112-114.

Steyn, P. 1966. A note on the breeding of *Camaroptera brevicaudata* (Cretzschmar). *Ostrich* 37:60-61.

Steyn, P. 1970. An exquisite nest. *African Wild Life* 24:337-340.

Tarboton, W.R. 1970. Nest and eggs of Burnt-necked Eremomela. *Ostrich* 41:212.

Took, J.M.E. 1959. Breeding of *Schoenicola brevirostris* in southern Rhodesia. *Ostrich* 30:138-139.

Vernon, F.J. & Vernon, C.J. 1978. Notes on the social behaviour of the Duskyfaced Warbler. *Ostrich* 49:92-93.

Cisticolas

Lynes, H. 1930. Review of the genus *Cisticola*. *Ibis* 6 (Series 12):1-673.

Masterson, A.N.B. 1972. The nests and eggs of Rhodesian 'Tailor-birds'. *Honeyguide* 69:21-25.

Penry, E.H. 1985. Notes on the breeding of *Cisticola brunescens* and *C. juncidis* in Zambia. *Ostrich* 56:229-235.

UEDA, K. 1984. Successive nest building and polygyny of Fan-tailed Warbler *Cisticola juncidis*. *Ibis* 126:221-229.
UEDA, K. 1989. Re-use of courtship nests for quick remating in the polygynous Fan-tailed Warbler *Cisticola juncidis*. *Ibis* 131:257-262.

Prinias and prinia-like Warblers

BENSON, C.W. 1946. A visit to the Vumba Highlands, Southern Rhodesia. *Ostrich* 17:280-296.
BROOKE, R.K. 1993. A case of apparent hybridization between Spotted and Blackchested Prinia *Prinia maculosa* and *flavicans*. *Ostrich* 64:137.
BROOKE, R.K. & DEAN, W.R.J. 1990. On the biology and taxonomic position of *Drymoica substriata* Smith, the so-called Namaqua Prinia. *Ostrich* 61:50-55.
MACLEAN, G.L. 1974. The breeding biology of the Rufouseared Warbler and its bearing on the genus *Prinia*. *Ostrich* 45:9-14.
PLOWES, D.C.H. 1972. The nest of the Redwing Warbler (*Heliolais erythroptra rhodoptera* (Shelly)). *Honeyguide* 69:34-35.
ROWAN, M.K. & BROEKHUYSEN, G.J. 1962. A study of the Karoo Prinia. *Ostrich* 33:6-30.
STEYN, P. 1966. Tawny-flanked Prinias *Prinia subflava* (Gmelin), utilizing Red Bishop *Euplectes orix* (Linnaeus) nests. *Ostrich* 37:195.

Flycatchers and Batises

BEASLEY, A.J. 1985. Black Flycatcher nesting in Red-headed Weaver's nest. *Honeyguide* 31:175-177.
BROEKHUYSEN, G.J. 1958. Notes on the breeding behaviour of the Cape Flycatcher *Batis capensis*. *Ostrich* 29:143-152.
FROST, S.K. 1990. Notes on the breeding of Marico and Pallid Flycatchers in the central Transvaal, South Africa. *Ostrich* 61:111-116.
HARRIS, T. & ARNOTT, G. 1988. *Shrikes of southern Africa*. Struik Winchester, Cape Town.
JAMES, H.W. 1922. Notes on the nest and eggs of *Stenostira scita* (Vieill.). *Ibis* 4 (Series 11):254-256.
LITTLE, J. DE V. 1964. Notes on the breeding behaviour of the Paradise Flycatcher. *Ostrich* 35:32-41.
MANSON, A.J. 1992. Biology of the White-tailed Crested Flycatcher. *Honeyguide* 38:6-11.
MYBURGH, N. & STEYN, P. 1966. Three unusual nests. *Bokmakierie* 18:99-100.
NÖLLER, R. 1975. Pririt Flycatcher fanning young. *Bokmakierie* 27:16.
PAGET-WILKES, G.D.B. 1976. An unusual Black Flycatcher's nest site. *Honeyguide* 87:27.
SAUNDERS, C.R. 1981. Paradise Flycatcher: two successive broods in one nest. *Honeyguide* 107/108:44.
SCOTT, J.A. 1984. Breeding behaviour of the Black-throated Wattle-eye. *Honeyguide* 30:72-74.
SKEAD, C.J. 1967. A study of the Paradise Flycatcher *Terpsiphone viridis* (Muller). *Ostrich* 38:123-132.
SKEAD, D.M. 1966. The Dusky Flycatcher *Muscicapa adusta* (Boie) in the southern Cape Peninsula. *Ostrich* 37:143-145.

Wagtails

NHLANE, M.E.D. 1990. Breeding biology of the African Pied Wagtail *Motacilla aguimp* in Blantyre, Malawi. *Ostrich* 61:1-4.
OATLEY, T. 1988. Waggi tales. *Bokmakierie* 40:120-121.
PIPER, S.E. 1989. Breeding biology of the Longtailed Wagtail *Motacilla clara*. *Ostrich* Suppl. 14:7-15.
PIPER, S.E. & SCHULTZ, D.M. 1989. Type, dimensionality and size of Longtailed Wagtail territories. *Ostrich* Suppl. 14:123-131.
SKEAD, C.J. 1954. A study of the Cape Wagtail *Motacilla capensis*. *Ibis* 96:91-103.
WINTERBOTTOM, J.M. 1964. Notes on wagtails of *Motacilla* southern Africa. *Ostrich* 35:129-141.

Pipits and Longclaws

MENDELSOHN, J. 1984. The Mountain Pipit in the Drakensberg. *Bokmakierie* 36:40-44.

Shrikes

HARRIS, T. & ARNOTT, G. 1988. *Shrikes of southern Africa*. Struik Winchester, Cape Town.
HODGSON, C. 1965. Bokmakierie in Rhodesia. *Honeyguide* 46:32-33.
STEYN, P. 1988. Whitecrowned Shrikes – an opportunity missed. *Bokmakierie* 40:35-36.
UYS, C.J. 1982. Fiscal Shrike uses same nest site for 14 years. *Promerops* 155:6.
WILLIAMS, J. 1989. Icterine Warblers feeding White-crowned Shrike chicks. *Honeyguide* 35:26.

Starlings

BROEKHUYSEN, G.J. 1951. Some observations on the nesting activities of the Red-winged Starling *Onychognathus morio morio* and especially the feeding of the young. *Ostrich* 22:6-16.
COOPER, J. & UNDERHILL, L.G. 1991. Breeding, mass and primary moult of European Starlings *Sturnus vulgaris* at Dassen Island, South Africa. *Ostrich* 62:1-7.
CRAIG, A. 1983. Co-operative breeding in two African starlings, Sturnidae. *Ibis* 125:114-115.
CRAIG, A.J.F.K. 1987. Co-operative breeding in the Pied Starling. *Ostrich* 58:176-180.
CRAIG, A.J.F.K. 1989. A review of the Blackbellied Starling and other African forest starlings. *Ostrich* Suppl. 14:17-26.
CRAIG, A.J.F.K., HULLEY, P.E. & WALTER, G.H. 1989. Nesting of sympatric Redwinged and Palewinged Starlings. *Ostrich* 60:69-74.
CRAIG, A.J.F.K., HULLEY, P.E. & WALTER, G.H. 1991. The behaviour of Palewinged Starlings, and a comparison with other *Onychognathus* species. *Ostrich* 62:97-108.
MITCHELL, C.S. 1976. Prolonged use of a nest and site by Red-winged Starlings. *Honeyguide* 86:35.
ROWAN, M.K. 1955. The breeding biology of the Redwinged Starling *Onychognathus morio*. *Ibis* 97:663-705.
SAUNDERS, C.R., SPARROW, R.L. & SPARROW, A. 1993. Mass breeding of Wattled Starlings in the lowveld. *Honeyguide* 39:198-200.
STEYN, P. 1976. Red-winged Starling: prolonged use of nest site. *Honeyguide* 87:34.
UYS, C.J. 1977. Notes on Wattled Starlings in the western Cape. *Bokmakierie* 29:87-89.
YOUNG, S. 1990. Birds with a talent for applied chemistry. *New Scientist* (December issue).

Oxpeckers

HENDERSON, G.H. 1953. Southern Red-billed Oxpecker *Buphagus erythrorhynchus caffer*. *Ostrich* 24:132.
MUNDY, P.J. & COOK, A.W. 1975. Observations of the Yellow-billed Oxpecker *Buphagus africanus* in northern Nigeria. *Ibis* 117:504-506.
STUTTERHEIM, C.J. 1982. Breeding biology of the Redbilled Oxpecker in the Kruger National Park. *Ostrich* 53:79-90.
STUTTERHEIM, C.J., MUNDY, P.J. & COOK, A.W. 1976. Comparisons between two species of oxpeckers. *Bokmakierie* 28:12-14.

Sugarbirds

BROEKHUYSEN, G.J. 1959. The biology of the Cape Sugarbird *Promerops cafer* (L.). *Ostrich* Suppl. 3:180-221.
BURGER, A.E., SIEGFRIED, W.R. & FROST, P.G.H. 1976. Nest-site selection in the Cape Sugarbird. *Zoologica Africana* 11:127-158.
DE SWARDT, D.H. 1992. Notes on the breeding biology of Gurney's Sugarbird. *Ostrich* 63:136-137.
DE SWARDT, D.H. & BOTHMA, N. 1991. Notes on the nest of Gurney's Sugarbird in the eastern Transvaal. *Birding in Southern Africa* 43:2-3.
RICHARDSON, D. 1990. Cape Sugarbird building nest in palm tree: symptom of abnormal predation pressure? *Promerops* 193:9.
SEILER, H.W. & PRYS-JONES, R. 1989. Mate competition, mate guarding, and unusual copulations of the Cape Sugarbird (*Promerops cafer*). *Ostrich* 60:159-164.
SKEAD, D.M. 1963. Gurney's Sugarbird *Promerops gurneyi* Verreaux in the Natal Drakensberg. *Ostrich* 34:160-164.
STEYN, P. 1973. A nest of Gurney's Sugarbird. *Honeyguide* 73:29-31.
WINTERBOTTOM, J.M. 1962. Breeding season of Long-tailed Sugarbird *Promerops cafer* (L.). *Ostrich* 33(2):77.

Sunbirds

BROEKHUYSEN, G.J. 1963. The breeding biology of the Orange-breasted Sunbird *Anthobaphes violacea* (Linnaeus). *Ostrich* 34:187-234.
BUCHANAN, D. & STEYN, P. 1964. The incubation and nestling periods of the White-breasted Sunbird (*Cinnyris talatala* A. Smith). *Ostrich* 35:65-66.
EARLÉ, R.A. 1982. Aspects of the biology and ecology of the White-bellied Sunbird. *Ostrich* 53:65-73.
HANMER, D.B. & MANSON, A.J. 1992. The Collared Sunbird in the Bvumba. *Honeyguide* 38:155-164.
HOWELLS, W.W. 1971. Breeding of the Coppery Sunbird at Salisbury, Rhodesia. *Ostrich* 42:99-109.
SCHMIDT, R.K. 1964. The Lesser Double-collared Sunbird *Cinnyris chalybeus* (Linnaeus) in the south-western Cape. *Ostrich* 35:86-94.
SKEAD, C.J. 1953. A study of the Black Sunbird *Chalcomitra amethystina amethystina* (Shaw). *Ostrich* 24:159-166.
SKEAD, C.J. 1962. A study of the Collared Sunbird *Anthreptes collaris* (Vieill.). *Ostrich* 33(2):38-40.
WILLIAMS, J.B. 1993. Nest orientation of Orangebreasted Sunbird in southern Africa. *Ostrich* 64:40-42.

White-eyes

BROEKHUYSEN, G.J. & WINTERBOTTOM, J.M. 1968. Breeding activity of the Cape White-eye *Zosterops virens capensis* Sundevall in the south-west Cape. *Ostrich* 39:163-176.
CLARK, A.R. 1969. The nest of the Yellow White-eye. *Ostrich* 40:137.
MASTERSON, A. 1970. Distraction by Yellow White-eye *Zosterops senegalensis*. *Honeyguide* 62:32.
MYBURGH, N. & STEYN, P. 1966. Three unusual nests. *Bokmakierie* 18:99-100.
SKEAD, C.J. & RANGER, G.A. 1958. A contribution to the biology of the Cape Province white-eyes (*Zosterops*). *Ibis* 100:319-333.

The Red-billed Buffalo Weaver, the White-browed Sparrow-weaver and the Sociable Weaver

BARTHOLOMEW, C.A., WHITE, F.N. & HOWELL, T.R. 1976. The thermal significance of the nest of the Sociable Weaver *Philetairus socius*: summer observations. *Ibis* 118:402-410.
BIRKHEAD, T.R., STANBACK, M.T. & SIMMONS, R.E. 1993. The phalloid organ of buffalo weavers *Bubalornis*. *Ibis* 135:326-331.
BROOKE, R.K. & HARRISON, J.A. 1992. An apparently unreported nest site of the Sociable Weaver. *Promerops* 206:14.
CLANCEY, P.A. 1970. Buffalo Weavers nesting on a windpump. *Ostrich* 41:262.
COLLIAS, E.C. & COLLIAS, N.E. 1978. Nest building and nesting behaviour of the Sociable Weaver *Philetairus socius*. *Ibis* 120:1-15.
COLLIAS, N.E. & COLLIAS, E.C. 1978. Cooperative breeding behaviour in the White-browed Sparrow Weaver. *Auk* 95:472-484.
EARLÉ, R.A. 1983. Aspects of the breeding biology of the Whitebrowed Sparrowweaver *Plocepasser mahali* (Aves:Ploceidae). *Navorsinge van die Nasionale Museum Bloemfontein* 4(7):177-191.
FERGUSON, J.W.H. 1988. Comparison of communication and signalling patterns of Whitebrowed Sparrowweavers and other gregarious Ploceid weavers. *Ostrich* 59:54-62.
KEMP, A. & KEMP, M. 1974. Observations on the Buffalo Weaver. *Bokmakierie* 26:56-58.
LEWIS, D.M. 1982. Cooperative breeding in a population of White-browed Sparrow Weavers. *Ibis* 124:511-522.
MACLEAN, G.L. 1973. The Sociable Weaver: Parts 1-5. *Ostrich* 44:176-190; 44:191-218; 44:219-240; 44:241-253 & 44:254-261.
VERNON, C.J. 1983. Aspects of the habitat preference and social behaviour of the White-browed Sparrow Weaver. *Honeyguide* 113:11-14.
WHITE, F.N., BARTHOLOMEW, G.A. & HOWELL, T.R. 1975. The thermal significance of the nest of the Sociable Weaver *Philetairus socius*: winter observations. *Ibis* 17:171-179.

Sparrows and the Scaly-feathered Finch

EARLÉ, R.A. 1986. Does intraspecific brood parasitism occur in the Cape Sparrow? *Bokmakierie* 38:70.
PLOWES, D.C.H. 1948. Bird life of Sekukuniland. *African Wild Life* 2:61-65.
ROWAN, M.K. 1966. Some observations on reproduction and mortality in the Cape Sparrow *Passer melanurus*. *Ostrich* Suppl. 6:425-434.
SUMMERS-SMITH, D. 1963. *The House Sparrow*. Collins, London.
SUMMERS-SMITH, D. 1983. The Great Sparrow. *Bokmakierie* 35:51-55.

Weavers

BERRY, H.H., ARCHIBALD, T.J. & BERRY, C.U. 1987. Breeding data on Chestnut Weavers in Etosha National Park, South West Africa/Namibia. *Madoqua* 15:157-162.
BRAINE, S.G. & BRAINE, J.W.S. 1971. Chestnut Weavers *Ploceus rubiginosus* breeding in South West Africa. *Ostrich* 42:299-300.
COLLIAS, N.E. & COLLIAS, E.C. 1970. Some experimental studies of the breeding biology of the Village Weaver *Ploceus (textor) cucullatus* (Müller). *Ostrich* Suppl. 8:169-177.
COLLIAS, N.E. & COLLIAS, E.C. 1971. Ecology and behaviour of the Spotted-backed Weaver in the Kruger National Park. *Koedoe* 14:1-27.
COLLIAS, N.E. & COLLIAS, E.C. 1984. *Nest building and bird behavior*. Princeton University Press, Princeton.
CRAIG, A. 1984. The Spectacled Weaver *Ploceus ocularis*, and monogamy in the Ploceinae. *Proceedings of the Fifth Pan-African Ornithological Congress* pp. 477-483.
CROOK, J.H. 1960. Nest form and construction in certain West African weaver-birds. *Ibis* 102:1-25.
HOWMAN, H.R.G. & BEGG, G.W. 1983. Nest building and nest destruction by the Masked Weaver *Ploceus velatus*. *South African Journal of Zoology* 18:37-44.
HOWMAN, H.R.G. & BEGG, G.W. 1987. Further observations on breeding behaviour of African Masked Weaver. *Honeyguide* 33:83-96.
LAYCOCK, H.T. 1979. Breeding biology of the Thick-billed Weaver. *Ostrich* 50:70-82.
SKEAD, C.J. 1947. A study of the Cape Weaver (*Hyphantornis capensis olivaceus*). *Ostrich* 18:1-42.
SKEAD, C.J. 1953. A study of the Spectacled Weaver (*Ploceus ocularius* Smith). *Ostrich* 24:103-110.
TALBOT, M.I. 1979. Masked Weavers *Ploceus velatus* Vieillot in ornamental cherry tree 1975-1978. *Honeyguide* 99:16-27.
VICTORIA, J.K. 1972. Clutch characteristics and egg discriminative ability of the African Village Weaverbird *Ploceus cucullatus*. *Ibis* 114:367-376.
WALSH, J.F. & WALSH, B. 1976. Nesting association between Red-headed Weaver *Malimbus rubiceps* and raptorial birds. *Ibis* 118:106-108.

Queleas, Bishops and Widows

ANDERSSON, M. 1982. Female choice selects for extreme tail length in a widowbird. *Nature* 299:818-820.
CRAIG, A.J.F.K. 1974. Reproductive behaviour of the male Red Bishop Bird. *Ostrich* 45:149-160.
EMLEN, J.T. 1957. Display and mate selection in the whydahs and bishop birds. *Ostrich* 28:202-213.
GRIMES, L.G. 1977. Nesting of the Red-headed Quelea *Quelea erythrops* on Accra Plains, Ghana. *Ibis* 119:216-220.
HORNBY, H.E. 1967. The breeding cycle of the White-winged Widow Bird *Euplectes albonotus*. *Ostrich* 38:5-10.
JONES, P.J. & WARD, P. 1976. The level of reserve protein as the proximate factor controlling the timing of breeding and clutch size in the Red-billed Quelea *Quelea quelea*. *Ibis* 118:547-576.
LACK, D. 1935. Territory and polygamy in a bishop bird *Euplectes hordeacea hordeacea*. *Ibis* 5 (Series 13):817-836.
MCLEAN, S. & TAYLOR, R.H. 1986. Redheaded Quelea breeding at Lake St Lucia. *Ostrich* 57:60-61.
SKEAD, C.J. 1956. A study of the Red Bishop Bird *Euplectes orix orix* (Linnaeus). *Ostrich* 127:112-126.
SKEAD, C.J. 1959. A study of the Redshouldered Widowbird *Coliuspasser axillaris axillaris* (Smith). *Ostrich* 30:13-21.
WARD, P. 1965. The breeding biology of the Black-faced Dioch *Quelea quelea* in Nigeria. *Ibis* 107:326-349.

Waxbills, Mannikins and other estrildid Finches

BARNARD, P. & MARKUS, M.B. 1990. Reproductive failure and nest site selection of two Estrildid finches in *Acacia* woodland. *Ostrich* 61:117-124.
BRICKHILL, N., HUNTLEY, B. & VORSTER, R. 1980. Observations on wild and captive Pied Mannikins. *Bokmakierie* 32:9-12.
COLAHAN, B.D. 1982. The biology of the Orange-breasted Waxbill. *Ostrich* 53:1-30.
GOODWIN, D. 1982. *Estrildid finches of the world*. British Museum (Natural History), London.
NUTTALL, R.J. 1992. Breeding biology and behaviour of the Quail Finch *Ortygospiza atricollis*. *Ostrich* 63:110-117.
SHILLINGLAW, N. 1977. Waxbills. *Bokmakierie* 29:104-106.
SKEAD, D.M. 1975. Ecological studies of four Estrildines in the central Transvaal. *Ostrich* Suppl. 11:1-55.
WOODALL, P.F. 1975. On the life history of the Bronze Mannikin. *Ostrich* 46:55-86.

Whydahs, Widow-finches and the Cuckoo Finch

BARNARD, P. 1989. Territoriality and the determinants of male mating success in the Southern African whydahs (*Vidua*). *Ostrich* 60:103-117.
BARNARD, P. & MARKUS, M.B. 1989. Male copulation frequency and female competition for fertilization in a promiscuous brood parasite, the Pin-tailed Whydah *Vidua macroura*. *Ibis* 131:421-425.
CHERRY, M.I. 1990. Tail length and female choice. *Tree* 5 (11):349-350.
DEAN, W.R.J. & VERNON, C.J. 1973. *Cisticola textrix*: a new host for *Anomalospiza imberbis*. *Ostrich* 44:86.
FRIEDMAN, H. 1960. The parasitic weaverbirds. *Smithsonian Institute United States National Museum Bulletin* 223:1-196.
PAYNE, R.B. 1994. The species of indigobirds in Zimbabwe. *Honeyguide* 40:78-86.
PAYNE, R.B., PAYNE, L.L. & NHLANE, M.E.D. 1992. Song mimicry and species status of the Green Widowfinch *Vidua codringtoni*. *Ostrich* 63:86-97.
SHAW, P. 1984. The social behaviour of the Pin-tailed Whydah *Vidua macroura* in northern Ghana. *Ibis* 126:463-473.
VERNON, C.J. 1964. The breeding of the Cuckoo-Weaver (*Anomalospiza imberbis* (Cabanis)) in Southern Rhodesia. *Ostrich* 35:260-263.
VERNON, C.J. 1975. Indigo birds at Victoria Falls. *Honeyguide* 82:41-42.
WICKLER, W. 1968. *Mimicry in plants and animals*. Weidenfeld & Nicolson, London.
WOODALL, P.F. 1972. Indigo finches. *Honeyguide* 70:31-32.

Canaries and Buntings

CUMMING, S.C. & STEYN, P. 1966. Observations on the breeding biology of the Rock Bunting *Emberiza tahapisi* Smith. *Ostrich* 37:170-175.
LORBER, P. 1973. Nest building of *Serinus atrogularis*. *Ostrich* 44:268-269.
MACLEOD, J.G.R. & STANFORD, W.P. 1958. Notes on *Poliospiza leucoptera*, at Somerset West, October/November 1957. *Ostrich* 29:153-156.
ROBSON, N.F. 1990. First recorded nest of Lemonbreasted Canary in the field. *Ostrich* 61:84-85.
SCHMIDT, R.K. 1982. The secret siskin. *Bokmakierie* 34:54-55.
SCHMIDT, R.K. 1982. The secret siskin (2). *Bokmakierie* 34:86.
SKEAD, C.J. 1948. A study of the Cape Canary (*Serinus canicollis canicollis*). *Ostrich* 19:17-44.
SKEAD, C.J. 1960. *The canaries, seedeaters and buntings of southern Africa*. Trustees of the South African Bird Book Fund, Cape Town.
VERNON, C.J. 1979. Mennell's Seedeater at Zimbabwe. *Honeyguide* 99:12-15.
WOLFF, S.W. & JACOBSEN, H.G. 1980. Cape Canary nest density. *Ostrich* 51:124.
WOODALL, P.F. 1971. The Cape Bunting round Salisbury. *Honeyguide* 67:25-26.

Index

Page numbers in **bold** indicate main references; page numbers in *italic* indicate photographs.

A

Accipiter 63
Acrocephalus 169, 172
Albatross, Black-browed *(Diomedea melanophris) 29*
Dark-mantled Sooty *(Phoebetria fusca)* 29, *29*
Wandering *(Diomedea exulans) 28*, 29-30
Yellow-nosed *(D. chlororhynchos) 11, 29*
albatrosses 11, **29-30**
Alethe, White-breasted *(Alethe fuelleborni)* 166
Aloe ferox 151
Andropadus 158
Apalis, Bar-throated *(Apalis thoracica)* 106, 169, 173
Black-headed *(A. melanocephala)* 169, 173
Chirinda *(A. chirindensis)* 169, 173
Rudd's *(A. ruddi)* 169, 173; nest *171*
Yellow-breasted *(A. flavida)* 169, *171*, 173
apalises 23, **169-73**
Aristida grasses 202
Aviceda 56
Avocet *(Recurvirostra avosetta)* 10, *16*, 86; nest *86*

B

Babbler, Arrow-marked *(Turdoides jardineii)* 105, 106, 156, *156*
Bare-cheeked *(T. gymnogenys)* 106, 156
Black-faced *(T. melanops)* 105, 156
Pied *(T. bicolor)* 105, 156; nest *156*
White-rumped (Hartlaub's) *(T. hartlaubii)* 106, 156
babblers 12, 23, 108, **156**, 157, 159
baobab tree 52, 57, 147, 201, *202*
Barbet, Black-collared *(Lybius torquatus) 9*, 135, 136, 137
Crested *(Trachyphonus vaillantii)* 135, *135*, 137, 139
Golden-rumped Tinker *(Pogoniulus bilineatus)* 135
Green (Woodward's)*(Stactolaema olivacea)* 135
Green Tinker *(Pogoniulus simplex)* 135
Pied *(Lybius leucomelas) 134*, 135, 137; nest *22*
Red-fronted Tinker *(Pogoniulus pusillus)* 135, *135*
White-eared *(Stactolaema leucotis)* 135, 137
Whyte's *(S. whytii)* 135
Yellow-fronted Tinker *(Pogoniulus chrysoconus)* 135
barbets 8, 12, 68, **135**, 136
nests 22; use by others 113, 122, 126, 128, 152, 181, 205
Bateleur *(Terathopius ecaudatus)* 8, 17, **60, 61**, 70, 201; nest *61*
Batis, Cape *(Batis capensis) 3*, 105, 106, *108*, *181*, 182
Chinspot *(B. molitor)* 106; nest *181*
Pririt *(B. pririt)* 106, 179, 182
batises 155, 171, **179-82**, 186
Bazaruto Island, Mozambique 124
Bee-eater, Böhm's *(Merops boehmi)* 124, 125, 137
Carmine *(M. nubicoides)* 11, *11*, 124, *124*, 125, 137
European *(M. apiaster)* 124, *124*, 125; nest *22*
Little *(M. pusillus)* 124, 125, 137; nest *125*
Olive *(M. superciliosus)* 124, 125, 137
Swallow-tailed *(M. hirundineus)* 124, 125, 137
White-fronted *(M. bullockoides)* 11, 12, 118, 124, 125, *125*, 137
bee-eaters 8, 10, 11, 12, 15, 19, **124-5**, 148
nests 22, 119 163
Betty's Bay 26
Bird Island 32, 95
Bishop, Fire-crowned *(Euplectes hordeaceus)* 214
Golden *(E. afer)* 214
Red *(E. orix) 12*, 105, 106, 214, *214*, 216; nest *216*
bishops 10, 14, 19, 23, 177, **213**, 218
Bittern, Dwarf *(Ixobrychus sturmii)* 35, 37, *37*
Little *(I. minutus)* 35, *36*, 37, *37*
bitterns 21, **35-7**
Blackcap, Bush *(Lioptilus nigricapillus)* **158-9**
Bokmakierie *(Telophorus zeylonus)* 185, 186, 187; nest *187*
Boubou, Southern *(Laniarius ferrugineus)*106
Swamp *(L. bicolor)* 106
Tropical *(L. aethiopicus)* 106, 188; nest *188*
boubous **185**
Bradypterus 169, 172
Bredasdorp 41, 45, 47
breeding cycle 13-17
Broadbill, African *(Smithornis capensis)* 23, 104, **140**, *141*
brood division: in coots 75, 78; in moorhens 78
brood parasites 21, 24; in bee-eaters 125; in coots 75; in Cuckoo Finch 24, 218, 220; in cuckoos 24, **104-8**, 125, 136, 157, 187, 218, 219, 220, 221; in ducks 49; in honeyguides 24, 136-7; in whydahs 24, 217, 219-20; in widow-finches 24, 217, 219; in widows 219-20
Bulbul, Black-eyed *(Pycnonotus barbatus)* 158, 159; nest *105*, 106
Cape *(P. capensis)* 106, 158, 159; nest *158*
Red-eyed *(P. nigricans)* 106, 158, 159
Slender *(Phyllastrephus debilis)* 158
Sombre *(Andropadus importunus)* 106, 158, 159; nest *158*
Stripe-cheeked *(A. milanjensis)* 158, *159*
Terrestrial *(Phyllastrephus terrestris)* 158; nest *158*
Yellow-bellied *(Chlorocichla flaviventris)* 158, 159
Yellow-streaked *(Phyllastrephus flavostriatus)* 158; nest *158*
bulbuls 20, 23, **158-9**
Bunting, Cabanis's *(Emberiza cabanisi)* 222
Cape *(E. capensis)* 220, 222; nest *222*
Golden-breasted *(E. flaviventris)* 106, 222; nest *221*
Lark-like *(E. impetuani)* 221, 222, *222*
Rock *(E. tahapisi)* 220-1, 222
buntings 23, **220-2**
Bustard, Kori *(Ardeotis kori)* 80; nest *79*
Ludwig's *(Neotis ludwigii)* 79; nest *79*
Stanley's *(N. denhami)* 79
bustards 21, **79-80**
Buteo 62, 63
Buttonquail, Black-rumped *(Turnix hottentotta)* 72
Kurrichane *(T. sylvatica)* 72; nest *72*
buttonquails 12, 21, **71-2**
Buzzard, Augur *(Buteo augur)* 62, 63; nest *62*
Forest *(B. trizonatus)* 62, *62*, 63
Jackal *(B. rufofuscus)* 62, 63; nest *62*
Lizard *(Kaupifalco monogrammicus)* 62, 63; nest *62*
buzzards **62-3**, 70
Bycanistes 130

C

Cainism 16, 19; in buzzards 63; in eagles 61; in goshawks 66; in gymnogenes 68; in pelicans 31; in sparrowhawks 66
camelthorn 201, 202
Canary, Black-eared *(Serinus mennelli)* 221
Bully *(S. sulphuratus)* 220
Cape *(S. canicollis)* 10, 220, 221, *221*, 222
Lemon-breasted *(S. citrinipectus)* 220, 221
Protea *(S. leucopterus)* 220
Streaky-headed *(S. gularis)* 221, 222; nest *220*
Yellow *(S. flaviventris)*: nest *222*
canaries 18, 23, **220-2**
Cape Point 34
Chaetops 157
Chaffinch *(Fringilla coelebs)* 220
Charadriiformes 81
Chat, Ant-eating *(Myrmecocichla formicivora)* 137, 147, 163, 164, 165; nest *165*
Arnot's *(Thamnolaea arnoti)* 164-5, *164*
Boulder *(Pinarornis plumosus)* 9, 105, 106, 108, **157**, 163; nest *157*
Buff-streaked *(Oenanthe bifasciata)* 163
Familiar *(Cercomela familiaris)* 140, 163-4, *163*, 165
Herero *(Namibornis herero)* 163, 165; nest *165*
Karoo *(Cercomela schlegelii)* 163; nest *21*
Mocking *(Thamnolaea cinnamomeiventris)* 147, 163, *163*, 164, 165
Mountain *(Oenanthe monticola)* 106, 163, *164*, 165
Sickle-winged *(Cercomela sinuata)* 163
Tractrac *(C. tractrac)* 163; nest *163*
chats 23, **163-5**
Chlorocichla 158
Chobe River 60
Chrysococcyx 104, 105, 106, 108
Cisticola, Ayres' *(Cisticola ayresii)* 175, 220
Black-backed *(C. galactotes)* 175
Chirping *(C. pipiens)* 175; nest *176*
Cloud *(C. tetrix)* 175, *175*, 220
Croaking *(C. natalensis)* 175, 176, 220
Desert *(C. aridula)* 175, 176, 220
Fan-tailed *(C. juncidis)* 10, 23, 175, 176, *176*, 220
Grey-backed *(C. subruficapilla)* 175
Lazy *(C. aberrans)* 176
Levaillant's *(C. tinniens)* 175, *176*, 220
Pale-crowned *(C. brunnescens)* 175, 176, 220
Rattling *(C. chiniana)* 106; nest *175*
Red-faced *(C. erythrops)* 175, 220; nest *175*
Singing *(C. cantans)* 175, 220
Tinkling *(C. rufilata)* 176
Wailing *(C. lais)* 137, 175
cisticolas 8, 136, 142, 174, **175-6**, 220; nests 23, 216
Clamator 104, 105, 106, 108
cobweb 144, 149, 150, 153, 155, 171, 173, 175 *176*, 181, 182, 185, 186, 187, 197, 200, 220, 221; *see also Stegodyphus*
Coliiformes 120
colonies, breeding in 8-9, 10-11; bee-eaters 124-5; bishops 213-14; buntings 222; Canary, Cape 221; cormorants 33-4, 35, 42; Darter 33-4, 35, 42; egrets 33, 35; flamingos 45-7; gannets 32; Grebe, Black-necked 27; gulls 35, 91, 93, 94, 95; herons 33, 35, 42; ibises 33, 35, 41, 42, 43-4; kites 54; martins 145, 148; mousebirds 120; nightjars 115; pelicans 31, 42; Penguin, Jackass 26, 30; petrels 29; plovers 84; pratincoles 89-90; queleas 213; Spoonbill, African 35, 42, 43; starlings 189-90; storks 35, 40, 42; swallows 145, 147; swifts 118-19; terns 93, 94, 95; vultures 41; weavers 201, 203-4, 207, 211-12; widows 213-14
colonies, roosting in: barbets 135; crows 151; estrildid finches 215; kites 55; sugarbirds 193
communal nests 9, 10; *see also* Weaver, Sociable
co-operative breeding 12; in babblers 12, 156; in barbets 12, 135; in bee-eaters 12, 125; in Chat, Ant-eating 163, 165; in Eremomela, Green-capped 174; in Finch, Scaly-feathered 206; in flufftails 77; in Flycatcher, Marico 179; in Hamerkop 39; in hoopoes 128-9; in hornbills 132-3; in kingfishers 12, 122, 123; in Lark, Spike-heeled 12, 144; in louries 103; in mousebirds 120; in oxpeckers 12, 192; in Robin, Karoo 166; in shrikes 12, 185, 187-8; in starlings 12, 190; in Thrush, Groundscraper 161; in Tit, Black 152-3; in weavers 12, 201-2; in woodhoopoes 12, 129-30
Coot, Red-knobbed *(Fulica cristata)* 75, 77-8, *78*; nest 22, 48, 49
coots **75-7**
Cormorant, Bank *(Phalacrocorax neglectus)* 33-4, *33*
Cape *(P. capensis)* 33-4, *33*
Crowned *(P. coronatus)* 33-4, *33*
Reed *(P. africanus)* 33-4, *34*
White-breasted *(P. carbo)* 33-4, *34*
cormorants 10, 21, **33-4**, 35, 42
Corvids 151-2
Coucal, Black *(Centropus bengalensis)* 109
Burchell's *(C. superciliosus)* 109, *109*
Coppery-tailed *(C. cupreicaudus)* 109
Green *(Ceuthmochares aereus)* 109
Senegal *(Centropus senegalensis)* 109, *109*
coucals 20, 23, 140, **109**
Courser, Bronze-winged *(Rhinoptilus chalcopterus)* 88
Burchell's *(Cursorius rufus)* 88
Double-banded *(Rhinoptilus africanus)* 13, 16, 17, *17*, 88, *88*, 89, 143
Temminck's *(Cursorius temminckii)* 88, *88*, 89
Three-banded *(Rhinoptilus cinctus)* 16, 88, 89
coursers 19, 21, **88-9**
courtship 13-14
Crake, African *(Crex egregia)* 75; eggs *75*
Baillon's *(Porzana pusilla)* 75
Black *(Amaurornis flavirostris)* 75, *75*
Corn *(Crex crex)* 75
Spotted *(Porzana porzana)* 75
Striped *(Aenigmatolimnas marginalis)* 75; nest *75*
crakes 21, **75-7**
Crane, Blue *(Anthropoides paradisea) 17*, 73, 74, *74*; nest 21
Crowned *(Balearica regulorum)* 73, 74; nest *73*
Wattled *(Grus carunculatus)* 22, 73-4, *73*

cranes **73**
crèche, formation of: in flamingos 46, 47; in Ibis, Sacred 43; in terns 95
Creeper, Spotted *(Salpornis spilonotus)* **155**, *155*, 185
Crimsonwing, Redfaced *(Cryptospiza reichenovii)* 215
Crombec, Long-billed *(Sylvietta rufescens)* 106, 169, *172*, 173
Red-faced *(S. whytii)* 169, 173
crombecs **169-73**
Crow, Black *(Corvus capensis)* 44, 105, 106, 151, 152; nest *152*
House *(C. splendens)*151, 152; nest *152*
Pied *(C. albus)* 70, 105, 106, 108, 151, 152
crows 24, 150, **151-2**; nests 21, 23, 48, 69, 70, *70*, 113, 114
Cuckoo, African *(Cuculus gularis)* 104, 105, 106, 150
Barred *(Cercococcyx montanus)* 104
Black *(Cuculus clamosus)* 104, 106
Diederik *(Chrysococcyx caprius)* 104, 105, 106, 108, 206, 212, 214
Emerald *(C. cupreus)* 104, 105, 106
Great Spotted *(Clamator glandarius)* 104, 105, 106, 108, 151
Jacobin *(C. jacobinus)* 104, 105, 106, 108, 158
Klaas's *(Chrysococcyx klaas)* 104, 105, 106, *108*, 197
Red-chested *(Cuculus solitarius)* *24*, 104, *104*, 106, 108, 157, 221; eggs *9*, 105, *105*, 167
Striped *(Clamator levaillanti)* 104, 105, 106, 108, 156
Thick-billed *(Pachycoccyx audeberti)* 104, 105, 106, 108
cuckoos 10, 15, 17, 20, **104-8**, 109; as a parasite 125, 136, 187, 218, 219, 220
Cuckoo-shrike, Black *(Campephaga flava)* 149, *149*
Grey *(Coracina caesia)* 149
White-breasted *(C. pectoralis)* 149; nest *149*
cuckoo-shrikes 23, **149**, 185
Cuculus 104, 105, 106, 108

D
Dabchick *(Tachybaptus ruficollis)* *22*, 27, *27*
Darter *(Anhinga melanogaster)* 21, **33-4**, *34*, 35, 42, 78
Dassen Island *26*, 31, *33*, 190
Dikkop, Spotted *(Burhinus capensis)* 20, 87, *87*
Water *(B. vermiculatus)* 87; nest *87*
dikkops 21, **87**
Dove, Blue Spotted *(Turtur afer)* 100
Cape Turtle *(Streptopelia capicola)* 101
Cinnamon *(Aplopelia larvata)* 100
Laughing *(Streptopelia senegalensis)* 100, *100*
Namaqua *(Oena capensis)* 100, 101, *101*; nest *21*
Red-eyed *(Streptopelia semitorquata)* 101
Tambourine *(Turtur tympanistria)* 100; nest *100*
doves 15, 16, 18, 20, **100-1**; nests 21, 161, 186
Dracaena fragrans 166, *167*
Drongo, Fork-tailed *(Dicrurus adsimilis)* 105, 106, 150; nest *150*
Square-tailed *(D. ludwigii)* 150; nest *150*
drongos 23, **150**
Duck, African Black *(Anas sparsa)* 47, 48, *48*, 49
Fulvous *(Dendrocygna bicolor)* 47, 48, 49
Knob-billed *(Sarkidiornis melanotos)* 39, 48, 49
Maccoa *(Oxyura maccoa)* *14*, 16, 20, 48, 49, *49*
White-backed *(Thalassornis leuconotus)* 47, *47*, 48, 49
White-faced *(Dendrocygna viduata)* 47, 49
Yellow-billed *(Anas undulata)* 47, 48
ducks 11, 15, 17, 21, 22, **47-9**

E
Eagle, African Fish *(Haliaeetus vocifer)* 10, 47, 58, 60-1, *60*
African Hawk *(Hieraaetus fasciatus)* 58, *58*, 59
Ayres' *(H. ayresii)* 59, 60
Black *(Aquila verreauxii)* 13, *15*, 19, 52, *57*, 60; nest 41, 57-8, 69
Black-breasted Snake *(Circaetus pectoralis)* 58, 60, *61*
Booted *(Hieraaetus pennatus)* 58, 59, 60, 61; nest *58*
Brown Snake *(Circaetus cinereus)* 58, 60
Crowned *(Stephanoaetus coronatus)* 10, 20, *59*, 60, 61, 63
Long-crested *(Lophaetus occipitalis)* 59
Martial *(Polemaetus bellicosus)* 9, 25, 59-60, 61; nest 21, 58, *58*, 68
Southern Banded Snake *(Circaetus fasciolatus)* 60
Tawny *(Aquila rapax)* 42, 47, 57, 58; nest *59*
Vulturine Fish *see* Vulture, Palm-nut
Wahlberg's *(Aquila wahlbergi)* 58-9, 61, 69, 114; nest *58*
Western Banded Snake *(Circaetus cinerascens)* 60
eagles 14, 52, 53, **57-61**, 63
eggs 15-17
Egret, Cattle *(Bubulcus ibis)* *37*, 44; nest *21*, 35, 44
Great White *(Egretta alba)* 35
Little *(E. garzetta)* 35
Slaty *(E. vinaceigula)* 37
Yellow-billed *(E. intermedia)* *13*
egrets 10, 11, 14, 18, 19, 33, **35, 37**; nests 21, 100
Elanus 54
Eremomela, Burnt-necked *(Eremomela usticollis)* 169, 174
Green-capped *(E. scotops)* 169, *172*, 173-4
Karoo *(E. gregalis)* 169, 173; nest 23, *172*
Yellow-bellied *(E. icteropygialis)* 106, 169, 173; nest *172*
eremomelas **169-74**
estrildid finches 23, **215-17**, 218, 219
Estrildidae 215
Etosha Pan 31, 45-7, 74
Euplectes 213, 214

F
Falco 69
Falcon, African Hobby *(Falco cuvierii)* 69
Lanner *(F. biarmicus)* 41, 69, *69*, 70
Peregrine *(F. peregrinus)* 8, 41, 69, *69*
Pygmy *(Polihierax semitorquatus)* 69, 70, *70*, 204
Red-necked *(Falco chicquera)* 69; nest *70*
Sooty *(F. concolor)* 13
Taita *(F. fasciinucha)* 69, *70*
falcons **69-70**
feathers, as nesting material 118
Finch, Cuckoo *(Anomalospiza imberbis)* 24, **218**, *219*, **220**
Cut-throat *(Amadina fasciata)* 216
Locust *(Ortygospiza locustella)* 215
Melba *(Pytilia melba)* *215*, 220
Quail *(Ortygospiza atricollis)* 215, *217*
Red-headed *(Amadina erythrocephala)* 204, 216
Scaly-feathered *(Sporopipes squamifrons)* **205-6**; nest 23, 113, *205*
Finch-lark, Black-eared *(Eremopterix australis)* 142, 144, *144*; nest *15*, 144
Chestnut-backed *(E. leucotis)* 142, 144, *144*
Grey-backed *(E. verticalis)* 142, 144; nest *144*
finch-larks **142-4**
Finfoot, African *(Podica senegalensis)* **78**, *78*
Firefinch, Blue-billed *(Lagonisticta rubricata)* *23*, *216*, 220
Jameson's *(L. rhodopareia)* 220
Red-billed *(L. senegala)* 220
firefinches 215
Flamingo, Greater *(Phoenicopterus ruber)* 45-7, *45*
Lesser *(P. minor)* 45-7, *45-6*
flamingos 10, 17, 20, **45-7**
Flufftail, Buff-spotted *(Sarothrura elegans)* 23, 75, 77; nest *77*
Red-chested *(S. rufa)* 75, 77, *77*
Streaky-breasted *(S. boehmi)* 75, 77; nest *77*
Striped *(S. affinis)* 75, 77, *77*; nest *77*
White-winged *(S. ayresi)* 75
flufftails 21, **75-7**
Flycatcher, Black *(Melaenornis pammelaina)* 179, 181; nest *179*
Blue-grey *(Muscicapa caerulescens)* 106, 179, *179*
Blue-mantled *(Trochocercus cyanomelas)* 182; nest *182*
Chat *(Melaenornis infuscatus)* 179, 181, 182; nest *179*
Dusky *(Muscicapa adusta)* 105, 106, 179, *179*, 221
Fairy *(Stenostira scita)* 179, 181-2, *182*
Fan-tailed *(Myioparus plumbeus)* 181
Fiscal *(Sigelus silens)* 181; nest *179*
Livingstone's *(Erythrocercus livingstonei)* 179, 181
Marico *(Melaenornis mariquensis)* 106, 165, 179, 181; nest *179*
Pallid *(M. pallidus)* 179, 181; nest *179*
Paradise *(Terpsiphone viridis)* 106, 179, *180*, 182
Vanga *(Bias musicus)* 179, 181
Wattle-eyed *(Platysteira peltata)* 179, *181*, 182
White-tailed *(Trochocercus albonotatus)* 182; nest *182*
flycatchers 171, 172, **179-82**, 186; nests 22, 23
Francolin, Cape *(Francolinus capensis)* *71*
Greywing *(F. africanus)* 72, *72*
Swainson's *(F. swainsonii)* 71, 72
francolins 15, 17, 18, 19, 21, **71-2**

G
Galium tomentosum, nesting material 21, *21*, 171
Gallinule, American Purple *(Porphyrula martinica)* 75, 77
Lesser *(P. alleni)* 78
Purple *(Porphyrio porphyrio)* *76*, 78, 81
gallinules 21, **75-8**
gamebirds 71-2
Gannet, Cape *(Morus capensis)* 10, *10*, 11, 16, 32, *32*
gannets 10, **32**
geese **47-9**
Glareolidae 89
Gonarezhou National Park, Zimbabwe 52
Goose, Egyptian *(Alopochen aegyptiacus)* 39, 47, *47*, 49
Pygmy *(Nettapus auritus)* 39, 47, 48, 49; nest *48*
Spur-winged *(Plectropterus gambensis)* 47, 48, 49
Gorongosa, Mozambique 166
Goshawk, African *(Accipiter tachiro)* 57, 63, 64, 66, 113; nest *64*
Dark Chanting *(Melierax metabates)* 63, 64, 66
Gabar *(Micronisus gabar)* 19, 63, 66, 156; nest 21, 64, *66*
Little Banded *(Accipiter badius)* 63, 64, *65*
Pale Chanting *(Melierax canorus)* *4*, 12, 63, 64, 66; nest *66*
goshawks 56, 62, **63-6**; nests *63*
Gough Island 29
Grassbird *(Sphenoeacus afer)* 169, *173*, 174
Grebe, Black-necked *(Podiceps nigricollis)* *14*, 27; nest *27*
Great Crested *(P. cristatus)* 27, *27*; nest *27*
grebes 22, **27**
guano collecting 32, 95
Guineafowl, Crested *(Guttera pucherani)* 72
Helmeted *(Numida meleagris)* *71*, 72, 121; nest *21*
guineafowl **71-2**; nests 21
Gull, Grey-headed *(Larus cirrocephalus)* 91, 93; nest *93*
Hartlaub's *(L. hartlaubii)* 10, 91, *92-3*, 94-5; nest *93*
Kelp *(L. dominicanus)* 10, 35, *91*, 93, 94; as avian predator 83, 95, 96
gulls 11, 15, 17, 18, 20, **91-3**; nests 21, 29
Gymnogene *(Polyboroides typus)* 19, 41, **68**, *68*, 119

H
Hadeda *see* Ibis, Hadeda
Hamerkop *(Scopus umbretta)* 9, 15, **38-9**, *38-9*; nest 21, 38-9, *38-9*, 41, 48, 63, 70, 114
Hardap Dam 31
Harrier, African Marsh *(Circus ranivorus)* 67, *67*
Black *(C. maurus)* *17*, 67, *67*
European Marsh *(C. aeruginosus)* 67
Montagu's *(C. pygargus)* 67
Pallid *(C. macrourus)* 67
harriers *17*, **67**
Hawk, Bat *(Macheiramphus alcinus)* **56-7**; nest *56*
Cuckoo *(Aviceda cuculoides)* **56-7**, 68; nest *56*
Heron, Goliath *(Ardea goliath)* 35
Green-backed *(Butorides striatus)* 35, 37; nest *35*
Grey *(Ardea cinerea)*: nest *37*
Purple *(A. purpurea)* 35, *35*
Rail *see* Bittern, Dwarf
Rufous-bellied *(Ardeola rufiventris)*; eggs *35*, 37
White-backed Night *(Gorsachius leuconotus)* 35, *35*
herons 10, 17, 19, 33, **35, 37**, 39, 42; nests 21, 172
Hluhluwe-Umfolozi Game Reserve 192
Honeyguide, Eastern *(Indicator meliphilus)* 136, 137
Greater *(I. indicator)* 136, *136*, 137
Lesser *(I. minor)* 135, 136, *136*, 137
Scaly-throated *(I. variegatus)* 136, 137
Sharp-billed *(Prodotiscus regulus)* 136, 137

Slender-billed *(P. zambesiae)* 136, 137, 201
honeyguides 17, 24, 135, **136-7**
Hoopoe *(Upupa epops) 18,* 19, 20, *22,* 106, **128-30**, 137
Hornbill, Bradfield's *(Tockus bradfieldi)* 130, *130*
Crowned *(T. alboterminatus)* 130, *131*
Grey *(T. nasutus)* 130
Ground *(Bucorvus leadbeateri)* 16, 19, 130, 132-3, *133*
Monteiro *(Tockus monteiri)* 130, 131, 132, *133*
Red-billed *(T. erythrorhynchus)* 130
Silvery-cheeked *(Bycanistes brevis)* 130, 131, 132
Trumpeter *(B. bucinator)* 130, 131, 132, *132*
Yellow-billed *(Tockus flavirostris) 22,* 130; nest *131*
hornbills 13, 19, 22, **130-3**
Hyliota, Mashona *(Hyliota australis)* 169, *169,* 171
Yellow-breasted *(H. flavigaster)* 169
hyliotas **169**, 171

I

Ibis, Bald *(Geronticus calvus)* 9, 41, 43-4, *43*
Glossy *(Plegadis falcinellus)* 43, 44; nest *43*
Hadeda *(Bostrychia hagedash)* 43, 44, 63; nest *43*
Sacred *(Threskiornis aethiopicus) 11,* 43, *43,* 44, 93, 95; nest *43*
ibises 10, 21, 33, 35, 41, 42, **43-4**
Ichaboe Island 34
Indicator 136, 137

J

Jacana, African *(Actophilornis africanus)* 12, 77, 80-1, *80-1,* 82
Lesser *(Microparra capensis) 16,* 80-1, *80*
jacanas 15, 22, **80-1**, 85
jackal-berry tree 51, 60

K

kapokbossie *(Eriocephalus) 21,* 153, 177
Kaupifalco 62
kelp 34
Kestrel, Dickinson's *(Falco dickinsoni)* 70
Greater *(F. rupicoloides)* 52, 69-70, 151
Grey *(F. ardosiaceus)* 39, 70
Rock *(F. tinnunculus)* 69
kestrels **69-70**
Kingfisher, Brown-hooded *(Halcyon albiventris)* 122, 123, 137
Giant *(Ceryle maxima)* 122, 123
Grey-hooded *(Halcyon leucocephala)* 122, 137; nest *123*
Half-collared *(Alcedo semitorquata)* 122, 123
Malachite *(A. cristata)* 122, 123; nest *123*
Mangrove *(Halcyon senegaloides)* 122
Pied *(Ceryle rudis)* 12, 122, 123, *123*
Pygmy *(Ispidina picta)* 122, *122,* 123, 137
Striped *(Halcyon chelicuti)* 122-3, *123,* 137, 192
Woodland *(H. senegalensis)* 122, 123, *123,* 137
kingfishers 12, 15, 18, **122-3**, 190; nests 22, 148
Kite, Black-shouldered *(Elanus caeruleus)* 54-5, *55,* 70
Yellow-billed (Black) *(Milvus migrans)* 54, *54,* 55, 68, 69
kites **54-5**, 56
Korhaan, Black *(Eupodotis afra)* 79
Black-bellied *(E. melanogaster)* 79
Blue *(E. caerulescens)* 79, 80
Karoo *(E. vigorsii) 10,* 16, 79, 80; nest *79*
Red-crested *(E. ruficrista)* 14, 79-80, *79-80*
Rüppell's *(E. ruepellii)* 79, 80
White-bellied *(Eupodotis cafra)* 79
korhaans 19, 21, 78, **79-80**
kransaasvoël *see* Vulture, Cape
Kruger National Park 60, 132, 133, 192, 210

L

Lake Magadi, Kenya 45
Lake Natron, Tanzania 46
Lake Ngami, Botswana 31
Lake St Lucia 31, 45, 47, 94, 213
Lambert's Bay 26, 32
Lanioturdus 186
Lark, Botha's *(Spizocorys fringillaris)* 142; nest *144*
Clapper *(Mirafra apiata)* 10, 142
Dune *(M. erythrochlamys)* 142, 144
Dusky *(Pinarocorys nigricans)* 142
Fawn-coloured *(Mirafra africanoides)* 142
Flappet *(M. rufocinnamomea)* 142
Gray's *(Ammomanes grayi)* 142
Karoo *(Mirafra albescens)* 142, *142*
Long-billed *(M. curvirostris)* 142; nest *144*
Melodious *(M. cheniana)* 142
Monotonous *(M. passerina)* 142
Pink-billed *(Spizocorys conirostris)* 142
Red *(Mirafra burra)* 142, *142,* 144
Red-capped *(Calandrella cinerea) 19*
Rudd's *(Mirafra ruddi)* 142
Rufous-naped *(M. africana)* 142
Sabota *(M. sabota)* 142
Sclater's *(Spizocorys sclateri)* 142, 143, *143,* 144
Short-clawed *(Mirafra chuana)* 142
Spike-heeled *(Chersomanes albofasciata)* 12, *18,* 142, 144; nest *23, 144*
Stark's *(Spizocorys starki)* 142; nest *144*
Thick-billed *(Galerida magnirostris)* 144
larks 9, 13, 14, 19, 23, **142-4**; *see also* Finch-larks
lichen 149, 150, 155, 171, 173, 182, 185, 220; *see also* 'Old man's beard'
Longclaw, Orange-throated *(Macronyx capensis) 184*
Pink-throated *(M. ameliae)* 184
Yellow-throated *(M. croceus)* 184
longclaws 23, **184**
Loriculus 102
Lourie, Grey *(Corythaixoides concolor)* 103, 113
Knysna *(Tauraco corythaix)* 103
Purple-crested *(T. porphyreolophus)* 103, *103*
Ross's *(Musophaga rossae)* 103
louries 15, 21, **103**
Lovebird, Black-cheeked *(Agapornis nigrigenis)* 102
Lilian's *(A. lilianae)* 102
Rosy-faced *(A. roseicollis)* 102, *102,* 204
lovebirds **102**

M

Macheiramphus 56
Magaliesberg 51
Makgadikgadi Pan 31, 45-7, *45-6,* 94, 98
Malgas Island, Saldanha Bay 26, 32
Mannikin, Bronze *(Spermestes cucullatus)* 216, 220
mannikins **215**
Marion Island 29, 30
Martin, Banded *(Riparia cincta)* 137, 146, 147-8
Brown-throated *(R. paludicola)* 137, 145; nest *22,* 135, 146, 147-8, *148*
House *(Delichon urbica)* 145, 146
Rock *(Hirundo fuligula)* 23, 146, *146,* 147, 148
martins 10, 22, 119, **145-8**
mating systems 11-12
Matobo Hills, Zimbabwe 58
Matobo National Park, Zimbabwe 192
Melierax 63, 66
Mercury Island 34
Micronisus 63
Milvus 54
Mirafra 142
Monticola 161
Moorhen *(Gallinula chloropus)* 78
Lesser *(G. angulata)* 78
moorhens 21, **75-8**, *78*
moss 161, 172, 173, 182, 183
Mossel Bay 41
Mount Gorongosa, Mozambique 150
Mousebird, Red-faced *(Colius indicus)* 120
Speckled *(C. striatus)* 120; nest *120*
White-backed *(C. colius)* 120, *120*
mousebirds 23, **120**
mud 13, 23, 130, 146-7, 161, 167, 191
Mynah, Indian *(Acridotheres tristis)* 106, 189, 190, 191

N

Namib Desert 142
Namib Desert Park 52
Namibornis 165
Ndumu Game Reserve 68
Neddicky *(Cisticola fulvicapilla)* 106, 137, 175, 220
nest-building 14-15
nest types 21-4
Nicator, Yellow-spotted *(Nicator gularis)* **158-9**, 186; nest *159*
Nightjar, Fiery-necked *(Caprimulgus pectoralis) 19,* 115, 116, *116*
Freckled *(C. tristigma)* 115, *115,* 116
Mozambique *(C. fossii)* 115, 116
Natal *(C. natalensis)* 115, 116
Pennant-winged *(Macrodipteryx vexillaria)* 115, 116; nest *115*
Rufous-cheeked *(Caprimulgus rufigena) 19,* 115, *115,* 116
nightjars 8, 15, 19, 21, **115-16**
Numididae 71
Nyika Plateau, Malawi 147

O

Okavango Delta 31, 37, 42, 60, 96, 114, 156, 171, 175, 176
'old man's beard' lichen *(Usnea)* 140, *140, 62, 150,* 150, 173, *194,* 211, 221
Oriole, African Golden *(Oriolus auratus)* 106, 150
Black-headed *(O. larvatus)* 150; nest *150*
European Golden *(O. oriolus)* 150
Green-headed *(O. chlorocephalus)* 150
orioles 23, **150**
Orthotomus 174
Osprey *(Pandion haliaetus)* **68**
Ostrich *(Struthio camelus)* 21, **25**, *25*; eggs 15, 17, *25,* 51
Owl, African Scops *(Otus senegalensis)* 111, *112,* 113, 138
Barn *(Tyto alba)* 9, 16, 39, 111, *111, 112-13, 204*
Barred *(Glaucidium capense)* 111, 114
Cape Eagle *(Bubo capensis) 113,* 114
Giant Eagle *(B. lacteus)* 19, 39, 112, 114, *114,* 201, 204
Grass *(Tyto capensis)* 111, 113; nest *111*
Marsh *(Asio capensis)* 111, 113, 114
Pearl-spotted *(Glaucidium perlatum)* 111, *112,* 113-14, 150
Pel's Fishing *(Scotopelia peli)* 112, 114; nest *114*
Spotted Eagle *(Bubo africanus) 6, 20, 113,* 114
White-faced *(Otus leucotis)* 113
Wood *(Strix woodfordii) 110,* 113
owls 19, 20, 21, 22, **111-14**, 115
Oxpecker, Red-billed *(Buphagus erythrorhynchus)* 192, *192*
Yellow-billed *(B. africanus)* 192
oxpeckers 12, 22, **192**
Oystercatcher, African Black *(Haematopus moquini)* 17, **83**, *83,* 96; nest 21, *83*

P

Pachycoccyx 104, 106
palm tree 118, 168; borassus 69; ilala 69; raphia 51
Parakeet, Rose-ringed *(Psittacula krameri)* **102**
Parrot, Brown-headed *(Poicephalus cryptoxanthus)* 102
Cape *(P. robustus)* 102
Meyer's *(P. meyeri)* 102, *102*
Rüppell's *(P. rueppellii)* 102
parrots 22, **102**, 103
Partridge, Chukar *(Alectoris chukar)* 71
Passer 205
passerines 12, 18, 165
Pelican, Pink-backed *(Pelecanus rufescens)* 9, 31, *31,* 42; nest 21, *31*
White *(P. onocrotalus)* 9, 31, *31,* 45, 47; nest 21, 31, *31*
pelicans 10, 19, **31**
Penguin, Adélie *(Pygoscelis adeliae)* 10, 11, 26, 30
Jackass *(Spheniscus demersus)* 9, 20, **26**, *26*
penguins 10
Petrel, Northern Giant *(Macronectes halli) 30*
Pintado *(Daption capense)* 29, *30*
Soft-plumaged *(Pterodroma mollis) 30*
Southern Giant *(Macronectes giganteus) 29*
petrels **29**
Petronia 205
Phasianidae 71
Phragmites reeds 211
Phyllastrephus 158
Piet-my-vrou see Cuckoo, Red-chested
Pigeon, Delagorgue's *(Columba delagorguei)* 100
Feral *(C. livia)* 69, 100
Green *(Treron calva)* 100, *100,* 101
Rameron *(Columba arquatrix)* 100, 101
Rock *(C. guinea)* 39, 100, 101; nest *100*
pigeons 16, 18, 21, **100-1**, 103
Pinarornis 157
Pipit, Grassveld (formerly Richard's) *(Anthus novaeseelandiae)* 184, *184*
Long-billed *(A. similis) 184*
Mountain *(A. hoeschi)* 184
Plain-backed *(A. leucophrys)*: nest *184*
Short-tailed *(A. brachyurus)* 184
Striped *(A. lineiventris)* 184
pipits 23, **184**
Pitta, Angola *(Pitta angolensis)* **140**; nest 23, *140*

Plover, Blacksmith *(Vanellus armatus)* *16*, 84, 85; nest *21*
Black-winged *(V. melanopterus)* 84, 85; eggs *84*
Chestnut-banded *(Charadrius pallidus)* 84; eggs *84*
Crowned *(Vanellus coronatus)* 17, 84, 85, 87
Kittlitz's *(Charadrius pecuarius)* *2*, 84, 85, 89; nest *85*
Lesser Black-winged *(Vanellus lugubris)* 84, 85
Long-toed *(V. crassirostris)* 84, 85; eggs *84*
Three-banded *(Charadrius tricollaris)* 84, 85; eggs *84*
Wattled *(Vanellus senegallus)* 84, 85, *85*
White-crowned *(V. albiceps)* 84, 85, 90; eggs *84*
White-fronted *(Charadrius marginatus)* 84, *84*, 85, 89
plovers 15, 16, 19, 21, **84-5**, 86, 88
Pochard, Southern *(Netta erythrophthalma)* 48, 49
Polihierax 69
polyandry 12; in buttonquails 72; in Coucal, Black 109; in Crake, Striped 75; in cuckoos 104; in goshawks 64; in Jacana, African 80-1; in Snipe, Painted 82
polygyny 11-12; in bishops 213-14; in bustards 79; in Cisticola, Fan-tailed 175; in cuckoos 104; in ducks 48; in Korhaan, Black 80; in mousebirds 120; in nightjars 115; in plovers 84; in quails 72; in Sunbird, Olive 196; in weavers 12, 201, 207; in Whydah, Pin-tailed 12, 218; in widow-finches 218-19; in widows 213
Potberg, S W Cape 52
Pratincole, Black-winged *(Glareola nordmanni)* 89
Red-winged *(G. pratincola)* 89-90; nest *89*
Rock *(G. nuchalis)* 16, 89, 90, *90*
pratincoles 10, 21, **89-90**
Prinia, Black-chested *(Prinia flavicans)* 177, *219*, 220
Roberts's *(Oreophilais robertsi)* 177, 178, *178*
Spotted (Karoo) *(Prinia hypoxantha/maculosa)* 106, 177, *177*; nest *23*
Tawny-flanked *(P. subflava)* 174, 177, 220; nest *177*
prinias 8, 23, **177-8**, 216, 220
prions **29-30**
Prodotiscus 136, 137
protea fluff 194, 197
purple-pod terminalias 52
Pycnonotus 158
Pytilia, Golden-backed *(Pytilia afra)* 215, 220

Q

Quail, Blue *(Coturnix adansonii)* 72
Common *(C. coturnix)* 72; nest *72*
Harlequin *(C. delegorguei)* 72
quails 15, 21, **71-2**
Quelea, Red-billed *(Quelea quelea)* 10, 213, *213*
Red-headed *(Q. erythrops)* 213
queleas 10, 23, **213**

R

Rail, African *(Rallus caerulescens)* 75
rails **75**, 78
Rallidae 75
Raven, White-necked *(Corvus albicollis)* 69, 114, **151-2**, *151*
Rhigozum trichonotomum bushes 178
Richard's Bay 89
Robben Island 10, 26, 71, 93
Robin, Bearded *(Erythropygia quadrivirgata)* 106, 166
Brown *(E. signata)* 168
Cape *(Cossypha caffra)* *104*, 105, 106, 166, 167; nest *166*
Chorister *(C. dichroa)* 105, 106, 166, 167
Gunning's *(Shephardia gunningi)* 166
Heuglin's *(Cossypha heuglini)* 105, 106, 166, *166*
Kalahari *(Erythropygia paena)* 106, 166
Karoo *(E. coryphaeus)* *12*, 106, 166, 167, *168*
Natal *(Cossypha natalensis)* 105, 106, 166, 167; nest *166*
Starred *(Pogonocichla stellata)* 105, 106, 166, 167, 168; nest 23, *166*, 167
Swynnerton's *(Swynnertonia swynnertoni)* 106, 166, 167, 168
White-browed *(Erythropygia leucophrys)* 106, 166, *167*, 168
White-throated *(Cossypha humeralis)* *9*, 106, 166, *167*
robins 23, **166-8**
Rockjumper, Cape *(Chaetops frenatus)* 157, *157*
Orange-breasted *(C. aurantius)* 157
rockjumpers 23, **157**
Rockrunner *(Achaetops pycnopygius)* 165, 169, 174; nest *173*
Roller, Broad-billed *(Eurystomus glaucurus)* 126, 127, *127*
European *(Coracias garrulus)* 126
Lilac-breasted *(C. caudata)* 126, 127
Purple *(C. naevia)* 126, *126*, 127
Racket-tailed *(C. spatulata)* 126, 127, *127*
rollers 22, **126-7**, 138, 190
Rondevlei Nature Reserve 37
Rostratulidae 82
Rynchopidae 94

S

Sandgrouse, Burchell's *(Pterocles burchelli)* 98
Double-banded *(P. bicinctus)* 98, *98*
Namaqua *(P. namaqua)* *17*, 98-9, *98-9*; nest *98*
Yellow-throated *(P. gutturalis)* 98; nest *98*
sandgrouse 8, 19, 21, **98-9**
Sandplover *see* Plover, Three-banded
Scolopacidae 82
seaweed 33, 34
Secretary Bird *(Sagittarius serpentarius)* 42, **50**, *50*, 70, 74
Seedcracker, Nyasa *(Pyrenestes minor)* 215
sexual dimorphism 14; in ducks 47; in egrets 14; in estrildid finches 215; in flufftails 75; in Jacana, African 81, 82; in kingfishers 122; in sandgrouse 98; in Snipe, Painted 14, 82; in weavers 14; in Whydah, Long-tailed 14; in widows 14; in Woodhoopoe, Scimitar-billed 14
shearwaters **29-30**
Shelduck, South African *(Tadorna cana)* 47, 48, 49
Shoebill *(Balaeniceps rex)* 38, 40
Shoveller, Cape *(Anas smithii)* 47, 49; eggs *48*
Shrike, Black-fronted Bush *(Telophorus nigrifrons)* 186
Brubru *(Nilaus afer)* 13, *19*, 185-6, 187, 188; nest *187*
Chestnut-fronted Helmet *(Prionops scopifrons)* 188
Crimson-breasted *(Laniarius atrococcineus)* 106, 185, 187, 188; nest *186*
Fiscal *(Lanius collaris)* 10, *10*, 105, 106, 161, 185, 187; nest *185*
Gorgeous Bush *(Telophorus quadricolor)* 186, *187*
Grey-headed Bush *(Malaconotus blanchoti)* 185, 186
Long-tailed *(Corvinella melanoleuca)* 185, *185*, 187, 188
Olive Bush *(Telophorus olivaceus)* 186, *187*
Orange-breasted Bush *(T. sulfureopectus)* 186
Puffback *(Dryoscopus cubla)* 106, 185, *186* 187, 188
Red-billed Helmet *(Prionops retzii)* 105, 106, 108, 188
Sousa's *(Lanius souzae)* 185; nest *185*
White-crowned *(Eurocephalus anguitimens)* 185, 187, 188, *188*
White Helmet *(Prionops plumatus)* 186, 187, 188, *188*
White-tailed *(Lanioturdus torquatus)* 185, 186, *187*, 188
shrikes 12, 19, 23, 159, **185-8**
siblicide *see* Cainism
Simon's Town 26
Siskin, Cape *(Serinus tottus)* 105, 106, 220, 221, 222, *223*
Drakensberg *(S. symonsi)* 220, 221
Skimmer, African *(Rynchops flavirostris)* 16, 21, 90, **94**, **96-7**, *97*
Skua, Sub-Antarctic *(Catharacta antarctica)* 29-30, *30*
skuas **29-30**
snake eagles 61
Snipe, Ethiopian *(Gallinago nigripennis)* 82; eggs *82*
Painted *(Rostratula benghalensis)* 12, *12*, 14, 82, *82*
snipes 21, **82**
Sparrow, Cape *(Passer melanurus)* 106, 113, 120, 147, 148, 205, 206; nest *23*, 206
Great *(P. motitensis)* 106, 205, 206
Grey-headed *(P. griseus)* 106, 137, 205-6; nest *206*
House *(P. domesticus)* 147, 205, 206; nest *206*
Yellow-throated *(Petronia superciliaris)* 137, 205, 206
Sparrowhawk, Black *(Accipiter melanoleucus)* 63, 64, *64*, 66, 68
Little *(A. minullus)* 63, *63*, 64, 66
Ovambo *(A. ovampensis)* 63, 64
Red-breasted *(A. rufiventris)* *18*, 63, 64; nest *21*, *63*
sparrowhawks 56, 62, **63-6**
sparrows 22, 23, **205-6**
Sparrow-weaver, White-browed *(Plocepasser mahali)* 102, **201-2**; nest *204*
sparrow-weavers: nests 23
spider, social *see Stegodyphus*
Spinetail, Bat-like *(Neafrapus boehmi)* 117, 119
Mottled *(Telacanthura ussheri)* 117, 118, *118*, 119
spinetails **117-19**
Spoonbill, African *(Platalea alba)* 11, *18*, 35, 42, **43-4**, *43-4*; nest 21, *44*
spoonbills 10, 41
St Croix Island 95
Starling, Burchell's *(Lamprotornis australis)* 106, *191*, 192
European *(Sturnus vulgaris)* 11, 21, 61, 189, 190, 191, 220; nest *189*
Glossy *(Lamprotornis nitens)* 106, 137, 190, 191
Greater Blue-eared *(L. chalybaeus)* 106, 190-1; nest *9*, *190*
Long-tailed *(L. mevesii)* 106, 191
Pale-winged *(Onychognathus nabouroup)* 106, 190, 191
Pied *(Spreo bicolor)* 106, 137, *189*, 190; nest 21, 119, 148
Plum-coloured *(Cinnyricinclus leucogaster)* 137, 189, 190, *190*, 191
Red-winged *(Onychognathus morio)* 106, 189, 191; nest *191*
Wattled *(Creatophora cinerea)* *13*, 113, 189, 190, 191; nest *189*
starlings **189-91**
Stegodyphus (social spider) 21, 63, 64, 66
Stilt, Black-winged *(Himantopus himantopus)* **86**; nest *86*
stilts 10
Stonechat *(Saxicola torquata)* 106, 163, 165, *165*
Stork, Abdim's *(Ciconia abdimii)* 40
Black *(C. nigra)* 40, 41-2, *41*, 69
Marabou *(Leptoptilos crumeniferus)* 40, 42, 47; nest *42*
Openbill *(Anastomus lamelligerus)* 40, 42
Saddlebill *(Ephippiorhynchus senegalensis)* 40, 42, *42*, 50
White *(Ciconia ciconia)* 40-1; nest *40*
Woolly-necked *(C. episcopus)* 40, *40*, 42
Yellow-billed *(Mycteria ibis)* 40-2, *40*
storks 10, 21, 35, 38, *40*, **40-2**
Sugarbird, Cape *(Promerops cafer)* 13, 193, *193*, 194, 199
Gurney's *(P. gurneyi)* 193, *193*, 194
sugarbirds 23, **193-4**
Sunbird, Black *(Nectarinia amethystina)* 106, 197; nest *198*
Blue-throated *(Anthreptes reichenowi)* 194
Bronze *(Nectarinia kilimensis)* 197
Collared *(Anthreptes collaris)* *15*, 106, 194, 197, *198*
Dusky *(Nectarinia fusca)* 106, 194
Greater Double-collared *(N. afra)* 106
Grey *(N. veroxii)* 194, 197; nest *196*
Lesser Double-collared *(N. chalybea)* *196*; nest *21*
Malachite *(N. famosa)* 105, 106, *195*, 197
Marico *(N. mariquensis)* 106
Neergaard's *(N. neergaardi)* 194; nest *194*
Olive *(N. olivacea)* 140, 194, 196, 197; nest *197*
Orange-breasted *(N. violacea)* 13, 194, 196, 197, 199, *199*
Purple-banded *(N. bifasciata)* 197; nest *194*
Scarlet-chested *(N. senegalensis)* *23*, 106, 137, 196, 197, *197*
Violet-backed *(Anthreptes longuemarei)* 197, *198*
White-bellied *(Nectarinia talatala)* 106, 197; nest *196*
sunbirds 23, 64, 105, 193, **194-8**
Swallow, Black Saw-wing *(Psalidoprocne holomelas)* 146, 148, *148*
Blue *(Hirundo atrocaerulea)* 8, 145, 146, 147; nest *146*
Eastern Saw-wing *(Psalidoprocne orientalis)* 146, 148
Greater Striped *(Hirundo cucullata)* *14*, 135, 137, 146, 165; nest *23*, *147*, 206
Grey-rumped *(H. griseopyga)* 145-6; nest 22, 146, 147, *148*
Lesser Striped *(H. abyssinica)* 123, 145, 146, *147*, 163, 164; nest *117*, *145*

Mosque *(H. senegalensis)* 146, 147
Pearl-breasted *(H. dimidiata) 23, 145*, 146, 147, 148
Red-breasted *(H. semirufa)* 137, 145, 146, 147, 148
South African Cliff *(H. spilodera)* 8, 145, 146, 147, 148; nest *147*
White-throated *(H. albigularis)* 137, 146, *146*
Wire-tailed *(H. smithii)* 145, 146, 148; nest *145*
swallows 10, 13, **145-8**; nests 21, 22, 23, 68, 117, 119, 135, 167, 205
Swan, Mute *(Cygnus olor)* 47
Swift, Alpine *(Apus melba)* 117, 118, 119, *119*
Black *(A. barbatus)* 117, 119
Bradfield's *(A. bradfieldi)* 117, 119
European *(A. apus)* 16, 117, 119
Horus *(A. horus)* 117, 118, 119; nest *118*
Little *(A. affinis)* 117, 118, 119, 123, 147, 205, 216
Mottled *(A. aequatorialis) 117*, 119
Palm *(Cypsiurus parvus)* 9, 16, 117, 118, *118*, 119
Scarce *(Schoutedenapus myoptilus)* 117
White-rumped *(Apus caffer)* 117, *117*, 118, 119, 147
swifts 10, 15, 21, **117-19**
Sylviidae 169

T

'tailor-birds' 174, 175
Tchagra, Black-crowned *(Tchagra senegala)* 186
Marsh *(T. minuta)* 186
Southern *(T. tchagra)* 186; nest *186*
Three-streaked *(T. australis)* 186, *186*
Teal, Cape *(Anas capensis)* 47, 49
Hottentot *(A. hottentota)* 47, 48-9
Red-billed *(A. erythrorhyncha)* 47
Tern, Caspian *(Hydroprogne caspia)* 10, 93, 94, *94*
Damara *(Sterna balaenarum)* 94, 95-6, *95*
Roseate *(S. dougallii)* 94, 95; nest *94*
Swift *(S. bergii)* 10, *17*, 93, 94-5, *94*; nest *94*
Whiskered *(Chlidonias hybridus) 1, 7*, 94, 96, *96*; nest 22, *22*
terns 13, 20, 21, **94-6**
territory, importance of 9-10
Thrush, Cape Rock *(Monticola rupestris)* 106, 161, 162; nest *162*
Collared Palm *(Cichladusa arquata)* 166, 167, 168; nest *168*
Groundscraper *(Turdus litsitsirupa)* 161, 162
Kurrichane *(T. libonyana)* 106, *160*, 161, 162
Miombo Rock *(Monticola angolensis)* 161, 162, *162*
Olive *(Turdus olivaceus) 17, 18*, 106, 161; nest *23*
Orange *(Zoothera gurneyi)* 161, 162; nest *162*
Rufous-tailed Palm *(Cichladusa ruficauda)* 166, 167; nest *168*
Sentinel Rock *(Monticola explorator)* 161
Short-toed Rock *(M. brevipes)* 161
Spotted *(Zoothera fischeri)* 161, *161*
thrushes 18, 140, **161-2, 166-8**, 193; nests 23, 100, 167, 181
Timaliidae 157
Tit, Ashy *(Parus cinerascens)* 152, 153
Black *(P. leucomelas)* 152-3, *153*
Cape Penduline *(Anthoscopus minutus)* 153, 154, *154*
Carp's Black *(Parus carpi)* 152, 153
Grey Penduline *(Anthroscopus caroli)* 153, 154
Northern Grey *(Parus griseiventris)* 152
Rufous-bellied *(P. rufiventris)* 152
Southern Black *(P. niger)* 137
Southern Grey *(P. afer)* 152, *152*
Tit-babbler *(Parisoma subcaeruleum)* 106, 169, *169*, 171
Layard's *(P. layardi)* 169, 171
tit-babblers **169**
tits 15, **152-4**; nests 21, 22, 64, 66
Tockus 130, 131, 132
Trogon, Narina *(Apaloderma narina)* 22, **121**, *121*
Turdidae 157
Turdus 161
Turnicidae 71, 72
Twinspot, Green *(Mandingoa nitidula)* 215, 216
Red-throated *(Hypargos niveoguttatus)* 218, 220
Tygerberg Zoo 41
Typha bulrushes 211

V

Van Wyksvlei 45
Viduidae 218
Vulture, Cape *(Gypos coprotheres)* 9, 10, 41, 51-2, 53, *53*
Bearded *(Gypaetus barbatus)* 51, 52, 53; nest *51*
Egyptian *(Neophron percnopterus)* 25, 51, 52, 53
Hooded *(Necrosyrtes monachus)* 51, 52, 53; nest *51*
Lappet-faced *(Torgos tracheliotus)* 50, 51, 52, 53, *53*, 69, 70, 113; nest *70*
Palm-nut *(Gypohierax angolensis)* 51, 52-3; nest *51*
White-backed *(Gyps africanus)* 51, 52, *52*, 53, 201
White-headed *(Trigonoceps occipitalis)* 51, 52; nest *53*
vultures 16, **51-3**

W

Wagtail, African Pied *(Motacilla aguimp)* 189
Cape *(M. capensis) 24*, 106, 183, *183*
Long-tailed *(M. clara)* 106, 183; nest *183*
wagtails 23, **183**, 184
Walvis Bay 31, 33-4
Warbler, African Marsh *(Acrocephalus baeticatus)* 169, *171*, 172
African Sedge *(Bradypterus baboecala)* 169, *171*, 172, 178
Barratt's *(Bradypterus barratti)* 169, 171
Barred *(Cameroptera fasciolata)* 169, 174
Bleating *(C. brachyura)* 106, 107, 137, 174, *174*; nest *105*
Brier *see* Prinia, Roberts's
Broad-tailed *(Schoenicola brevirostris)* 169, 172
Cape Reed *(Acrocephalus gracilirostris)* 169, *170*, 172
Cinnamon-breasted *(Euryptila subcinnamomea)* 169, 174; nest *173*
Cinnamon Reed *(Acrocephalus cinnamomeus)* 169, 172
Greater Swamp *(A. rufescens)* 169, 172
Icterine *(Hippolais icterina)* 188
Knysna *(Bradypterus sylvaticus)* 169, *169*, 171, 172
Moustached *(Melocichla mentalis)* 169, 174
Namaqua *(Phragmacia substriata)* 177-8, *178*
Red-winged *(Heliolais erythroptera)* 177
Rufous-eared *(Malcorus pectoralis)* 177, 178, *178*
Stierling's Barred *(Camaroptera stierlingi)* 169, 174
Victorin's *(Bradypterus victorini)* 17, 169, *169*, 171, 172, 174
Yellow *(Chloropeta natalensis)* 169, 172; nest *171*
Yellow-throated *(Seicercus ruficapillus)* 169, 171, 172-3; nest 23
warblers 23, 136, **169-74**
Waterberg Plateau, Namibia 59
Waxbill, Black-cheeked *(Estrilda erythronotos)* 215
Blue *(Uraeginthus angolensis)* 14, 216; nest *216*
Common *(Estrilda astrild) 14*, 215, 218, 220; nest *217*
Grey *(E. perreini)* 216
Orange-breasted *(Sporaeginthus subflavus)* 214, 216, *216*, 220
Swee *(Estrilda melanotis)* 220
Violet-eared *(Uraeginthus granatinus)* 220
waxbills **215-16**
Weaver, Brown-throated *(P. xanthopterus)* 207, 212
Cape *(Ploceus capensis)* 106, 207, 211; nest *24*, 179, 200-1, *207, 210*
Chestnut *(P. rubiginosus)* 207, 212, 216
Forest *(P. bicolor)* 207, 208, 211, 212, 216; nest *211*
Golden *(P. xanthops)* 106, 207, *208*, 211
Lesser Masked *(P. intermedius)* 106, 207, *210*, 211, 212; nest *208*
Masked *(P. velatus)* 105, 106, 207, *207*, 210, 212; nest *216*
Olive-headed *(P. olivaceiceps)* 207, 208, 211
Red-billed Buffalo *(Bubalornis niger)* 52, 102, 114, **201**, 216; nest *202*
Red-headed *(Anaplectes rubriceps)* 207-8, 212; nest *24*, 181, 211, *211*
Sociable *(Philetairus socius)* 9, 10, 12, 16, 70, **201-4**, 212; nest 21, 59, 102, 114, 135, 163, **203-4**, *203*, 216
Spectacled *(Ploceus ocularis)* 106, 207, *209*, 211, 212
Spotted-backed (Village) *(P. cucullatus)* 15, 106, 207, 210, 211
Thick-billed *(Amblyospiza albifrons)* 23, 24, 207, 208, 211, 212, *212*
Yellow *(Ploceus subaureus)* 106, 207, 211, *212*
weavers 10, 12, 14, 15, 68, **207-10**, 218, 220; nests 21, 24, 60, 153, 206, 214, 216
Wheatear, Capped *(Oenanthe pileata)* 163, *163*, 164, 165
White-eye, Cape *(Zosterops pallidus)* 17, *20*, 200-1, *201*
Yellow *(Z. senegalensis)* 136, 137, 200, *200*, 201
white-eyes 23, **200-1**
Whydah, Broad-tailed Paradise *(Vidua obtusa)* 219, 220
Paradise *(V. paradisaea)* 219, 220
Pin-tailed *(V. macroura)* 9-10, 12, 14, 218, *218*, 219, 220
Shaft-tailed *(V. regia)* 219, 220
whydahs 24, 217, **218-19**
Widow, Long-tailed *(Euplectes progne)* 214, 218
Red-collared *(E. ardens)* 214
White-winged *(E. albonotatus)* 105, 106
Yellow-rumped *(E. capensis)* 10, 105; nest 214
Widow-finch, Black *(Vidua funerea)* 220
Green *(V. codringtoni) 218, 220*
Purple *(V. purpurascens)* 220
Steel-blue *(V. chalybeata)* 220
Violet *(V. incognita)* 218
widow-finches 24, 217, **218-19**
widows 13, 14, 19, **213**, 218; nests 23, 177, 216
Woodhoopoe, Red-billed *(Phoeniculus purpureus)* 12, 128, 129, 130, 137, 190, 201
Scimitar-billed *(P. cyanomelas)* 14, 128, 129, *129*, 130, 137
Violet *(P. damarensis)* 128
woodhoopoes 19, 20, 22, **128-30**
Woodpecker, Bearded *(Thripias namaquus)* 138
Bennett's *(Campethera bennettii)* 137, 138
Cardinal *(Dendropicos fuscescens)* 137, *137*, 138
Golden-tailed *(Campethera abingoni) 22*, 137
Ground *(Geocolaptes olivaceus 22, 106, 119, 137-8*, 138
Knysna *(Campethera notata)* 137
Little Spotted *(C. cailliautii)* 137
Olive *(Mesopicos griseocephalus)* 137, 138, *139*
Speckle-throated *(Campethera scriptoricauda)* 137
woodpeckers 8, 68, 113, **137-9**, 155; nests 22, 122, 126, 128, 152, 181, 205, 216
Wryneck, Red-throated *(Jynx ruficollis)* 20, 22, *137-9, 138*

Z

Zoothera 161

LIST OF SUBSCRIBERS

Sponsors' Edition

Steve Bales

Simon Barlow

E. Bertelsmann SC

Jim Gerard Paul Broekhuysen

D.A. Hawton

M.F. Keeley

Robin Moser

Carel A. Nolte

Kathleen Satchwell

Collectors' Edition

Giaco Angelini
J.A. Ardington
N.J. Arkell
Mrs Elma-Louise Banks
Alex A. Barrell
Don Barrell
P.A. Becker
Dr & Mrs R.M.F. Berard
Ian & Sue Bishop
& Kerry Chambers
John E. Bishop
David K. Bond
Joan S. Bortnick
Professor M.C. Botha
N.R.G. Brunette
André Claassen
Vicki & Nigel Colne
Grant Cornish-Bowden
Jagger Cornish-Bowden
Kevin H. Culverwell
Dr P. Deminey
Jalal & Kulsum Dhansay
W.R. Doepel
Caroline & Grant Donaldson

J.P. du Plessis
S.G. Edwardes-Evans
Lesley Gail England
Fernwood Press (Pty) Limited
Eugene & Lalie Fourie
Dr Leslie Frankel
E.S.C. Garner
Francis & Karin Garrard
Amanda Gazendam
Robert Gee
Saul Goldblatt
Lydia Gorvy
Pat & Karin Goss
Trish Henchoz
John K. Hepburn
Heinz Heuser
James William Alistair Hill
Mr & Mrs Rupert Horley & Family
Rosemary Faith Hundleby
C.R. Hunting
L.R. Hunting
M.S. Hunting
Ismail Hussain
M.J. Hyde

Mr & Mrs Benjamin Jonsson
Hilary & Graham Killerby
Albert E. Kuschke
Graham Leslie
Hilton & Louise Lissack
Nigel Marven
James McLuskie
Ian McPherson
Robin Moser
Natal Portland Cement
(Pty) Limited
Jan & Elizabeth Nel
P.J.D. Nel
L-M. Nicholls
D.J. Opperman
Ian Outram
John & Jeannine Pearse
Petersfield Nurseries
Dr & Mrs D.G.C. Presbury
Shaun V. Price
Heather Rahn
Dr Ludger Reißig
Douglas Stuart Roberts
Dr Colin R. Saunders

Lorraine Shalekoff
Ilse Shuttleworth
D.T. Smith
Johanna Soper
Isolde & David Stegmann
Peter Steyn
Stocks Housing (Pty) Limited
Pieter & Pam Struik
William Taylor
Charles & Igna Tregoning
Jean Turck
Margarete Unite
G.A. Upfill-Brown
J.A. Janse van Rensburg
N.P. Janse van Rensburg
W. van Rÿswÿck
Paul van Schalkwyk
O. von der Lancken
Peter A. Watt
Roy & Lorraine Webber
Mr J.B.M. Weeple
J.D. Weyers
Peter & Jeffe Williams
J.A. Windell

STANDARD EDITION

A.C. Loide Abbott
John Abbot
ABC Bookshop
Rob & Eveline Abendanon
Doug & Bo Acheson
Pete & Sharon Acheson
Dr D.J.J. Ackermann
Dr J.J. Ackermann
Mike Adams
Peter Adams
Stephan Albat
Ivonne & Oswald Albers
Albi J. Alberts
M. Alberts
Albertus Delport-biblioteek
Margaret Alexander
Terry M. Allan
David & Ann Alston
David P. Ambrose
Mike & Monica Amm
Annelise Andersen
Mr & Mrs D.J. Anderson
Mark & Tania Anderson
S.A.G. Anderson
J.R. Angove
Graham Anthony
Ian & Anne Anthony
A.M. Archer
N.J. Arkell
Mark & Ruth Armitage
D.M. Armstrong
F.I. Armstrong
Brian Askew
Bernice R. Aspoas
Roelf & Jill Attwell
H.L. Aucamp
John & Margie Austin
Yvonne A. Austin
Ann & David Ayre
Colin & Julia Baker
Paoletta Baker
David & Linda Baldie
Gina Baldo
D.N. Bantjes
J. Barker
Simon Barlow
Simon & Michele Barlow
Carl Barnard
Bert Barnes
Bill & Leila Barnes
Eric & Margaret Barnes
Don Barrell
Mrs Suzanne Barrow
Kevin Barry
David Basckin
Malcolm & Peta Anne Basford
Allan Batchelor
David & Cathy Bath
E.J.M. Baumbach
Darryl L. Baxter
Aidan Beard
Professor T.V.R. Beard
Peter Becker
Edward Beesley, Illovo Beach
Annette & Hans Beetge
D.W. Beghin
Kathy Beling
Keith, Lesley, Trevor & Stuart Bell
Beatrice Bellegoni-Sarzana
Mrs P.N. Bellingan
J. & J.P. Bello
Graham & Heather Benfield
David & Salomé Bennett & Family
P.J. Bennett
Mrs Leslie Berger
C.H. Berman
Drs H. & Z. Bernitz

John Bethune-Williams
Professor Gerhard & Isolde Beukes
H.J. Beunk
Danie & Ina Bezuidenhout
M.A. Billing
A.W. Birkholz
R.C. Birkholz
Peter & Ruth Blackwell
K. Blatherwick
Anthony G. Bloomer
Neil & Wendy Bloy
Arryn Blumberg
Ms T. Bodbijl
Richard Booth
Rhoda Booyens
E.L. Booysen
Peter & Brenda Borchert
Rick Bosch
Professor Elizabeth Boshoff
Johan Boshoff
Barbara Bosman
Mrs E. Bosman
Theo & Varina Bosman
André J. Botha
Charles & Julia Botha
Frikkie Botha
Rudolph Botha
Zelda Botma
Geoff & Jeanette Bowers-Winters
Bill Boyd
D.G. Boyes
John Bradfield
Dieter & Susan Brandt
Harry & Sadie Braun
The Brenthurst Library
Dr David N. Brereton
C.J. & Honor Breyer-Menke
Terry Briceland
Murray John Bridgman
D.J. Bridle
Marie Brink, Misty Mountain Farm
Mr P.F. Brink
J. Hylton Briscoe
Cheryl Brocklehurst
Frances & Anthony Broekhuizen
Luc Broes
J-J. Brossy
S.M. Broster
R.J. & M.N. Brown
Tanya Browne
Mr W.G.R. & Mrs D.M. Brückner
Mrs J.M. Brunette
Buck & Diana Buchanan
Pat & Mike Buchel
Gerald & Karen Bullen
John & Penny Burchmore
Nino & Karin Burelli
J.E. Burger
Dr A.J. Burgess
Ernie & Rose Buric
Nanette Burkheiser
E.E. & N.R. Burnett
Rob & Wendy Burnett
John & Tracey Burnton
Chris, Andre & Deon Buys
J.M. Caddick
Rupert George Calcott
Alan & Joanne Calenborne
Karen Cameron
Virginia Cameron
Picard Camille
Mr & Mrs B.H. Campbell
Mrs D.J. Campbell
Cape Nature Conservation
Cape Technikon Library
Dario Cappelli
Margaret W. Carbis

Harold Carless
Robert Guy Carr
J.J. Carstens
David & Sal Carter
Louise & Randle Carter
Margaret Cary
S.V.R. Casserly
Mr A.W. Castleman
A.E. Caudle
Alan Cave
Charles Cawood
G.A. Cawood
M. Cazzavillan
R. Cemernjak
Dawie & Margot Chamberlain
P.N. Chamberlain
Samantha Chandler
J.B. Chaplin
M.J. & T.A. Chapman
Colleen & Jeremy Chennells
Dr Jettie Chipps-Webb
Ms Corinna Chiu
Eugénie Chopin
P.J. Cillié
Japie Claassen
Lawrence & Petronella Clark
Mrs V. Clarke
Elizabeth A. Clayton
R.J. Clinton
Bruce Cloete
Steve, Raine & Choni Cloete
Dr P.G. Close
Peter & Sally Cobbold
Roger S.V. Cockram
Dorothy Cody
Johann & Steph Coetzee
M.J.R. Coetzee
Nico Coetzee
Mrs D.A.H. Coetzer
K. Cohen
Brian D. Colahan
Des & Naureen Cole
Pam Cole
Derek Coley
V.A. Coley
Mrs Dawn Colin
John & D. Collard
Jean Collier
L. Compton-James
Carol Coney
Margôt Connor
T.W. Connor
Ian & Elaine Conry
Garth Martin Cook
Doug & Nola Cooke
Marene Coote
Monica Corder
Pietro Corgatelli
F.G. Cornish
Roy & Margie Corrans
Mrs J.S. Cotterrell
C.F.G. Cottino
Dr Graham Coupland
C.L. Cousins
S.M. Cove
Ron Cowley
Professor R.M. Cowling
Graham & Jillian Cox
I.A. Cox
W.L. Cox
Anton Coy
Llewellyn & Nellien Crewe-Brown
Duncan Cromarty
Hannes en John Cronjé
M.J. Crosby
Marge & Joe Crouch
Shannon S. Culverwell

Toni B. Culverwell
Richard & Stephanie Cunliffe
Paul Cunningham
Derek Custers
S. Dal Col
Dr S.K. Dantu
André Darmont, Belgium
J.C. Davel
B. Davies
H.K. Davies
Herb & Elissa Davies
Dave & Lise Day
Chris M. de Beer
D. de Beer
Kobus & Elizabeth de Beer
Richard & Maureen de Beer
Sandra Fay de Coning
J.A. de Gier
Julian & Jean de Jager
M.C. de Klerk
Mr & Mrs J.D. de Lange
Mr Gustav C. de Muelenaere
Manuel José de Sousa Pita
H.A.C. de Villiers
J.C. (Kay) de Villiers
Dr Jake de Villiers
Jan de Vos
Peter & Dyanne de Vos
Dr W.N. de Vos
J.A. de Waal
Blanche de Wet
K.N. de Wet
C.P. de Wit
Mrs F.W. Deacon
Barbara Mabel Deeks
Clinton Deeks
C.J. & M.J. Dekker
Dr Hennie Dekker, Menlopark
Michael & Ingrid Dennill
Michael C.M. Denny
R. Desenclos
Kevin Deutschmann
Helen Dewar
Sara Dewar
Martin, Lisa & Vincent Di Bella
Leicester Dicey
Hendrik & Coba Diederiks
Patricia & Peter Dinkelmann
S.M. Dippenaar
V. Estelle Dippenaar
J.G. Ditchfield
Karen V. Dixon
Peter Dodds
Ellen Doidge
Jennifer Dianne Don
Patrick J.K. Donaldson
Terry Donnelly
Seamus & Veronique Donoghue
Dr A.S. Donohoe
Graeme Dott
Rodney Douglas
Maggie Douwes
G.H. Dreyer
Jeanine Dreyer
I.M. McK. Drummond
Iain Drummond
Mike Drury
Barry du Plessis
Dup & Maggie du Plessis
Mr & Mrs René du Plessis
Johann du Preez
Pierre du Preez
Francois du Randt
A.F. du Toit
C.J.F. (Cathy) du Toit
Charl du Toit
Fanus & Dalene du Toit

Gys du Toit
P.F. du Toit
Pierre du Toit
W.T.F. (Will) du Toit
Dr F.B.W. Ducasse
Frieda Duckitt
Patrick Duigan
Graham Dumbrill
Mr & Mrs C.A.F. Dunn
B.A. Dürr
Marc Durr
Andrè & Nanette Düsterhöft
John & Sue Duthie
Rienk Duursma
Keith & Greer Dyer
J.A. Dykes
Joan Dykhouse
Mike Dykhouse
Mrs I.G. Eales
R.D. Eddie
Gary Ronayne Edwards
G.T.S. Eiselen
Pierre Eksteen
Vaughan Elliott
Geoffrey D. Ellis
Roger Ellis
Mrs E.L. Els
Mark & Cheryl Eltringham
Barry Emberton
Janet & Ian Emott
Schalk Engelbrecht
Engineering Fabricators C.C.
Céleste Enslin
Liesel Erasmus
Einhard & Dagmar Erken
P.W. Erlangsen
Carla & Felix Ernst
ESKOM
Mike & Biddy Evans
J.M. Fairbairn
Colleen Fairlamb
Joe & Gwyn Faller
Pauline Farquhar
Gavin & Antoinette Faulds
John & Muffy Featherstone
Mr & Mrs Nigel Fernsby
Khakie Ferreira
Meeding & Barendien Ferreira
Yvonne Pyne-James & Peter Ferrett
Liz & Alex Fick
Dr Dawn Finlay
Margie Firer
First National Bank
J.H. Fisher
Heribert Flach
S. Foggo
D.J. Fölscher
Malcolm Foster
Beverley Anne Fourie
Eugene & Lalie Fourie
Jannes & Helena Fourie
M.C. Fourie
Dr Rowan Fourie
Lenie Franken
Candy & John Fraser
Erica Fraser
Ian Cameron Fraser
Reuben Albert Freed
Bill Freund
Friends of Pilanesberg Society
N.B.R. Fuller
Malcolm Funston
Paul Funston
R. & J. Furlong
John Furniss
Erich M. Gaertner
Gametrackers Botswana

Mervyn & Allison Gans
Adrian Gardiner
John Gardner
David & Niki Garratt
Jerry Garrett
Rendall Garrett
Jaenine Geldhof
Arlene Georgeson
J.S. Gericke Library, University of Stellenbosch
Brian & Susan Germond
R. Gettliffe
Barrie & Pam Gibson
Robin Gill
I.F.G. Gillatt
Peter Ginn
Viv & Mary Goddard
Vivien Goldsack
Ian A.D. Gordon
James & Trish Gordon Lennox
T.P. & G.R.N. Gordon-Cumming
Dr G.J.M.R. Gorter
Jerry & Marilyn Gosnell
Dr K.C. Goulding
Christoph Grabherr
A.R. Gregory
Uwe Gressmann
Ben & Rosemary Griesel
R. Griesel
Bob & Marlene Griffin
Dr Gwyn Griffiths
Charmaine Grobbelaar
N. Grobbelaar
Jors & Mariet Grobler
John Groenewald
Siegfried J. Gross
Joan & Peter Grossett
M.L. Guittard
Dave Gunn
A.H. Gurnell & Family
Jeff Guy
David Hakime
Mr & Mrs D.A. Hall
Dorothy G. Hall
R.S. & B.A. Hall
Andrew Halsted
Philippa Halsted
Thomas Halsted
Joseph J. Hamman
Dr Christopher A. Hammond
Rolf Hangartner
Rudolf & Sonia Hanni
Gill & Paul Hardingham
Robert H. Harm
Jan Harmse
Angela & Bill Harrison
Marcelle Harrison
S.G. Harrison
J.O.C. Hart
R.A. Harvey
Chris Hatton
Hatty Family Trust
Maureen & Henk Havinga, The Netherlands
Neville Haworth
Margie & Barry Hawthorne
D.A. Hawton
C.A.S. Hayne
Mr E.F. Hayward
G.D. Hazell
Dr & Mrs B.W. Hellberg
Leigh David Hen-Boisen
Craig N. Hepburn
Glynn & Anne Herbert
Frank Herholdt
Hermanus Botanical Society
Marjorie M. Heron
M. Heyns
John Hickman
Rowley & Moreen Hide
D.E. Hill
Dave & Goldie Hill
Dennis & Margaret Hill
Mr & Mrs E.H.O. Hoal
B.C. Hodge
John Hodgson
Neville Hoets
Hanneke Hoffman
Gordon Holtshausen
Janine Honiball
J.R.L. Hoog
Mrs D. Hooper
Jeremy & Sally Horne
J.N. & S.E. Horsley
Jim & Anne Horton
W. Howells
Craig & Claerwen Howie
Irralie Howland
Tim & Paula Howse
Bishop Norman Hudson
Philip Huebsch
David & Carol Hughes
John F. Hume
Peter Humphrey
J.T. Hund
Ian B. Huntley
Erika Huntly
Robin & Joanie Hutchinson
Patrick & Julie Hutchison
Doug & Jane Hutson
John Huxter
Nan Hyslop
Drs Ronald & Pauline Ingle
B. Irving-Smith
Hilary Isherwood
K. & C. Iuel
L.J. Jacobi
Brig. D.J.D. Jacobs
Adriaan & Jenny Jacobsz
Ron James & Family
J.A.C. Jankowitz
Elodie Janovsky
R. & K. Jaschek
Marty Jasper
R.G. Jeffery
Richard Jennett
Mrs I. Jessnitz
Mrs Shirley Jex
Johannesburg Zoological Gardens
Ralf Johannsen
John Voelcker Bird Book Fund
Athol & Dalene Johnson
Dr D. Johnson
Mr & Mrs P.A.A. Johnson
Linda Jones
Peter Jones
Shirley Jones
Stephen E. Jones
I.D. Jonker
Louis A. Joos-Vandewalle
Abri Jordaan
Jean Jordaan
Petré Jordaan
Inez Jordan
Sharon Joss
Dr Johan J. Joubert
S.C.J. Joubert
Mrs E.F. (Dinky) Kahn
Hillel M. Kahn
Sidney H. Kahn
Gillian Karstaedt
Mike & Leigh Kay
Professor Derek Keats
Aliki & Clive Kelly
R.E.H. Kemp
Rodney Kenyon
Mike & Marion Kerby
Mrs D.B. Kilpin
Jean Aveline Kimble
Clive S. King
Dudley King
Leslie S. King
Dr Michael J. King
Michael W. King
Donald B. Kinross
Mr & Mrs I.S. Kirby
Richard A. Kirk
Freda Kirschner
Keith Edward Kirsten
Ralph & Lorna Kirsten
Timothy L. Klapwijk
Dick Klein
Dr F.A. Klein
Dawie & Sarieta Kleynhans
J.D. Klinck
George Kluckow
Erica Knickelbein
Cicely Knight
David Garth Knott
Henry Kok
O.B. Kok
Claire & Holger Kolberg
Volker Konrad
Marilyn B. Korck
Carol Kramer
Peter Roderick Kraunsoe
Nora Kreher
Brian David Kriedemann
Mr & Mrs K.P. Krog
Franci & Mariaan Krone
Margarita Krusche
Hajo Kullmann
Jocelyn Kuper
P. de W. la Grange
B.A. Laing
Nic E. Lambrechts
Nina Landman
Ian R. Lang
Christine Langenegger
Inge Langenhorst
Wallie Latham
John & Moira Lauderdale
F. Laurens
M. Lawrenson
Gavin & Dawn Lawrie
R.M. Lawrie
Anne Lawson
Eckaard le Roux
Emma & Reinhard le Roux
Lawrence, Marlene, Paul & Rachel Le Roy
Tony & Sue Leask
Wicus Leeuwner
John & Thelma Legg
J. Lello
Christopher Lenferna
John Lennard
Letaba Arts & Crafts
Raymond Lévêque
I.M.A. Lewis
Jean M. Lewis
N.A.L. Lexander
Mrs Sabine Liebherr
Edward B.L. Lightbody
Lionel Lindsay
Lorna Linford
Lex Liston
Ian Hamilton Little
Brian & Mary Lloyd
Barry & Judi Loader, Geelong, Australia
Lynton & Iris Lockwood-Hall
Margaret Lombaard
Martin & Desirée Long
Henk & André Loots
Pat Stuart Lorber
Gill & Rupert Lorimer
Merle Lötter
M.M. Loubser
Betty Louw
David & Patsy Louw
John W. Louw
Peter Louw
L.J. Louwrens
Max Lowe
Shëlagh M. Lubbock
Christopher & June Lucas
Elizabeth Ludick
Mrs B.M.J. Ludorf
The Luyt Family
Dianne Lyall
Ann & Peter MacDonald
G.A. MacDonald, Falcon College, Zimbabwe
Mr Andrew A. Mackay
C.A. MacKenzie
Ian MacKenzie
Ken MacKenzie
O.J. MacKenzie
Gordon Lindsay Maclean
Peter Ian MacQuilkan
Madikwe River Lodge
Mahenye Safari Lodge, Zimbabwe
Mike & Lyn Mair
Yakoob M. Makda
Christian Manciot
John & Bev Manning
Cecilia Manson
Deon Marais
Eugene Marais
Hennie & Anna Marais
Dr S.J. Marais
Athol Marchand
Dr Paul Marchand
Mr & Mrs J.H. Maree
Austin & Cathie Markus
Bill & Olga Marsay
Ronald Erling Marthinusen
D.M. Martin
Ian Martin
Robbie & Lyn Massey-Hicks
Matkovich & Hayes
Nigel & Leslye Matthews
Maggie & Tom Matthis
Ann Mawer
Harry Mays
Lionel Mazabow
Bruce & Margaret McBride
Heather K. McBurnie
Ian McCall
Kevin McCann
M.P. McCarthy
Tony & Monica McClean
R.A. McClelland
Pauline McClure
B. McCormick
Kevin & Loreen McDonald
Ron McDonald
T. McDonald
Peter McEwen
Stephen McGregor
Grant M. McIlrath
Dr B.M. McIntosh
Candace E. McIntosh
Sean McKeag
Beryl McMenamin
S.E. McNeill
Lesley Meise
Charles Melck
Carl Wilhelm Josef Menne
William Grant Menzies
Raleigh Llewellyn Meredith
Tim Metelerkamp
Yvonne Meter
Eric Meyer
Phil & Bertha Meyer
Pierre, Irene & Julie-Mari Meyer
Quartus en Elize Meyer
Dr N.T. Michau
A.B. Micklethwait
Dr C.F. Middelberg
Mrs H.A. Middelberg
Belinda & Tony Miek
Anita & Heimo Mikkola
Ria Milburn
G.A. & N. Millar
Dawn & Ron Miller
Mrs Rosemary Miller
Keith Milne
Mrs L. Milton
Dr W.N. Minnaar
David B. Mitchell
Gloria G. Mockford
G.S. & Z. Moffat
Mrs M.A. Moffitt
Lorraine Mollentze
Ara Monadjem
Tony Montinaro
Megan Mooney
Charles W. Moore
Dr William Moore
Heather Carruthers Morawski
Barak Morgan
Michael J. Morgan
L.M. Morris
Bruce Morrison
Maria Morrison
Doug & Terri Morton
P.D. Morum
Robin Moser
C.J. Mostert
D. Mostert
Laurie Dykhouse Moult
Ros & Alan Muirhead
Peter & Hester Müller
W.O.A. Müller
Dr Peter J. Mundy
Neil Munro
Mégan Sarah Musiker
Reuben Musiker
R.M. Murray
Koos Myburgh
W.J. Mylrea
David Nabarro
Mr & Mrs Peter D.R. Nairn
Namibrand Nature Reserve
R.C.M. Napier
Dr J.C. Nathoo
Cornelia Naudé
Simon James Naylor
Paul A. Neal
Bruce & Cilla Nel
Dr Hans Jurie Nel
Jan & Elizabeth Nel
W.A.J. Nel
Woody Nel
Hein Nell
Edgar Nelson
F.A. Neuhoff
T.M. & Mrs J.W. Newsome
Geoff, Lynne & Douglas Nichols
Pam & Bill Nicol
Phillip & Helena Nieuwoudt
George & Cherrie Nisbet
Rosemary Nisbet
Kim Nixon
Martin & Erica Northam-Brown
G. Nürnberger
Eric Eugene Nutt
S.C.V. O'Brien
Peter K. O'Connor
Bruce O'Ehley
David & Margaret O'Reilly
Niël Oberholzer
Francois en Estelle Odendal
Johann Oelofse
T. Ogilvie Thompson
Lida & Peter Oldroyd
Dr P.G. Olivier
Willie & Sandra Olivier
Peggy & Andre Oosthuizen
Gert J. Opperman
Dr H.J.L. Orford
W.A. Orrock
S.M. Osborn
Out of Africa Guest Lodge
Cora & Cyril Ovens
Dr Rainer Pampel
Dee & Colin Paterson-Jones
Andrew J.S. Peace
Andrew Peacock
Cynthia Peacock
Geoff & Jenny Peatling
Anne & Anthony Peepall
P.J.W. Pelle

Ernst J.V. Penzhorn
Johnny Pereira
Wendy Perks
Jeffry Perlman
Nicholas Persse
Kathleen Petersen
Professor H.C. Petrick
R.L. Phillips
P. Clifton Piek
Dr A.J. en Mev Pienaar
Herbert & Cynthia Pienaar
Pietermaritzburg: Parks & Recreation
Mr D.G. Pieterse
Judy Pitout
Doug & Lynn Pitter
John Platter
Rosemarie Ploger
Joy Pocklington
Kevin & Glynis Podmore & Family
C.F. Pohl
Mr & Mrs Philip Pohl
Fritz & Karin Potgieter
Jeanne A. Potgieter
Johann & Rosa-Marje Potgieter
Caroline M.R. Potts
Sharon Preddy
George & Sally Prentice
Coenie & Elma Pretorius
Mark Pretorius
Henriëtte Prince
Renate & Joachim Prinsloo
Myles Baker Protheroe
Dr O.W. Prozesky
Jennifer Pudifin
Uwe Putlitz
Liz Quarmby
Lorraine & Roger Raab
E.S. Rabie
Gail Ractliffe
P. Rademan
Adrian Rademeyer
Benny & Joe Rainsford
Linda Ralphs
Victor Ralphs
Irene Rasmussen
Brian & Carol Ratcliffe
Stan Ratcliffe
David & Anne Rattle
Dave, Margot & Christine Raulstone
Leon & Sue Rautenbach
Patti & Bernard Ravnö
Mr & Mrs P.S. Raymond
Suzette M. Raymond
Christine Read
Johan Rebel
Nadia Rebelo
Dr Anita Redelinghuys
Graham & Joan Reed
Donald Reid
Mr & Mrs J.F. Reid
Lionel Renwick
Johan Repsold
G.C. Retief
Sally Reunert
Joan Reynolds
Steve Reynolds
Rhône-Poulenc Agrichem SA (Pty) Limited
Richards Bay Minerals
Dave & Corlia Richardson
Pam Richardson
Wavency & Allan Richardson
Morelle & Bob Rickards
G.A. Rissik
Pilmann & Mara Ritz
Giovànni Diègo Riva
Colleen & Norman Roberts
Douglas Stuart Roberts
Pierre Roberts
David Charles Robertson
C.H. Robinson
D.L. Robinson
Helmer & John Robinson
John Robinson
R.L. Robinson
Cedric Roché
Brian & Tami Roos
Professor C.J. Roos
Ursula Rose
Ralph Rosen
Johan Rosiers
Mrs E. Rothwell
Sean Kennedy Routledge
John Roux
Vic Roux
Neville Rowe
John Rowntree
Y. Rundle
Charles & Sheryl Rust
Robert A. Rutemoeller
Marc Ryder
Errol & Julienne Sackstein
M.J.V. Samuels
G.H. Sauer
Ken & Glynis Saunders
H.L. Schaary
Con en Jacqueline Schabort
Elsa Schaffer
Rudy W. Schats
Margaret Elizabeth Schleyer
S.J. Schoeman, Pretoria North
Gordon & Marilyn Scholtz
Marjonne Carla Schönefeld
E.M. Schoute-Vanneck
Martin Erich Schroda
Albert Schultz
Michael Schwarz
J.A. Scott
Jonathan Scott
R.W.S. Scott
Shannon Scott-Ronaldson
W.G. Scurr
Allen & Ann Sealey
P.W. St L. Searle
Ron & Maureen Searle
Y.H. Searle
Mary Seely
Gavin Selfe
Julia Selley
G. & L. Sellick
Frederick R. Sellman, London
Roy Sellmann
Robert, Kate, Andrew & James Semple
Ingrid E.F. Shaul
Andrew, Stuart & Joseph Short
Sylvia & Bernard Shull
Hennie Siebert, Lynnwood Manor, Pretoria
Hanno Sierts
J.F.K. Sievers
I.R. Simms
R.G. Simms
Audrey Linda Jean Simpson
George Simpson
Mrs P.E. Simpson
Ann & Jim Sinclair
Singita Game Lodge
Gordon & Gail Sink
D. & H. Skawran
Joan & Hugo Slabbert
Ernest Slome
Ansie & Dennis Slotow
Jennifer Slotow
Louise Smit
Bradley K. Smith
Errol & Rosemary Smith
Keith & Dorothy Smith
Paul & René Smith
Michiel & Morag Smuts
Peter Smuts
Corrie Smuts-Steyn, Botswana
Charles Edward Snell
Daleen Snyman
Trevor Snyman
Mrs Annemarie Solms
Derek Solomon
John & Carrie Solomon
Sondelani Game Lodge
S. Soroczynski
Michelle & Natalie Southwood
Derek William Spencer
Michael Springer
Archie Sprott
H.G. Squires
Rosemarie St. Clair
Diane & Ivor Stack
Wald & Lilla Stack
Harold & Paddy Stainbank
Malcolm Dering Stainbank
Nicholas Stamatiadis
Norman C. Standring
June Stannard
Eurydice Stanton
Ines & Basil Stathoulis
Elizabeth Steenkamp
Jan & Ann Stekhoven
Jim & Anne Stenhouse
Menno en Krysia Stenvert
Coral Stephens
Bruce (J.W.B.) Stewart
Mrs A. Steyn
Charl Steyn
David J. Steyn
Lynne Junius Steyn
Laetitia Steynberg
Allen & Jenny Stidworthy
A.D. Stoltz
Dr Susan Stott
F.J. Strauss
J.I. Strauss
B. & C. Street
Liz & Peter M. Street
Beatrice Strivens
Johannes & Leonore Strobos
Henriette Stroh
M.E. Struwig
M.S. Strydom
P.C. Stunden
R. Summers
Andrew C. Sutherland
Ivan H. Swaffield
Mark S. Swaffield
Lieben Swanevelder
Caroline Mildred Swarts
Mr & Mrs T.G. Sweeney
Barry & Janet Sweet
Catherine, Caylynne, Cherylee & Clive Symes
Dr Andrew Tarr
J.E.S. Taylor
Vincent & Pam Taylor
William Taylor
G.J. Teasdale
Technicare (Pty) Limited
Colin & Ann Tedder
Arlene Temlett
Pierre-Jean Ternamian
A.D. Theron
Annette E. Theron
Dr Walther Thiede
Gaby Thinnes
F. Thomas
Jim Thomson
Keith Thomson
Dr A.S. Thornton
Mr D. Thorrington-Smith
Mervyn Todd
C.A. Tomaselli
Antony Tooley
Craig Tooley
Raymond Topp
Angel Tordesillas
Treehaven Waterfowl Trust
Roy Trendler
Mr D.D. Trent
John & Jill Tresfon
Professor Elizabeth Triegaardt
Jill Trollip
Stanley Trollip
Allen & Ingrid Tucker
B.A. Tucker
Colin K. Tucker
Mrs D.M. Tucker
Mr A.S. Tuckett
I.P. & M.E. Turner
James W.D. Turner
Dawn Tyler
Joe, Cindy, Jessica and James Tyrrell
L.G. Underhill
Don & Irma Uytenbogaardt
Ismail Ebrahim Vahed
Norma van Aarde
Benno van Assen
Dr A.C. van Bruggen
André L. van den Berg
Hettie van den Bergh
Aat & Thea van der Dussen
Dick & Liz van der Jagt
C.C. van der Merwe
Christo van der Merwe
Corni van der Merwe
François van der Merwe
Johann en Marié van der Merwe en Seuns
Dr Roelof van der Merwe
Paul & Gerda van der Reyden
Arnon van der Spuy
Johann van der Spuy
Andries & Sarie van der Walt
Shirley & Arthur van der Westhuizen
B.J. van Dyk
Madel van Dyk
Jill van Eck
J. van Eyk
Anne & Errol van Greunen
Hentie & Annelie van Jaarsveldt
H. van Kerken
Dr Karel M.A. van Laeren
M.S. van Loggerenberg
D.J. van Niekerk
Hendrik van Niekerk
Jacqueline van Niekerk
Pieter & Jess van Niekerk
David & Diana van Nierop
Mrs T.A.M. van Ogtrop
Gert C. van R. van Oudtshoorn
Gary van Rensburg
M. & M.A. van Rijswijck
Willem van Rÿswÿck (Jnr)
Suzanne van Rÿswÿck
Theo van Schaik
Johann L. van Schalkwyk
Paul van Schalkwyk
R. van Schalkwyk
Dr M.C.E. van Schoor
Dr Paul van Staden
Joyce T. van Wyk
Marius & Lenette van Wyk
Helm & Gillian van Zijl
M.H. Veldman
Graham Veldsman
Susan en Vellies Veldsman
Chris Venter
A.E. Vercueil
Leo Vereycken
Paul C. Verleye
C.T. Vermaak
G. Vermoter
H.J. Vickery
Audrey Viljoen
Coral & John Vinsen
Chris, Marietjie & Tiaan Visser
R.L. Vivier
Etienne & Susann Vlok
Willie & Joan Vlok
Rina Vogel
E.A. von Bonde
Peter von Broembsen
O. von der Lancken
Eric von Glehn
Joy von Korff
A.P. Vorster
Lambert Vorster
Dr & Mrs P. Vorster
Unus & Lorna Vos
Pat & Jim Wallace
Dave & Tania Wallis
Terry Walls
Bryony Walmsley
Dr D.H. Wang
Martine Ward
J.M. Wardhaugh
Llewellen & Marianne Wardle
A.F. & S.J. Watermeyer
Professor Ian B. Watt
Cynthia Weaver
Gill Leach & David Weaver
Alan Weaving
Derick & Ethné Webb, King Williams Town
Roy & Lorraine Webber
Dr Arno Weber
Masterman Carl Weber
Fiona Welch
Heino, Zelda, Karl, Andrea & Bernd Wellmann
J.N.H. Wells
M.C. Went-Schultz
Melanie Wes
Jennie Wessels
Wessel P. Wessels
Tim & Sheila West
Mrs J. Weys
Jean White
Mrs I. Whitelock
John W. Whittaker
Vernon Whittal
Colin Whittle
Alan & Jill Whyte
Dr M.K. Wickham
Anthony Haig Pohl Wienand
K.D. Wienand
Dietmar & Lesley Wiening
M.E. Wiid
R.C. Wiid
David & Marjory Wijnberg
Jan Wilkens
Lesley Willers
Mrs Jacko Williams
Peter M. Williams
Nigel Willis
Jean Wilson
Judy & Gavin Wilson
Rod & Wendy Wilson
Graham Winch
Isobel Windsor
B.C. Winkler
M.A. Winkler
Roy & Val Wise
Hilda & Jürgen Witt
Harald Witt
R.E.H. Wohlberg
Anton Wolfaardt
Corinne Wolpe
B. Nigel Wolstenholme
Zoë E. Woodcock
G.W. Woodland
Mrs Wooten
D.O. Wright
The Wykeham Collegiate
Dr Ina Wykowski
Gavin Yell
Ross Young
Thomas Youngleson
Jake & Heather Yule
A.H. Zeederberg
Dorothy Zerbst
Dr Albert Zinn
Maurice Zunckel